For Frederick Mosteller and Judith Tanur, with thanks and great regard: The vision and creative commitment that brought us the first *Guide* in 1971 still continues to influence students and inspire teachers 30 years later.

CONTENTS

Essays Classified by Data Source   vi
Essays Classified by Statistical Method   ix
Preface   xv
Introduction: Learning from Data by David S. Moore   xvii

PART ONE
PUBLIC POLICY AND SOCIAL SCIENCE   1

Statistics in the Courtroom: *United States v. Kristen Gilbert* • *George Cobb & Stephen Gehlbach*   3

The Anatomy of a Preelection Poll • *Edward C. Ratledge*   19

Counting and Apportionment: Foundations of America's Democracy • *Tommy Wright & George Cobb*   35

Evaluating School Choice Programs • *Jennifer Hill*   69

Designing National Health Care Surveys to Inform Health Policy • *Steven B. Cohen & Trena M. Ezzati-Rice*   89

PART TWO
SCIENCE AND TECHNOLOGY   103

Monitoring Tiger Prey Abundance in the Russian Far East • *Ken Gerow, Dale Miquelle, & V. V. Aramilev*   105

Predicting the Africanized Bee Invasion • *James H. Matis & Thomas R. Kiffe*   119

Statistics and the War on Spam • *David Madigan*   135

Should You Measure the Radon Concentration in Your Home? • *Phillip N. Price & Andrew Gelman*   149

Statistical Weather Forecasting • *Daniel S. Wilks*   171

Space Debris: Yet Another Environmental Problem • *David R. Brillinger*   183

PART THREE
BIOLOGY AND MEDICINE   195

Modeling an Outbreak of Anthrax • *Ron Brookmeyer*   197

The Last Frontier: Understanding the Human Mind • *William F. Eddy & Margaret L. Smykla*   211

Leveraging Chance in HIV Research • *Charles M. Heilig, Elizabeth G. Hill, & John M. Karon*   227

Statistical Genetics: Associating Genotypic Differences with Measurable Outcomes • *Rongling Wu & George Casella*   243

DNA Fingerprinting • *Bruce S. Weir*   255

How Many Genes? Mapping Mouse Traits • *Melanie Bahlo & Terry Speed*   271

PART FOUR
BUSINESS AND INDUSTRY   291

To Catch a Thief: Detecting Cell Phone Fraud • *Diane Lambert & José C. Pinheiro*   293

Reducing Junk Mail Using Data Mining Techniques • *Richard D. De Veaux & Herb Edelstein*   307

Improving the Accuracy of a Newspaper: A Six Sigma Case Study of Business Process Improvement • *Ronald D. Snee*   323

Assuring Product Reliability and Safety • *Necip Doganaksoy, Gerald J. Hahn, & William Q. Meeker*   339

Randomness in the Stock Market • *Richard J. Cleary & Norean Radke Sharpe*   359

Advertising as an Engineering Science • *William Kahn & Leonard Roseman*   373

PART FIVE
HOBBIES AND RECREATION   391

Baseball Decision Making by the Numbers • *Hal Stern*   393

Predicting the Quality and Prices of Bordeaux Wines • *Orley Ashenfelter*   407

Index   425

ESSAYS CLASSIFIED BY DATA SOURCE

Available Data

Statistics in the Courtroom: *United States v. Kristen Gilbert, Cobb & Gehlbach*   3

The Anatomy of a Preelection Poll, *Ratledge*   19

Predicting the Africanized Bee Invasion, *Matis & Kiffe*   119

Statistics and the War on Spam, *Madigan*   135

Should You Measure the Radon Concentration in Your Home? *Price & Gelman*   149

Statistical Weather Forecasting, *Wilks*   171

To Catch a Thief: Detecting Cell Phone Fraud, *Lambert & Pinheiro*   293

Reducing Junk Mail Using Data Mining Techniques, *De Veaux & Edelstein*   307

Assuring Product Reliability and Safety, *Doganaksoy, Hahn, & Meeker*   339

Randomness in the Stock Market, *Cleary & Sharpe*   359

Baseball Decision Making by the Numbers, *Stern*   393

Predicting the Quality and Prices of Bordeaux Wines, *Ashenfelter*   407

Samples

The Anatomy of a Preelection Poll, *Ratledge*   19

Predicting the Africanized Bee Invasion, *Matis & Kiffe*   119

Should You Measure the Radon Concentration in Your Home? *Price & Gelman*   149

Space Debris: Yet Another Environmental Problem, *Brillinger*   183

Leveraging Chance in HIV Research, *Heilig, Hill, & Karon*   227

Improving the Accuracy of a Newspaper: A Six Sigma
Case Study of Business Process Improvement, *Snee*   323

Surveys

The Anatomy of a Preelection Poll, *Ratledge*   19

Counting and Apportionment: Foundations of
America's Democracy, *Wright & Cobb*   35

Evaluating School Choice Programs, *Hill*   69

Designing National Health Care Surveys to Inform
Health Policy, *Cohen & Ezzati-Rice*   89

Should You Measure the Radon Concentration
in Your Home? *Price & Gelman*   149

Leveraging Chance in HIV Research,
Heilig, Hill, & Karon   227

Advertising as an Engineering Science, *Kahn & Roseman*   373

Experiments

Evaluating School Choice Programs, *Hill*   69

The Last Frontier: Understanding the Human Mind,
Eddy & Smykla   211

Statistical Genetics: Associating Genotypic
Differences with Measurable Outcomes, *Wu & Casella*   243

How Many Genes? Mapping Mouse Traits,
Bahlo & Speed   271

Assuring Product Reliability and Safety,
Doganaksoy, Hahn, & Meeker   339

Advertising as an Engineering Science, *Kahn & Roseman*   373

Observational Studies

The Anatomy of a Preelection Poll, *Ratledge*   19

Designing National Health Care Surveys to Inform
Health Policy, *Cohen & Ezzati-Rice*   89

The Last Frontier: Understanding the Human Mind,
Eddy & Smykla   211

DNA Fingerprinting, *Weir*   255

Reducing Junk Mail Using Data Mining Techniques,
De Veaux & Edelstein   307

Improving the Accuracy of a Newspaper: A Six Sigma
Case Study of Business Process Improvement, *Snee*   323

Assuring Product Reliability and Safety,
Doganaksoy, Hahn, & Meeker   339

Field Studies

Monitoring Tiger Prey Abundance in the Russian
Far East, *Gerow, Miquelle, & Aramilev*   105

Predicting the Africanized Bee Invasion, *Matis & Kiffe*   119

Statistics and the War on Spam, *Madigan*   135

ESSAYS CLASSIFIED BY STATISTICAL METHOD

Probability and Probability Distributions

The Anatomy of a Preelection Poll, *Ratledge*   19

Predicting the Africanized Bee Invasion, *Matis & Kiffe*   119

Statistics and the War on Spam, *Madigan*   135

Statistical Weather Forecasting, *Wilks*   171

Space Debris: Yet Another Environmental Problem,
Brillinger   183

Modeling an Outbreak of Anthrax, *Brookmeyer*   197

How Many Genes? Mapping Mouse Traits,
Bahlo & Speed   271

To Catch a Thief: Detecting Cell Phone Fraud,
Lambert & Pinheiro   293

Assuring Product Reliability and Safety, *Doganaksoy,
Hahn, & Meeker*   339

Baseball Decision Making by the Numbers, *Stern*   393

Experimental Design

The Last Frontier: Understanding the Human Mind,
Eddy & Smykla   211

Sampling

The Anatomy of a Preelection Poll, *Ratledge*   19

Designing National Health Care Surveys to Inform
Health Policy, *Cohen & Ezzati-Rice*   89

Leveraging Chance in HIV Research,
Heilig, Hill, & Karon   227

Percents and Rates

Designing National Health Care Surveys to Inform
Health Policy, *Cohen & Ezzati-Rice*   89

Leveraging Chance in HIV Research,
Heilig, Hill, & Karon   227

Advertising as an Engineering Science, *Kahn & Roseman*   373

Tables and Graphs

Statistics in the Courtroom: *United States v. Kristen
Gilbert, Cobb & Gehlbach*   3

The Anatomy of a Preelection Poll, *Ratledge*   19

Designing National Health Care Surveys to Inform
Health Policy, *Cohen & Ezzati-Rice*   89

Predicting the Africanized Bee Invasion, *Matis & Kiffe*   119

Statistical Weather Forecasting, *Wilks*   171

Space Debris: Yet Another Environmental Problem,
Brillinger   183

Modeling an Outbreak of Anthrax, *Brookmeyer*   197

Leveraging Chance in HIV Research,
Heilig, Hill, & Karon   227

Statistical Genetics: Associating Genotypic Differences
with Measurable Outcomes, *Wu & Casella*   243

How Many Genes? Mapping Mouse Traits,
Bahlo & Speed   271

Improving the Accuracy of a Newspaper: A Six Sigma
Case Study of Business Process Improvement, *Snee*   323

Assuring Product Reliability and Safety,
Doganaksoy, Hahn, & Meeker   339

Advertising as an Engineering Science, *Kahn & Roseman*   373

Estimation

The Anatomy of a Preelection Poll, *Ratledge*   19

Evaluating School Choice Programs, *Hill*   69

Predicting the Africanized Bee Invasion, *Matis & Kiffe*   119

Statistics and the War on Spam, *Madigan*   135

Should You Measure the Radon Concentration
in Your Home? *Price & Gelman*   149

Space Debris: Yet Another Environmental Problem,
Brillinger   183

Modeling an Outbreak of Anthrax, *Brookmeyer*   197

Leveraging Chance in HIV Research,
Heilig, Hill, & Karon   227

How Many Genes? Mapping Mouse Traits,
Bahlo & Speed   271

To Catch a Thief: Detecting Cell Phone Fraud,
Lambert & Pinheiro   293

Assuring Product Reliability and Safety,
Doganaksoy, Hahn, & Meeker   339

Baseball Decision Making by the Numbers, *Stern*   393

Hypothesis Testing

Statistics in the Courtroom: *United States v. Kristen
Gilbert, Cobb & Gehlbach*   3

Evaluating School Choice Programs, *Hill*   69

Statistics and the War on Spam, *Madigan*   135

The Last Frontier: Understanding the Human Mind,
Eddy & Smykla   211

Statistical Genetics: Associating Genotypic Differences
with Measurable Outcomes, *Wu & Casella*   243

How Many Genes? Mapping Mouse Traits,
Bahlo & Speed   271

Reducing Junk Mail Using Data Mining Techniques,
De Veaux & Edelstein   307

Decision Analysis

Should You Measure the Radon Concentration
in Your Home? *Price & Gelman*   149

To Catch a Thief: Detecting Cell Phone Fraud,
Lambert & Pinheiro   293

Reducing Junk Mail Using Data Mining Techniques,
De Veaux & Edelstein   307

Improving the Accuracy of a Newspaper: A Six Sigma
Case Study of Business Process Improvement, *Snee*   323

Advertising as an Engineering Science, *Kahn & Roseman*   373

Baseball Decision Making by the Numbers, *Stern*   393

Bayesian and Empirical Bayesian Analysis

Statistics and the War on Spam, *Madigan*   135

Should You Measure the Radon Concentration
in Your Home? *Price & Gelman*   149

Regression

Evaluating School Choice Programs, *Hill*   69

Monitoring Tiger Prey Abundance in the Russian
Far East, *Gerow, Miquelle, & Aramilev*   105

Should You Measure the Radon Concentration
in Your Home? *Price & Gelman*   149

Statistical Weather Forecasting, *Wilks*   171

Randomness in the Stock Market, *Cleary & Sharpe*   359

Advertising as an Engineering Science,
Kahn & Roseman   373

Predicting the Quality and Prices of Bordeaux Wines,
Ashenfelter   407

Time Series

Space Debris: Yet Another Environmental Problem,
Brillinger   183

Modeling an Outbreak of Anthrax, *Brookmeyer*   197

Randomness in the Stock Market, *Cleary & Sharpe*   359

Forecasting and Prediction

Predicting the Africanized Bee Invasion, *Matis & Kiffe*   119

Statistics and the War on Spam, *Madigan*   135

Statistical Weather Forecasting, *Wilks*   171

Space Debris: Yet Another Environmental Problem,
Brillinger   183

Modeling an Outbreak of Anthrax, *Brookmeyer*   197

Reducing Junk Mail Using Data Mining Techniques,
De Veaux & Edelstein   307

Assuring Product Reliability and Safety, *Doganaksoy,
Hahn, & Meeker*   339

Randomness in the Stock Market, *Cleary & Sharpe*   359

Advertising as an Engineering Science, *Kahn & Roseman*   373

Modeling

Counting and Apportionment: Foundations
of America's Democracy, *Wright & Cobb*   35

Predicting the Africanized Bee Invasion, *Matis & Kiffe*   119

Space Debris: Yet Another Environmental Problem,
Brillinger   183

Modeling an Outbreak of Anthrax, *Brookmeyer*   197

To Catch a Thief: Detecting Cell Phone Fraud,
Lambert & Pinheiro   293

Reducing Junk Mail Using Data Mining Techniques,
De Veaux & Edelstein   307

Assuring Product Reliability and Safety,
Doganaksoy, Hahn, & Meeker   339

Randomness in the Stock Market, *Cleary & Sharpe*   359

Baseball Decision Making by the Numbers, *Stern*   393

Predicting the Quality and Prices of Bordeaux Wines,
Ashenfelter   407

Transforming Data

Monitoring Tiger Prey Abundance in the Russian
Far East, *Gerow, Miquelle, & Aramilev*   105

Predicting the Africanized Bee Invasion, *Matis & Kiffe*   119

Statistical Quality Control

Improving the Accuracy of a Newspaper: A Six Sigma
Case Study of Business Process Improvement, *Snee*   323

Assuring Product Reliability and Safety,
Doganaksoy, Hahn, & Meeker   339

PREFACE

"The most important science in the whole world, for upon it depends the practical application of every other science and of every art. The one science essential to all political and social administration, all education, all organization based on experience, for it only gives results of our experience."

Florence Nightingale

It has been more than 30 years since the first edition of *Statistics: A Guide to the Unknown* was published and more than 15 years since the publication of the third edition. These previous editions made an important contribution to the statistics literature by providing a collection of essays that highlighted the role of statistics in a wide variety of applications and in a way that was accessible and of interest to a general audience.

In the intervening years, many new and challenging problems have emerged, and the need for statistical methods is perhaps greater now than ever. In the spirit of the previous editions, this fourth edition seeks to demonstrate the impact of statistics through an entirely new collection of essays that illustrate how statistics can provide an organized way of learning from data and how the ensuing knowledge can be used to address important social, environmental, and economic problems.

We have assembled 25 essays from many different fields, with the goal of creating a collection that demonstrates the impact and wide applicability of statistics. We hope that we have achieved a volume that promotes the discipline of statistics by illustrating its power in addressing important problems and that the reader will find these essays both informative and engaging.

Our hope is that SAGTU will continue to be useful as a supplementary text for introductory statistics courses. To assist instructors who would like to use SAGTU in this way, we have provided two tables following the table of contents that classify the essays by data collection method and by the statistical tools referenced in the essay.

Many people contributed to the success of this edition. First and foremost, we thank all the distinguished researchers who made time to write essays for this edition. We owe a great debt to them for their time, insight, and contribution to statistics education. Our thanks also go to David Moore for crafting a fine introduction to the collection. There are others deserving of our thanks as well: Larry Weldon for his thoughtful comments on the essays and for writing most of the study questions that appear at the end of each essay; Judith Tanur for her priceless advice and support; Frederick Mosteller and the editorial boards of the previous SAGTU editions for providing such successful models for this edition; and our editor at Duxbury Press, Carolyn Crockett, for her good humor and unwavering support.

Each essay was reviewed by several students, who provided comments and suggestions for revision, and we thank them for their work: Mathew Bowyer, Julia Busso, Marissa Deffebach, Dawn Eash, Breanne Henkelman, Katherine Ianiro, Mary Joynt, Brad Kaplan, Jordan Keene, Amanda King, Michael Kolkowski, Maya Markowitz, Katie Pesicka, Rebecca Russ, Tierra Stimson.

We also thank the American Statistical Association for its support of this project, which included providing the funding necessary for the book to be produced in color. And, our thanks go also to the Sloan Foundation for providing the original funding for the first edition of this book.

Finally, I would like to express my deepest gratitude to the members of the editorial board. It was a pleasure to work with such a supportive and responsive group of individuals. Each member of this eminent group contributed many, many hours to this project and their commitment and dedication to this project was extraordinary.

Roxy Peck for the SAGTU Editorial Board

INTRODUCTION

Learning from Data

DAVID S. MOORE

Purdue University

———— ❧ : : ❧ ————

What genes are active in a tissue? Answering this question can unravel basic questions in biology, distinguish cancer cells from normal cells, and distinguish between closely related types of cancer. To learn the answer, apply the tissue to a "microarray" that contains thousands of snippets of DNA arranged in a grid on a chip about the size of your thumb. As DNA in the tissue binds to the snippets in the array, special recorders pick up spots of light of varying color and intensity across the grid and store what they see as numbers.

What's hot in popular music this week? SoundScan (part of Nielsen Media Research) knows. SoundScan collects data electronically from the cash registers in more than 14,000 retail outlets, and also collects data on download sales from websites. When you buy a CD, the checkout scanner is probably telling SoundScan what you bought. SoundScan provides this information to Billboard Magazine, MTV, and VH1, as well as to record companies and artists' agents.

Should women take hormones such as estrogen after menopause, when natural production of these hormones ends? In 1992, several major medical organizations said yes. In particular, women who took hormones seemed to reduce their risk of a heart attack by 35% to 50%. The risks of taking hormones appeared small compared with the benefits. But in 2002, the National Institutes of Health declared these findings wrong. Use of hormones after menopause immediately plummeted. Both recommendations were based on extensive studies. What happened?

DNA microarrays, SoundScan, and medical studies all produce data (numerical facts), and lots of them. Using data effectively is a large and growing part of most professions. Reacting to data is part of everyday life. That's why statistics is important:

Statistics is the science of learning from data.

The essays in this book illustrate learning from data in many settings. To get started, here are some comments on how we learn.

WHERE THE DATA COME FROM MATTERS

What's behind the flip-flop in advice offered to women about hormone replacement? The evidence in favor of hormone replacement came from a number of *observational studies* that compared women who were taking hormones with others who were not. But women who choose to take hormones are very different from women who do not: they are richer and better educated and see doctors more often. These women do many things to maintain their health. It isn't surprising that they have fewer heart attacks.

Large and careful observational studies are expensive, but are much easier to arrange than careful *experiments*. Experiments don't let women decide what to do. They assign women to either hormone replacement or to dummy pills that look and taste the same as the hormone pills. The assignment is done by a coin toss, so that all kinds of women are equally likely to get either treatment. Part of the difficulty of a good experiment is persuading women to agree to accept the result—invisible to them—of the coin toss. By 2002, several experiments with women of different ages agreed that hormone replacement does *not* reduce the risk of heart attacks.

Observational studies just observe; they don't try to change behavior. We can learn from observational studies how chimpanzees behave in the wild, or which popular songs sold best last week, or what the public thinks about the president's performance. But to learn whether some act *causes* a change in behavior—for example, whether taking hormones causes a reduction in the risk of heart attacks—observational studies are a poor choice. Experiments that directly compare different actions—for example, taking a hormone pill and taking a dummy pill—are designed to help us learn about cause and effect.

The most important information about any statistical study is how the data were produced.

Observation versus experiment is just one aspect of how data are produced. SoundScan can't get data from every music-selling source in the United States. Its data come from a large *sample* of sellers. When downloading music became popular, SoundScan had to add online sellers in order to keep its sample representative of all sellers. As CDs are increasingly sold in untraditional places such as Hallmark card stores and Starbucks, SoundScan faces more challenges.

The opinion poll that tells us that 55% of the public approves of the president's performance faces challenges more severe than Sound-Scan's, because there are so many of us whose opinions are part of public opinion. Professional polls use *random samples.* That is, they let impersonal chance choose the sample of people they talk with. That's a great idea—it allows rich and poor, black and white, Democrat and Republican the same chance to respond. You might contrast the professional approach with just gathering a sample by stopping people at shopping malls. Shopping mall patrons aren't "the public." They are more likely to be either teens or retired. They are less likely to be really poor. And the poll-taker may not want to question the unshaven hulk in the muscle shirt. Random samples avoid the kinds of favoritism that a mall survey produces.

But even with a random sample, we must watch the details. Almost all polls create random samples by dialing telephone numbers at random. That misses the few people without phones (and often leaves out Alaska and Hawaii to hold down cost), but that's not the big problem. Answering machines, never at home, don't want to talk with someone who might be a telemarketer—the big problem is that many of the people in the random sample don't respond. Most polls don't announce their rates of nonresponse, because the truth would be embarrassing. It appears from studies by groups such as the Pew Research Center that roughly two-thirds of the initial sample typically fails to respond to a telephone survey. Do you think the nonresponders are different from the responders? If so, don't take the poll results too seriously.

Some sample surveys are more trustworthy. Government surveys, such as the monthly Current Population Survey (CPS) that produces the unemployment rate and much other information, have much higher rates of response. Only about 6% or 7% of the households chosen at random for the CPS sample don't respond. The Bureau of Labor Statistics, unlike pollsters, makes its response rates public. Knowing the details increases our confidence in the findings.

Before you trust the results of a statistical study, ask about details of how the study was conducted.

YOU CAN OBSERVE A LOT JUST BY WATCHING

Yogi Berra's famous saying is a motto for learning from data. A few carefully chosen graphs are often more instructive than great piles of numbers. Consider the outcome of the 2000 presidential election in Florida.

Elections don't come much closer: after much recounting, state officials declared that George Bush had carried Florida by 537 votes out of almost 6 million votes cast. Florida's vote decided the election and made George Bush rather than Al Gore president. Lawsuits followed, and the Supreme Court upheld the result. Legal and political issues aside, Figure 1 displays a graph that plots votes for the third-party candidate Pat Buchanan against votes for the Democratic candidate Al Gore in Florida's 67 counties.

What happened in Palm Beach County? The question leaps out from the graph. In this large and heavily Democratic county, a conservative third-party candidate did far better relative to the Democratic Party candidate than in any other county. The points for the other 66 counties show votes for both candidates increasing together in a roughly straight-line pattern. Both counts go up as county population goes up. Based on this pattern, we would expect Buchanan to receive around 800 votes in Palm Beach County. He actually received more than 3400 votes. That difference determined the election result in Florida

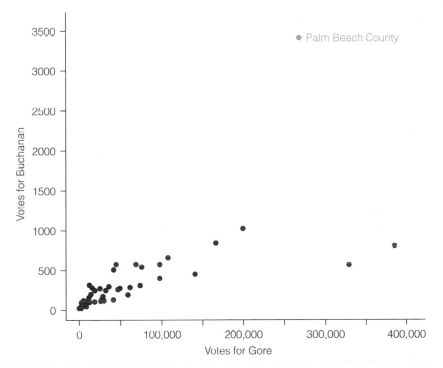

FIGURE 1 Votes for Pat Buchanan versus votes for Al Gore in Florida's 67 counties

and in the nation. All this from a simple graph. Once you have data in hand, the first rule of data analysis is:

Always plot your data.

The graph demands an explanation. It turns out that Palm Beach County used a confusing "butterfly" ballot in which candidate names on both left and right pages led to a voting column in the center. It would be easy for a voter who didn't look carefully to cast a vote for Buchanan when a vote for Gore was intended. The graph is convincing evidence that this in fact happened, probably more convincing than the complaints of voters who (later) were unsure where their votes ended up.

Plotting the data and thinking about the plots is the start of learning from data, but only the start. Issues such as the effect of hormone replacement on women's health are too complicated to be settled by looking at graphs. The women studied vary in age, race, bad habits such as smoking, good habits such as regular exercise, and so on. The variation among women will overwhelm the effect of taking hormones unless we can find a way to see through the variation.

FIGHTING THE CURSE OF VARIATION

Why don't opinion polls interview just one person? Because not all people have the same opinions. Why can't we just compare the long-term health of two women, one with hormone replacement and one without? Because individuals vary even when the two individuals are the same in sex, age, race, income, previous health history, and so on. Accounting for variation, and making sure that variation among individuals doesn't obscure important overall patterns, is a big reason why we need the science of statistics, not just a quick look at the data.

Statisticians have two main strategies for overcoming variation. First, *take enough observations* so that the effects of variation are "averaged out." A poll of 50 people, if repeated, may give quite different proportions who approve the president's performance. A poll of 2500 people will almost always give close to the same result, because the variation caused by choosing people at random disappears as we choose more and more people. It's just like tossing a coin—one, or 10, or even 50 tosses give quite variable percentages of "heads," but thousands of tosses give close to half "heads" every time. In fact, another big reason to use random samples is that probability, the same mathematics that describes coin tosses, describes how random samples behave. That larger samples are less variable isn't just common sense—it's a mathematical fact. The mathematics of probability allows statisticians to say how large a sample we need to reduce the variation in results to whatever level we want.

The second way to fight variability is to *measure characteristics that explain variation among individuals*. If we know the age, race, income, and health history of a woman, we can use this information to predict her future health. If we do this for all the women in a study of hormone replacement, it becomes easier to see the effect of hormone replacement because we can remove variation explained by the things we measured.

These strategies can reduce variation in outcomes, but they don't produce certainty. Experimental findings that hormone replacement has few benefits and some risks trump observational studies that show benefits, but we can't be absolutely sure that the experimental findings are right. There remains some risk that by bad luck the dummy-pill group received healthier women than the hormone group. So, statistical findings are always uncertain. The laws of probability again come to our rescue: we can attach to our findings a statement of just how uncertain they are, and we can design our studies to make the remaining uncertainty as small as we may wish. Opinion polls, for example, give not only the percentage of the sample who support the president but also a "margin of error" that describes the uncertainty in applying the sample result to the wider universe of all adults. Saying how much variation remains belongs with strategies for reducing variation in the statistician's toolkit for dealing with variation.

Variation is everywhere. Individuals vary. Repeated measurements on the same individual vary. The science of statistics provides tools for dealing with variation.

Applying the laws of probability, removing variation due to characteristics we have measured, and saying how much uncertainty remains are technical matters—that's why there are statisticians. It's important to realize that this technical magic can't remedy bad data production. The mathematics of probability doesn't apply to a haphazard shopping mall sample, and it doesn't account for nonresponse in a random sample. The observational studies that seemed to show beneficial effects from hormone replacement did measure and adjust for many other variables, but still came up with the wrong answer. Did these studies fail to make some essential measurement? Were the women they studied not typical in some unknown way? We don't know. We do know that experiments comparing two groups formed at random are the gold standard for understanding the effect of an intervention such as hormone replacement.

STATISTICS: A GUIDE TO THE UNKNOWN

Data enlighten; they shed light in dark places. We are interested in whether the president has the support of the country and whether unemployment is dropping. Women need to know whether the benefits of hormone replacement outweigh the risks. We may be surprised to learn from SoundScan's cousin, BookScan, that classics such as Jane Austen's *Pride and Prejudice* outsell even the best sellers of a few years ago—*Pride and Prejudice* sold 110,000 copies in 2002, and that doesn't include students who had to buy the book. Biologists were surprised to learn that the pattern of data from a microarray can distinguish two types of leukemia that seem identical to the usual tools of medicine.

The ideas and methods of statistics guide us in using data to explore the unknown: how to produce trustworthy data, how to look at data (starting with graphs), and how to reach sound conclusions that come with an indication of just how confident we can be. These are three important aspects of statistical science: data production, data analysis, and inferring conclusions from data. We have seen brief examples of each. The essays in this book illustrate in more detail the reach of these statistical ideas, from counting tigers to reducing junk mail. Every application of statistics has its special features—tigers and junk mail have little in common—but common ways of thinking illuminate all.

PUBLIC POLICY AND SOCIAL SCIENCE

⁂ : : ⁂

Statistics in the Courtroom: *United States v. Kristen Gilbert*

George Cobb & Stephen Gehlbach

The Anatomy of a Preelection Poll

Edward C. Ratledge

Counting and Apportionment: Foundations of America's Democracy

Tommy Wright & George Cobb

Evaluating School Choice Programs

Jennifer Hill

Designing National Health Care Surveys to Inform Health Policy

Steven B. Cohen & Trena M. Ezzati-Rice

STATISTICS IN THE COURTROOM

United States v. Kristen Gilbert

GEORGE COBB

Mount Holyoke College

STEPHEN GEHLBACH

University of Massachusetts, School of Public Health and Health Sciences

———— ❊ : : ❊ ————

A NURSE ACCUSED

By the mid-1990s, Kristen Gilbert had been working for several years as a nurse at the Veteran's Administration (VA) hospital in Northampton, Massachusetts. For a time, she had been one of the nurses the others most often looked up to as an example of skill and competence. She had established a reputation for being particularly good in a crisis. If a patient went into cardiac arrest, for example, she was often the first to notice that something was wrong. She would sound a "code blue," the signal that brought the aid of the resuscitation team. She stayed calm, and she knew how to give a shot of the stimulant epinephrine, a synthetic form of adrenaline, to try to restart a patient's heart. Often the adrenaline did its job, the heart began to beat again, and the patient's life was saved.

Lately, though, other nurses had become increasingly suspicious that something was not right. To some, it seemed that there were too many codes called, too many crises when Gilbert was on the ward. Over time, the suspicions became stronger. Several patients who went into arrest died, and to some of the staff, the number of deaths was a sinister sign. An investigation was launched. Although an initial report by the VA found that the numbers of deaths were consistent with the patterns at other VA hospitals, the suspicions of the staff remained. Eventually, after additional investigation, including a statistical analysis by one of us (Gehlbach), Assistant U.S. Attorney William Welch convened a grand jury in 1998 to hear the evidence against Gilbert. Welch accused her of killing several patients by giving them fatal doses of heart stimulant, and he wanted her indicted for multiple murders.

Kristen Gilbert was the mother of two young children. Although she was divorced, she had been dating a male friend for some time. She had a steady job, one that paid reasonably well, and her skill as a nurse was generally recognized. What could possibly motivate her to commit the murders that she was now suspected of? These were not "mercy killings"; the victims in Welch's indictment were not old men or in poor health but were middle-aged, and healthy enough that their deaths were unexpected. Welch argued that Kristen Gilbert did have reasons for her actions. She liked the thrill of a crisis, she needed the recognition that came from her skillful handling of a cardiac arrest, and, especially, she wanted to impress her boyfriend, who also worked at the hospital.

Part of the evidence against Gilbert dealt with her motivation, part of it came from the testimony of coworkers about her access to the epinephrine she was accused of using in the alleged murders, and part came from a physician who testified about the symptoms of the men who had died. Taken together, this evidence was certainly suggestive, but would it be convincing? No one had seen Gilbert give fatal injections, and although the patients' deaths were unexpected, the symptoms could have been considered consistent with other possible causes of death. It turned out that a major part of the evidence against Gilbert was statistical.

HYPOTHESIS TESTING I

A key question for the grand jurors was this: Was it true that there were more deaths when Kristen Gilbert was working? Not just one or two extra deaths—one or two could easily be due just to coincidence—but enough to be truly suspicious? If not, there might not be enough evidence to justify bringing Gilbert to trial. On the other hand, an answer of yes would call for an explanation, and enough other evidence pointed to Gilbert to make an indictment all but certain.

The prosecutors recognized that the key question about excess deaths was one that could only be answered using statistics, and so they asked Stephen Gehlbach, who had done the statistical analysis of the hospital records, to present a summary of the results to the grand jury. In what follows, we will present you with a similar summary of the statistical evidence. As you read through the summary, imagine yourself as one of the grand jurors. Do you find the evidence strong enough to bring Gilbert to trial?

The statistical substance involves hypothesis testing, a form of reasoning that uses probability calculations to decide whether or not an observed outcome should be regarded as so unusual—so extreme—that it qualifies as a "scientific surprise." The logic and interpretation of hypothesis testing is fundamental to a lot of work in the natural and social sciences, important enough that anyone serious about understanding how science works should understand this form of reasoning. Unfortunately, in many statistics courses, the logic of hypothesis testing is taught at the same time as some of the probability calculations that you need for particular applications, and the details of the computations tend to eclipse the underlying logic. Part of the challenge facing Gehlbach was to make the logic clear to the grand jury without going into the details of the calculations.

GEHLBACH'S TESTIMONY TO THE GRAND JURY

Dr. Gehlbach's testimony was delivered orally, with Gehlbach in the witness stand, talking to the members of the grand jury. The next several paragraphs summarize three parts of Gehlbach's testimony, a first part about the pattern of deaths, by shift and by year, on the medical ward where Gilbert worked; a second part about variability and p-values; and a third part about a statistical test for whether the pattern linking the excess deaths to Gilbert's presence on the ward was too extreme to be regarded as due to ordinary, expectable variability.

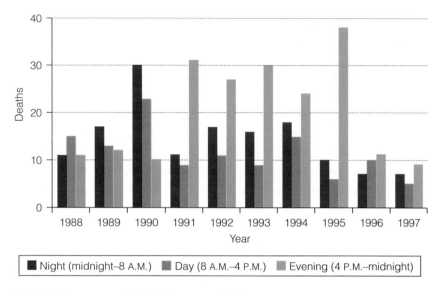

Night (midnight–8 A.M.) ■ Day (8 A.M.–4 P.M.) ■ Evening (4 P.M.–midnight)

FIGURE 1 The pattern of deaths, by year and shift

The summaries don't use the exact words from the grand jury testimony, but they cover some of the same substance.

Part One: The Pattern of Deaths

Imagine that at this point in Gehlbach's testimony, the jurors are looking at a graph like the one in Figure 1:

> Dr. Gehlbach: "The graph you see shows data from the VA hospital where Kristen Gilbert worked. Each set of three bars shows one year's worth of data, starting in 1988 and running through 1997. Within each set of three bars, there is one bar for each shift. The left bar is for the night shift, midnight to 8 A.M.; the middle bar is for the day shift, 8 A.M. to 4 P.M.; and the right bar is for the evening shift, 4 P.M. to midnight. The height of each bar tells how many deaths there were on a shift for the year in question.
>
> "Now look at the pattern from one year to the next. For the first two years, '88 and '89, the bars are short, showing roughly ten deaths per year on each shift. Then, there is a dramatic increase. For the years 1990 through 1995, there is one shift in each set of three with 25 to 35 deaths per year.

Then for the last two years, the bars are all short again, a bit under ten deaths per year on each shift.

"How does this pattern fit with Kristen Gilbert's time at the VA? It turns out that Ms. Gilbert began work on Ward C, the medical ward, in March of 1990, and stopped working at the VA in February of 1996. Looking at the deaths by year, the pattern tracks Ms. Gilbert's work history: small numbers of deaths in years when she didn't work at the VA, and large numbers when she was there.

"We can learn more by looking at the different shifts. You'll notice that in each of the years that Ms. Gilbert worked on Ward C, one of the three shifts always shows more deaths than the other two. For five of these six years, 1991 through 1995, it's the evening shift that stands out. During these five years, Ms. Gilbert was assigned to the evening shift.

"What about the exception, 1990? That year it is the night shift, not the evening shift, that stands out as having an unusually large number of deaths. Well, it turns out that 1990 was also an exception for Kristen Gilbert's work history. That year she was assigned not to the evening shift but to the night shift.

At this point in the argument, there is a clear pattern associating Gilbert's presence with excess deaths. However, in principle the pattern *might* be nothing more than the result of ordinary, expectable variation. The goal of a statistical test in this situation would be to determine whether the numbers of excess deaths were too extreme to be accounted for by such variation. In order to prepare the jurors to think about a statistical test here, Gehlbach first explained the basic ideas in a more familiar context.

Part Two: Variability and *p*-values

Dr. Gehlbach: "To understand the idea of a statistical test, think about tossing a coin. How can you decide whether there's something suspicious about a set of 10 coin flips? Ordinarily, we expect a coin to be fair, which means there's a 50-50 chance of heads. This is our hypothesis, the starting point of our reasoning. If you flip 10 times, and the coin is fair, then on average you'd expect five heads to show up. But you know that you might get six, seven, or even eight heads. Things vary, and sometimes the variation is due just to chance.

"Now suppose you got 10 heads in 10 flips. Is that result extreme enough to be suspicious? How extreme an outcome do we need before we should doubt our hypothesis that the chance of heads is 50-50?

"To answer this question, statisticians compute a p-value: start with the hypothesis—a 50-50 chance of heads—and compute the probability of six heads, or seven heads, and so on. It turns out that the probability of at least six heads in 10 flips is about 0.38. This means that 38% of the time when you make 10 flips with a fair coin, you'll get at least six heads. If something happens 38% of the time, there's nothing surprising or suspicious about it.

"For seven heads, the p-value is about 0.17: about 17% of the time you do 10 flips of a fair coin, you'll get seven heads or more. So, seven out of 10 isn't really surprising, either.

"If you get nine heads in 10 flips, however, that's unusual. If the coin is fair, then you're unlikely to get a result that extreme. The probability, or p-value, for nine or more heads is only about 0.01, or 1%.

"For 10 out of 10, the p-value is about 0.001, or one in 1000. This is a result so extreme that you'd almost never get it from a fair coin. If you saw me pull a coin out of my pocket, flip it 10 times, and get heads every time, you'd be justified in thinking there's something going on besides just chance variation.

"That's how statisticians use a p-value. If we see a result with a really low p-value, then either we've seen a really rare outcome or else the hypothesis we used to compute the p-value must be wrong.

"In many medical trials—testing whether antihistamines relieve your symptoms of allergies, or things like that—we compute a p-value assuming the drug has no effect, and a probability of one out of 100 is unusual enough, and the evidence would be considered strong enough, to conclude that the medicine actually worked.

GILBERT PRESENT	DEATH ON SHIFT?		Total
	Yes	No	
Yes	40	217	257
No	34	1350	1384
Total	74	1567	1641

TABLE 1 The basis of the statistical test

Note: The table is based on the following data:

Number of days	547
Number of shifts	1641
Number of deaths	74
Deaths per shift	0.045
Shifts with KG present	257
Expected number of deaths	11.59
Observed number of deaths	40

Now, with the basic logic out on the table, it was time to present a formal test. What follows is just one focused part of the set of tests Gehlbach actually presented.

Part Three: A Statistical Test

At this point, Gehlbach showed the jury data like that shown in Table 1.

Dr. Gehlbach: "The table summarizes records for the 18 months leading up to the end of February 1996. (That February was the month when Ms. Gilbert's coworkers met with their supervisor to express their concerns; shortly after that, Ms. Gilbert took a medical leave.) With 547 days during the period in question, and three shifts per day, there were 1641 shifts in all. Out of these 1641 shifts, there were 74 for which there was at least one death.

"Now think of each shift as like a coin flip, with a death on the shift if the coin lands heads. The fraction of shifts with a death is $\frac{74}{1641}$ or 0.045. This means that out of every 100 shifts, you would expect four and one-half shifts, or 4.5%, with at least one death. It's like tossing a coin, one toss per shift, with a probability of 0.045 that the toss lands heads.

"Now let's look just at the shifts when Ms. Gilbert worked. There were 257 of these. If the deaths distributed themselves like coins landing heads, we'd expect between 11 and 12 of these shifts to experience a death, because 4.5% of 257 is 11.6.

"What does the record show? As you can see from the table, there were in fact not 11 or 12 shifts with a death, but 40. How extreme is 40? Could you get a number like that just from chance variation, or is 40 really suspicious? To answer that, we compute a p-value.

"Assume that the 257 shifts that Ms. Gilbert worked behaved like coin tosses, with a chance of heads equal to 0.045. What is the probability of 40 or more deaths? The p-value turns out to be less than one in 100 million. In other words, it is virtually impossible to get as many as 40 shifts with deaths from ordinary, chancelike variation.

GILBERT ON TRIAL

The grand jury found the evidence persuasive and indicted Gilbert. Because the VA hospital is legally the property of the federal government, Gilbert would stand trial in federal district court on four counts of murder and three additional counts of attempted murder. The question of jurisdiction was important because although the state of Massachusetts has no death penalty, Gilbert was facing a federal indictment, governed by federal rather than state laws, and Assistant U.S. Attorney Welch decided to ask for the death penalty. Kristen Gilbert would be on trial for her life.

Before the trial got under way, the judge, Michael A. Ponsor, had to rule on whether the jury should be allowed to hear the statistical evidence. On the one hand this seems like a no-brainer. After all, if the evidence was an important part of what was presented to the grand jury, if it was appropriate for them to hear, and if they found it compelling, what could possibly be wrong with letting the trial jury hear the same testimony? On the other hand, a counterargument might be that allowing the statistical evidence would just lead to the unhelpful distraction of "dueling experts." The court system allows expert testimony when the evidence involves specialized technical or scientific issues that go beyond what members of the jury would ordinarily be familiar with. The purpose of the experts is to provide explanations of the science or of the technical facts involved, along with the appropriate conclusions. In other words, they help the jury understand the evidence better—and the U.S. Supreme Court has set guidelines aimed at making sure that unscientific testimony is not admitted. The goal is to help ensure that the verdict will be scientifically sound. Nevertheless, attorneys sometimes say that if there is expert testimony on one side, the other side hires another expert who will disagree, and the jury, rather than think through the explanations, will simply ignore it all. One expert cancels the other. Although this view may be overly cynical, no doubt it does have a basis in fact.

Rather than rely on the crude strategy of dueling experts, Gilbert's defense attorneys asked the other of the two of us (Cobb) to prepare a written report for the judge summarizing the reasons why it would not be appropriate for the new jury, the trial jury, to hear the same evidence that Gehlbach had presented earlier to the grand jury. In the next several paragraphs, you will read a summary of the main points in that report. This time, put yourself in the position of Judge Ponsor. Do you find these points persuasive? Would you have allowed the jury to hear the statistical evidence or not?

HYPOTHESIS TESTING II

So far, in the Gehlbach testimony, the interpretation of hypothesis testing has focused on what it is that a tiny p value *does* tell you. It tells you that the observed result is too extreme to be explained as due to chancelike variation. This was exactly the relevant issue for the grand jury: Were there so many excess deaths when Gilbert was present as to be suspicious in the eyes of science? The clear answer was yes. In the Cobb report, the focus was on things that tiny p-values do *not* tell you. Unfortunately for people who need to understand hypothesis testing, these invalid conclusions are a constant temptation. They seem to make sense intuitively, but they are wrong, and so they have great potential to mislead the unwary. This potential for logical mischief was the basis for the defense team's request that Judge Ponsor not allow Attorney Welch to present statistical evidence to the trial jury.

COBB'S REPORT TO JUDGE PONSOR

Leaving aside a variety of secondary technical issues, the Cobb report made three main points. One of them was to agree with the bottom line conclusion in Gehlbach's testimony. The other two dealt with two limitations on what you can learn from a tiny p-value.

Point One: The Defense and Prosecution Statisticians Agree

As mentioned earlier, often the two experts who provide testimony on scientific evidence disagree. However, that was *not* what happened in the Gilbert case. Cobb's report *agreed* with Gehlbach's testimony before the grand jury. We both thought the pattern linking Gilbert's presence on the ward with excess deaths was far too strong to be regarded as mere coincidence due to chancelike variation. We both thought, too, that in the absence of any innocent explanation for the pattern, the association was more than strong enough to justify the indictment. Why then, shouldn't the trial jury hear the testimony? To answer that question we proceed with Cobb's other two points.

Point Two: Association Is Not Causation

The grand jury and the trial jury have quite different decisions to make as they weigh the evidence, and the difference is closely tied to what a *p*-value does and does not tell you. The grand jury had to decide whether or not Gilbert should stand trial. Was there enough suspicion to justify the expense to the government and the psychological burden on Gilbert to hold what promised to be a long and expensive trial? A grand jury does not have to decide guilt or innocence beyond a reasonable doubt. For them, the standard is much lower. They are simply asked to determine whether the level of suspicion is high enough. This is precisely the kind of question that logic of hypothesis testing is designed to answer. In statistics, and in science generally, the bar is set quite high for what deserves to be considered strong suspicion, typically a *p*-value of 0.05 or 0.01. A low *p*-value establishes suspicion by *ruling out chance variation* as an explanation. Notice that a low *p*-value does not *provide* an explanation. It doesn't say, "Here. This is the reason for the excess deaths." What it says is much more limited: "Whatever the explanation may be, you can be quite confident that it is *not* mere chance variation."

The trial jury isn't asked to decide whether the facts look suspicious. By the time a case comes to trial, the decision about suspicion has already been made. The trial jury is asked to decide the reason for the suspicious facts. Were the excess deaths caused by Gilbert giving fatal injections? Or were there enough uncertainties that the cause could not be determined beyond a reasonable doubt? Because a low

p-value cannot tell you about cause, the Cobb report argued, the statistical evidence was not an appropriate part of the evidence for the trial jury.

But wait. Isn't statistical evidence used all the time to draw conclusions about cause? Doesn't the FDA use statistics to decide whether a particular medication will cause a disease to go away, or at least cause its symptoms to go away, or whether that same medication will cause side effects? Didn't scientists use hypothesis testing to decide, for example, that antihistamines can relieve the symptoms of allergies? If hypothesis testing can tell us about cause in these situations, why not in the Gilbert case also?

The answer involves what some statisticians consider to be the single most important contribution that statistics has made in the last hundred years: an understanding of the difference between an observational study and a randomized experiment. The statistical analysis in the Gilbert case is based on observational data; in the studies used to decide such things as whether taking aspirin lowers the risk of heart attacks, the data come from randomized experiments. The distinction here is so important that it is worth pausing to take a look at it in more detail.

AN IMPORTANT DISTINCTION: OBSERVATIONAL STUDIES VERSUS RANDOMIZED EXPERIMENTS

A famous study from the early research on smoking and health illustrates why observational studies can be misleading about cause and effect. Look at the death rates from that study, as shown in Table 2, and notice what the "obvious" conclusion would be. The sample sizes in this study were huge, and the tiny p-values conclusively rule out chance variation as an explanation for the differences among the three groups. Taking the numbers at face value would leave us with the conclusion that cigarette smoking carries only a miniscule risk, but that pipes and cigars are highly dangerous.

Nonsmokers	20.2
Cigarette smokers	20.5
Cigar and pipe smokers	35.3

TABLE 2 Death rates per 1000 people per year

Nonsmokers	20.3
Cigarette smokers	28.3
Cigar and pipe smokers	21.2

TABLE 3 Death rates, adjusted for age, per 1000 people per year

To avoid this logical trap, you need to recognize that a low *p*-value, by itself, does not prove a cause-and-effect relationship; it only eliminates chance as one of the possible causes. For this study, there was another cause at work behind the scenes: age. The non-smokers, on average, were 54.9 years old; the cigarette smokers only 50.5; and the cigar and pipe smokers were 15 years older, 65.9. Because the researchers had this additional information, they were able to use statistical methods to adjust for the effect of the "lurking variable," age. The adjusted death rates, illustrated in Table 3, are in line with what we have come to expect.

For our purposes, the key point is this: With an observational study, you can never know for certain whether your numbers look the way they do for the reasons you know about or whether, instead, there are hidden causes at work. With a randomized experiment, the groups being compared are created using randomness, or chance. If the group sizes are large enough, the randomization process evens out all possible influences that might make one group different from another. The beauty and power of the randomization is that it evens out all unwanted influences, including the ones you don't know about.

Consider how randomization worked for an influential study of aspirin and heart attacks. Back in the 1980s, researchers began a huge study involving 21,996 physicians across the United States. All of them had volunteered to take part. Some were older; some were younger. Some were overweight; others were not. Some exercised regularly; others didn't. Cholesterol levels varied from low to very high. In short, there were many influences, both known and unknown, that caused big differences in the risk of heart attack. To ensure that all these influences would even out, the researchers used a chance device to assign each physician to one of two groups. Those assigned to the treatment group took a daily pill that actually contained aspirin. Those assigned to the placebo group also took a daily pill, one that was identical to the other pill except that it contained no aspirin.

The results of the study were striking. Even before the study was supposed to have ended, preliminary p-values were so low that the scientists in charge of the research decided to call an end to the experiment so that the physicians in the placebo group could start taking the aspirin if they wanted to.

In a nutshell, here is the key difference between the experiment and the observational study. In the observational study, when a low p-value ruled out chance variation as an explanation for the differences in death rates, it was not clear what was causing those rates to differ. The difference in rates might have been caused by differences in smoking habits, might have been caused by differences in age, and might have been caused by any of a number of other influences. The p-value alone was of no help in deciding the cause. In the experiment, one possible cause—aspirin—was singled out for investigation, and the experiment was carefully designed to eliminate all other possible causes, apart from chancelike variability. When a statistical test was then able to rule out chance variability, there was only one possible explanation left, the cause that was singled out for study. Observational studies can be useful; after all they did play a large part in making the link between smoking and cancer, but one must work hard to eliminate the many uncontrolled factors as possible explanations for the observed association. Randomized experiments make it much easier to draw causal conclusions.

The data in the Gilbert trial was observational. To make it an experiment, Gilbert's presence on the ward would have had to be assigned using a chance device to decide which shifts she worked. Because there was no experiment, the tiny p-value, though it ruled out chance as an explanation for the excess deaths, did not rule out other possible explanations. In his ruling, Judge Ponsor gave a hypothetical example that would produce similarly damaging statistical evidence: Suppose that on a shift when Gilbert was present, a boiler had accidentally exploded and killed several dozen patients. Such an accident could lead to a tiny p-value showing that Gilbert's presence was associated with a very high number of deaths, but it would not be evidence of her guilt. As statisticians often say, "Association is not causation." The temptation, especially when the evidence of association is strong and there is a plausible explanation, is to conclude that the test provides evidence that the explanation is right. With observational data, such a conclusion would be based on false logic.

Point Three: The Prosecutor's Fallacy

The Cobb report pointed out a second, closely related temptation that is present with hypothesis testing. The *p*-value is a conditional probability, computed by assuming that a result is due to chancelike variation. It summarizes logic that goes as follows: "If the cause is just random variation, then the extreme result is very unlikely. We got an extreme result. Therefore, it is not reasonable to think that random variation is the cause." Notice that this logic says nothing about other causes. If there was a boiler explosion, for example, the extreme result would not be at all surprising.

Now look at how slippery the logic can get if you're not careful: "Suppose Gilbert is not guilty, and that the deaths behave in a chancelike way, like coin tosses. Then the probability is less than one out of 100 million that you would see so may excess deaths on Gilbert's shifts." (Correct.) It's a quick jump to the following shorter version: "If Gilbert is innocent, then it would be almost impossible to get so many excess deaths." (Also correct.) And then, "With this many excess deaths, the chance is less than one in 100 million that Gilbert is innocent." (*Not* valid.) This kind of "reasoning" is so tempting, and so common, that it has become known to statisticians as the prosecutor's fallacy. Because the false logic beckons so seductively, it is often used as the basis for arguing, as the Cobb report did, that the statistical evidence was likely to be misinterpreted by the jury in a way that favored the prosecution and was therefore prejudicial.

CONCLUSION

Judge Ponsor ruled that the statistical evidence should not be allowed at trial. Nevertheless, the other, nonstatistical evidence proved to be enough to convince the jury, and after many days of deliberation, Gilbert was convicted on three counts of first-degree murder, one count of second-degree murder, and two counts of attempted murder. After a penalty phase of the trial, the jury voted 8-4 for a death sentence, and because the vote was not unanimous, Gilbert's life was spared. She is now serving a sentence of life in prison without possibility of parole.

The statistical analysis that uncovered the pattern linking Gilbert's presence to the excess deaths was an essential part of the process that brought her to justice. The two juries that Gilbert faced, and their different roles in our system of justice, illustrate neatly the proper interpretation of hypothesis testing. First, a small *p*-value *does* allow you to rule out chancelike variability as a plausible explanation for an observed pattern. It tells you that the observed pattern is so extreme as to qualify as a surprise in the eyes of science. Second, if your data are observational, a small *p*-value does *not* tell you what has *caused* the surprise. Association is not causation. Inferences about cause are much more straightforward with a randomized experiment.

ADDITIONAL READINGS

Cameron, J. B. (2001). "Gilbert Guilty of Four Murders." *Northampton (MA) Daily Hampshire Gazette*, March 15, 2001.

DeGroot, M. H., S. E. Fienberg, and J. B. Kadane. (1994). *Statistics and the Law*. New York: John Wiley.

Finkelstein, M. O. (2001). *Statistics for Lawyers*, 2nd ed. New York: Springer-Verlag.

Gastwirth, J. L. (Ed.). (2000). *Statistical Science in the Courtroom*. New York: Springer-Verlag.

———. (1998). *Statistical Reasoning in Law and Public Policy*. San Diego: Academic Press.

Good, P. I. (2001). *Applying Statistics in the Courtroom*. Boca Raton, Fla.: CRC Press.

Zeisel, H., and D. H. Kaye. (1997). *Prove It with Figures*. New York: Springer-Verlag.

QUESTIONS

1. Explain why Figure 1 (along with the shift pattern of Ms. Gilbert) suggests that Ms. Gilbert may be guilty of excess deaths on the medical ward.

2. Why was the evidence from Figure 1 (along with the shift pattern of Ms. Gilbert) not conclusive evidence that Ms. Gilbert was guilty of the excess deaths? Suggest an explanation that could have caused the association without the unusual activity of Ms. Gilbert.

3. What is the relevance of the coin-tossing story to the trial of Ms. Gilbert?

4. Cobb argued that a jury would likely fall into the "prosecutor's fallacy." What is it, and why was the defense concerned about it?

THE ANATOMY OF A
PREELECTION POLL

EDWARD C. RATLEDGE

University of Delaware

———————— ❋ : : ❋ ————————

Everyone loves a race. The race may involve stock cars at the Daytona 500, thoroughbreds at the Kentucky Derby, athletes at the Olympics, or teams chasing a championship. Many in this country favor the political race, a race in which people are observers and participants at the same time.

Each of these races almost always ends with a winner. However, many exciting events, both positive and negative depending on your point of view, occur along the way to the finish line. In addition, progress is measured along the way. Progress may mean elapsed time, win-loss records, or lengths ahead of pursuers. In races other than those settled on Election Day, progress is readily observed on television (live or prerecorded) or routinely reported in your favorite morning newspaper. While there are certainly many opportunities to observe political races as they unfold, the observations will not produce accurate assessments of each candidate's likely performance on Election Day. The political poll, if well executed, provides that information. It joins the stopwatch at the racetrack and the scorekeeper at the ballpark in measuring the progress of the race.

An election, though usually categorized as a race, is more like a tug-of-war between the two major parties. The eligible voters from each party are of equal strength and may stand on the sidelines or join the fray. They are also able to return to the sidelines or even switch sides at will. They also may support a third-party candidate. Each candidate hopes that more than 50% of those willing to take part will choose his or her team. The preelection poll attempts to accurately measure the preferences of the voters as the battles of personalities and ideas are waged.

There are two types of polls conducted prior to an election. Private polls are taken by each candidate's organization. The results will influence the candidate's strategy and influence the decision as to whether the candidacy is even viable. Details of the polls are rarely made public, although you may hear a candidate say, "Our polls show we are in good shape." Public polls are provided by news media and other organizations as a public service to provide an overview of the races that otherwise would be missing or distorted by public statements about private polls (and to sell newspapers). In the final analysis, the quality of the information provided by these public sources depends on how well statistical design and analysis were utilized.

The roles statisticians play in an election poll, which are described in the remainder of this essay, are briefly summarized here. To accurately assess the current preferences of the people who will vote on Election Day, statisticians must first identify the likely voters. Second, because there is not time, money, or need to talk to each and every likely voter, statisticians must decide how many potential voters need to be contacted. Third, statisticians need to decide how much effort will be made to reach each person in the sample. Fourth, statisticians must ask questions that accurately assess both the likelihood of voting and the voters' preferences among the candidates. All of these decisions will influence the accuracy of the poll.

WHO WILL VOTE

A great deal of care is taken in deciding whose opinion is to be measured. Early in the election cycle, opinions of all adults may be of particular importance in influencing the political dialogue. Later, during the year of the election, statisticians and pollsters narrow their target to the adults who are likely to vote.

The objective of most public political polls is to measure the voters' preferences among the candidates at a point in time—that is, if the election were held today. Because statisticians are interested in people who will actually vote, the actual voters must be sifted from those who are eligible to vote. (The U.S. Census Bureau reported that 60% of citizens voted in 2000. In the 1998 congressional election 42% of citizens voted.) The more accurately the actual voters can be identified, the closer the poll results will approximate Election Day results.

No one can determine precisely who will go to the polls on Election Day. Voter participation is influenced by factors such as the weather, quality of the candidates, activity of the political parties, and a variety of other factors too numerous to mention. Statisticians want to narrow the focus to exclude as many nonvoters as possible. At the same time, the possibility of excluding potential voters must be minimized. Ideally, all voters should have an equal chance of being selected in the sample. The success of this process may substantially impact the accuracy of the poll. Several avenues are available to statisticians.

1. *All adults.* First, all adults could be polled. This approach may yield interesting information about the preferences of the population in general, but it is unlikely to accurately represent the preferences of actual voters on Election Day unless voters and nonvoters are very much alike.

2. *Registered voters.* Second, lists can be obtained that show individuals who are currently registered to vote. (Most, if not all, state election commissioners routinely provide registration lists to political parties and news organizations. They also sell the information in a variety of forms to individuals and other organizations.) Some statisticians use this strategy, especially as Election Day nears and registrations are nearly complete. The major difficulty is identifying the telephone numbers for the registered voters. Some recent research suggests that surveys of registered voters may be more accurate than those that search for likely voters later in the election cycle. While this approach will clearly exclude unregistered adults, the percentage of registered voters who will participate in the election can be quite high for some types of elections. According to the U.S. Bureau of Census, between 85% and 90% of registered voters will participate in presidential elections.

3. *Voters in prior elections.* Third, registration lists will usually indicate who participated in the last several elections. If so,

another strategy is to poll only those who voted in the last election or perhaps the last presidential election. People who have moved into the state or have only recently become of voting age will be excluded. Surely at least some of those people will vote, as will some of those who didn't vote in the previous election.

4. *Complex screening.* Fourth, more complex screening questions can be used in the poll. This approach begins with all adults and asks questions about voting, voting history, registration status, and their plans for voting in the coming election in order to identify likely voters. The way the screening questions are designed and delivered varies widely between polling organizations, and each feels that its method is the best.

One thing is certain. Polls of adults, registered voters, and likely voters are all of interest for different reasons but are not equivalent. When looking at the results of a poll in the newspaper or viewing them on television, an individual should make sure to note which population was polled.

HOW MANY INTERVIEWS

Some people discount poll results because "they never call me." Others say "they call me every night at dinner time." Still others complain that 1200 people could not possibly represent the opinions of 150 million individuals. All of these are relevant issues when deciding the sample size and how to contact that sample.

When poll results are reported in the newspaper or on television, a statement regarding the poll's margin of error ($\pm X$ percent) will accompany the results. In general, the margin of error tends to be between 3% and 5%. The margins of error reported in the organization's report reflect only a particular kind of error called sampling error. If statisticians drew several samples of the same size during the same time period and used identical interviewing procedures, the results for the samples would never be identical. Assuming the polls were done perfectly, variations between the samples would arise from sampling error. Statisticians control the amount of sampling error in polls by choosing appropriate sample sizes that are large enough to make the results appear credible.

To arrive at an appropriate sample size for the new poll, statisticians need to employ one of the many weapons in the statistical arsenal: the well-known binomial distribution. We all have flipped a coin or two in our lives and know that there are only two possible outcomes: heads or tails. (We will exclude the possibility that it lands on its edge.) If you flip the coin 10 times and count the number of heads that come up, the proportion of heads out of the 10 flips can be calculated. If the coin is flipped 10 times on several occasions, sometimes five heads will occur and other times four or six heads will come up. On rare occasions none will appear. Assuming it is a "fair coin"—that is, not one employed by a magician—the sampling distribution for the number of heads obtained will follow the binomial. This gives statisticians a reliable way of measuring sampling error.

This is very much akin to the problem statisticians face with polls where the proportion of adults voting for Candidate A is the desired result. Assuming that the sample is what statisticians call random—that is, one in which each potential voter has the same chance of inclusion in the sample—then the number of voters in the sample planning to vote for Candidate A behaves according to the binomial distribution. When the sample size is large, as it is in today's polls, the binomial distribution approaches an even more familiar normal distribution.

To use these statistical tools in setting the appropriate sample size, statisticians look at the races in which they are interested and try to determine the extent to which voters are divided. Polls taken early in the race will also provide evidence of this split. The importance of this decision is illustrated in Figure 1. If likely voters were split 50-50 between two candidates, statisticians would use a sample of more than 1067 to achieve a margin of error of ±3%. If, on the other hand, the voters were split 90-10, then a sample size of about 384 would achieve the same level of accuracy. As the variability decreases moving from left to right on the curve, the sample size also decreases. However, as the desired margin of error decreases from the bottom curve to the top curve, sample sizes increase dramatically. This has a profound effect on the time and budget required for polling. Thus, statistical design techniques are crucial to both statisticians and the clients.

Once the design of a poll is completed, the sample must be drawn. At the outset statisticians realize that every individual of voting age in the United States (excluding institutionalized people such

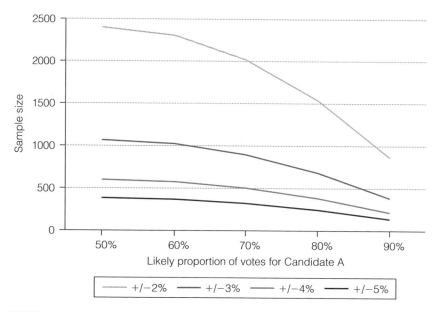

FIGURE 1 Sample size, margin of error, and variability

Source: Center for Applied Demography & Survey Research, University of Delaware

as prisoners) should have an equal chance of being selected. "Likely voters" will be a subset of that population. Most preelection polls are still conducted by telephone, although experiments with web surveys surfaced during the 2000 election. About 95% of the population has a telephone at home. This excludes households without a telephone or those that have only cell phones (2% don't have access to a landline) from the sample and thus represents a compromise with the principle of equal chance of selection.

Since many households have unlisted numbers, drawing the sample from databases of listed numbers would further compromise the sample. To avoid most of the problem, pollsters create telephone numbers that are a combination of the first five digits of numbers found in the database and two randomly generated digits. This process is called random digit dialing and is guaranteed to include (and irritate) some people with unlisted numbers.

Some households contain only a single adult, which does not pose a problem. What happens to those that contain two or more adults? Should the interview be taken from the person who answers the telephone? The approach generally taken is to find out how many adults are in the household and randomly select one for the interview. There are several methods to achieve this, but all aim to give each voter an equal chance of being selected.

From this discussion, it should be clear that there are many challenges that confront statisticians in achieving a sample that reasonably reflects the views of that elusive universe of likely voters. All of these issues need to be weighed when statements about the margin of error are made.

CONDUCTING THE POLL

Statisticians are now armed with their statistical design and an exquisitely drawn sample of telephone numbers (including households with listed and unlisted numbers, disconnected numbers, modems, and government offices). By this time they will have already prepared the survey instrument (questionnaire) for use by the interviewers. This instrument will almost certainly contain more questions than "If the election were held today, would you vote for candidate Jones or candidate Smith?" In addition to that type of question, there likely will be questions about the current condition of the country and whether one views each candidate positively or negatively. If the candidate is an officeholder, questions about job performance will be included.

More important, questions designed to find out if one is a "likely voter" will be asked. Some of those questions will be direct and others will be indirect. As was mentioned earlier, the scheme for finding this crucial subset of people will vary from statistician to statistician. Finally, questions about one's age, race, religion, income, state/county of residence, and an assortment of other demographics will be gathered so that the data can be sliced and diced in a variety of ways. Preelection polls are always seen through a prism, and there is always some facet that can be found to produce a result favorable to almost every candidate.

Response Rate

One of the more contentious aspects of preelection polling is the amount of effort expended by statisticians to reach the selected sample. Most people are not sitting by their telephones waiting for an interviewer to call. They are busy and are becoming busier as life becomes more hectic. Statisticians must decide when the sample household is called and how many times it is called in an attempt to complete an interview. The constraints are time and money. These constraints may well affect the desire to give every voter an equal chance of being part of the completed sample.

Most preelection polls are completed within two or three days. During that time fewer than five attempts will usually be made to contact someone at a particular telephone number. Calls will generally be attempted in both the daytime and the evening. Some polls are conducted on Sundays and Mondays to improve the chance of reaching those who have unusual work hours. Additional telephone numbers are then added to the original sample to replace those that were not reached in the allowed time.

Non-Sampling Error

The sampling error mentioned above is not the only source of error in polls! In addition to those households that were "never reached," there are some people who simply will not answer the statistician's questions regardless of the interviewer's professionalism. In fact, it may be hard to believe, but there are even a few who threaten to call the police or hurl a few choice epithets. These people are nicely classified as "refusals." Since statisticians will never know the opinions of the "never reached" or the "refusals," they are not represented in the sample and further compromise the equal chance of selection into the sample. Now, it may be that they are exactly like those people who have participated and the sample is not "biased" in some way. It is one more risk that statisticians and candidates take when they draw conclusions about the entire population from the sample. Short "overnight" polls tend to reach less than 30% of the targeted sample, while slightly longer polls may reach up to 40%.

Academic organizations engaged in preelection polling have the luxury of extending a poll over several weeks to reduce the "never reached" and "refused" households. This is accomplished by making as many as 15 calls at a wide variety of times and days of the week to reach someone. In addition, "refused" households may be called again in an attempt to obtain the interview. This usually results in more of the original sample being completed. Since the time period is extended, the timeliness of the poll is sacrificed and trends may not be as obvious. These polls will generally reach 60% or more of the original sample. While the statistician still does not know if the non-participants are substantively different from the respondents, the chances are probably less than with polls reaching 30% of the original sample. This issue is still being researched, and many would contend that increases in the response rate do not necessarily improve the estimates that much. True nonresponders are difficult, if not impossible, to coax into the study and are likely to remain recalcitrant independent of the methodology used.

Additional sources of non-sampling error include inadequate or inappropriate sampling frames, poorly designed questions, and faulty survey execution.

PREELECTION POLL EXAMPLES

This section contains four examples, each describing the result of a series of polls of Delaware voters for a year 2000 election. The four examples illustrate different features of polling.

Statistical Dead Heat: 2000 Delaware Senate Election

A "dead heat" in an election poll occurs when two candidates receive the same number of votes. When the difference in votes between candidates is less than the margin of error of the poll, news anchors frequently proclaim that the race is a "statistical dead heat." A statistical dead heat is not a tie; it simply indicates that the level of confidence in the observed polling results is not high enough to be certain which candidate would win if the voting occurred today. Even though a statistical dead heat is ambiguous about the winner, the outcome of Candidate A receiving more votes than Candidate B is more likely when the voters truly favor Candidate A and less likely when the voters truly favor Candidate B. Thus, it is better to lead than to follow.

An example of a statistical dead heat occurred in a U.S. Senate race in 2000, when two of Delaware's political heavyweights went head-to-head for a critical seat that had an impact on control of the U.S. Senate. They were the two-term incumbent governor Tom Carper and the long-time incumbent senator Bill Roth. Both were widely known and respected in this small state. Also, they usually won their offices by large margins. Few would hazard a guess as to who would win.

	Feb 25	May 17	Jul 29	Sep 5	Sep 20	Oct 24
Likely Voters	466	336	365	382	335	651
Margin of Error	4.5%	5.2%	5.0%	4.9%	5.2%	3.8%

TABLE 1 Poll Details

Source: Center for Applied Demography & Survey Research, University of Delaware

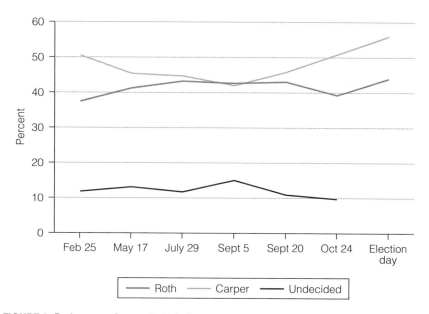

FIGURE 2 Roth versus Carper for U.S. Senate

Source: Center for Applied Demography & Survey Research, University of Delaware

In 2000, the Center for Applied Demography and Survey Research of the University of Delaware conducted six preelection polls as a public service. The studies used random digit dialing, random selection of the respondent among eligible respondents, and up to 15 callbacks. On average the polls were conducted during a 10-day period. The date of the polls reported represents the center of the interviewing period. The sample sizes and estimated margins of error are found in Table 1.

Figure 2 shows the results from six different polls taken during 2000. (The estimates shown are simply the proportion of likely voters that selected a particular candidate or remained undecided.)

With the exception of the first and last polls, the two veterans were in what the local papers called a "statistical dead heat." And aside from the September 5 poll, however, Governor Carper was always ahead, albeit slightly. The most interesting result was between the September 20 and the October 24 polls, when Carper's lead jumped from 3% to 11%. This shift was not due to a change in the undecided voters. Apparently, 4% of those supporting Senator Roth had switched to Governor Carper. Some attribute the drop to two health-related campaign incidents in the late stages of the race that made Senator Roth's age (79) a more significant factor. An alternative explanation from the Carper campaign suggests it was an effective media campaign focused on independent voters during the late stages of the race. Preelection polls are not designed to show that a particular action "caused" a change in the polls results. For example, Senator Roth gained slightly (2%) among Republican voters and lost support slightly (3%) among Democratic voters. Both changes were within the margin of error (which is larger for subgroups of voters due to the correspondingly smaller number of voters from a subgroup represented in the poll). In contrast, support among independent voters fell by 15%. At the same time, support for Senator Roth among voters with ages under 65 decreased by 7%, while support among voters aged 65 and older increased by 11%. In other words, there is evidence to support either of these "causes" for the change in voter preferences. On Election Day ("the only poll that counts") the actual results for the U.S. Senate race mirrored those found in the poll (Figure 2).

Heavily Contested Primary Election:
2000 Delaware Governor Election

The Delaware governor's race in 2000 illustrates the impact of a heavily contested primary election on an ensuing election. In the race for governor, incumbent lieutenant governor Ruth Ann Minner squared off against John Burris, who was a well-known public figure in both business and political circles, although not then a current officeholder. Burris, the Republican challenger, also had to contend with Judge William Swain Lee in a heavily contested primary, which was not settled until September. The primary complicated the polling since respondents had to deal with two potential opponents for Lieutenant Governor Minner. In addition, the undecided voters

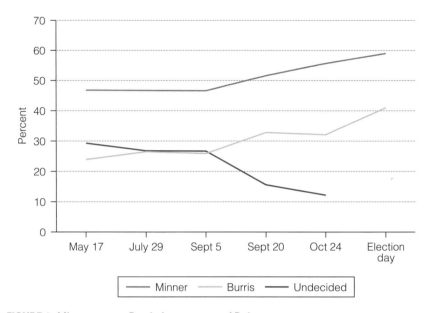

FIGURE 3 Minner versus Burris for governor of Delaware

Source: Center for Applied Demography & Survey Research, University of Delaware

were substantially greater than those observed in the senatorial race. In spite of this, the polls showed Lieutenant Governor Minner maintaining about a 20-point lead throughout the race (see Figure 3), and she won by 18 points on Election Day to become Delaware's first female governor.

Relatively Unknown Candidates:
2000 Delaware Lieutenant Governor Election

Both Delaware races for senate and governor involved well-known individuals, and the undecided vote at the end was close to 10%. How will estimates by the poll compare to actual results when the combatants are relatively unknown? The Delaware race for lieutenant governor pitted John Carney (D) against Dennis Rochford (R). While both men were politically active, their name recognition was relatively low. As a result, the pool of undecided voters was larger than that for either candidate in all but the last poll.

Figure 4 shows that Carney was leading among likely voters that had an opinion on the race. It also shows a widening gap between the

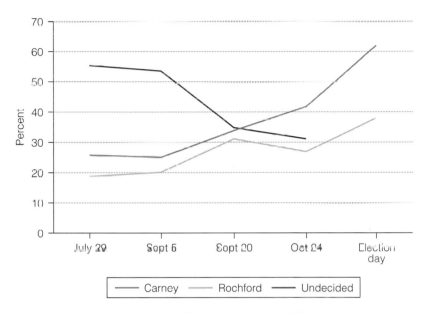

FIGURE 4 Carney versus Rochford for lieutenant governor of Delaware

Source: Center for Applied Demography & Survey Research, University of Delaware

two candidates two weeks before the election. Still, 31% of the voters were undecided. Carney received 62% of the vote on Election Day. He also was the preferred candidate for 61% of those who expressed a preference in the last preelection poll. Even with nearly one-third of the sample undecided, the poll could reasonably predict the likely winner. Carney consistently had more than 52% of those voters having a preference through all the polls.

Change in Front-runner: 2000 U.S. Presidential Election

The 2000 presidential race in Delaware between George W. Bush and Albert Gore provides an interesting contrast to those presented thus far. In each of the other races, the front-runner won. This was not the case in Bush v. Gore, as is shown in Figure 5. During the early going, George Bush was ahead in this poll and in the national polls. The chart shows a major shift during August. A successful Democratic convention that included a solid performance by Gore is generally given credit for this major shift. Bush narrowed the gap in the national polls but failed to regain his earlier advantage in Delaware. In the final vote, almost the entire undecided vote went to Gore.

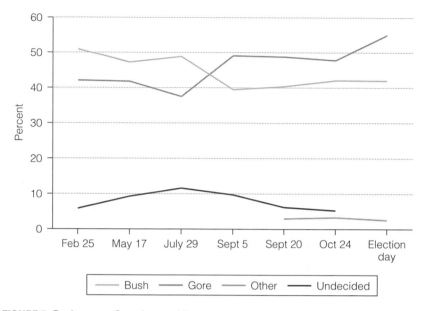

FIGURE 5 Bush versus Gore for president

Source: Center for Applied Demography & Survey Research, University of Delaware

Other Uses of Polling

The candidate preference issue is only one of the many interesting topics that are explored with preelection polls. Geographic differences across a state or the entire country can be fascinating. Certainly new polling data sparks interest and discussion (even arguments) within a broad spectrum of the general population including voters, nonvoters, and those not eligible to vote. Without a doubt, candidates would not leave home without them.

CONTRIBUTIONS OF STATISTICS

More than 327,000 Delawareans voted in the 2000 election and yet polls with sample sizes varying from 450 to 750 were able to provide a good snapshot of the race. More important, the actual result in each of the three regional races was reasonably close to that of the final poll. Since election polls are one of the few types of opinion research where an actual result is available to compare with poll results, statisticians must be quite confident in their methods and techniques if they are to avoid embarrassment. During election season, there will be occasions when more than one poll within the same time period will cover a particular race. If the results are substantially different, and they can be if the methods vary significantly, there will be much shouting and finger-pointing. Fortunately, those who use the concepts and tools provided by statistics are well served by the decades of research and practice that formed this discipline.

It should now be apparent that in election polling, several issues are paramount. First, defining and locating the population of interest (adults, registered voters, likely voters) is of the utmost importance. This is easier said than done and has required great skill and ingenuity. Second, a design must be used that will allow statisticians to answer the questions of interest with a precision that is appropriate. Third, a sample of sufficient size to meet the requirements of the design must be drawn. It must be drawn in a way that preserves adherence to the proposition that everyone in the population of interest has an equal chance of inclusion. Fourth, an instrument must be developed that can elicit reliable responses from the participants and even identify those who are being deceptive. Fifth, as many individuals of the original sample as is humanly possible must be interviewed to provide a true picture of the population from which they were drawn. Sixth, the data must be analyzed and reported in a way that accurately represents the views of those interviewed. Understanding the role of statistical methods in each task allows consumers to judge the strength and limitations of a particular poll.

REFERENCES

Green, D. P., and A. S. Gerber. (2003). "Enough Already with Random Digit Dialing: Using Registration-Based Sampling to Improve Pre-Election Polling." Paper presented at the American Association for Public Opinion Research annual meeting, Nashville, Tenn., May 2003.

Keeter, Scott, et al. (2000). "Consequences of Reducing Non-response in a National Telephone Survey." *Public Opinion Quarterly* 64, no. 2 (Summer): 125–48.

U.S. Bureau of Census. (2002). "Voting and Registration in the Election of 2000." February. Washington, D.C.

QUESTIONS

1. What are the different roles of private polls and public polls?

2. Why is it desirable to have each voter have an equal chance of being included in the sample? Why is this difficult to achieve in practice?

3. If the rolls of early-registered voters contain 90% of ultimate voters, is a survey of the early-registered voters likely to be unbiased?

4. What kind of survey error can be controlled by using an adequate sample size? What kinds of errors cannot be controlled in this way?

5. Does the appropriate sample size depend on the population size in political polls? Explain.

6. Use Figure 1 to approximate the accuracy of a sample size of 500 in which the proportion of votes is actually split 20%–80%.

7. What assumptions are made for Figure 1 to accurately predict accuracy?

8. Why is it important to make repeated attempts to contact individuals selected for a telephone survey? How might the number of attempts to contact provide useful information for anticipating nonresponse bias?

9. In Table 1 what causes the "margin of error" to vary?

10. Is the margin of error suggested by Table 1, about 5 percentage points or less, consistent with the switch in percentages for Carper and Roth, from more that 13 percentage points in favor of Carper on February 25, to a slight favor of Roth on September 5?

11. Was the change noted in question 10 due to the emerging choice of the undecided voters?

12. What likely causes are there for the switch in poll results between September 5 and Election Day?

13. In the survey results portrayed in Figure 3, the undecided vote dropped markedly between September 5 and September 21. Did the undecided vote favor Burris or Minner?

14. In the survey results portrayed in Figure 4, the undecided vote dropped markedly between September 5 and September 21. Did the undecided vote favor Carney or Rochford?

COUNTING AND APPORTIONMENT

Foundations of America's Democracy

TOMMY WRIGHT

U.S. Census Bureau

GEORGE COBB

Mount Holyoke College

———— ❀ ∶ ∶ ❧ ————

My political ideal is democracy. Everyone should be respected as an individual, but no one idealized.   Albert Einstein

The 2002 headline ran, "Supreme Court to Weigh Utah Census Case Wednesday." At stake was a seat in the U.S. House of Representatives; a court victory would bring Utah an extra seat, raising that state's total from three to four seats. Because the total number of House seats has been fixed by Congress at 435 since 1911, Utah could gain a seat only if some other state were to lose one, and, according to the methods used for allocating seats, that other state would be North Carolina.

Why North Carolina? What rules does the Census Bureau use to allocate seats in the U.S. House of Representatives? The answer is a simple one. On November 15, 1941, President Franklin Roosevelt signed the bill for apportionment (following the 1940 census) to be based on the *method of equal proportions,* which has been used since. More recently, the U.S. Supreme Court has articulated the closely related principle of "one person, one vote" when speaking of the drawing of boundaries within the various states to determine which districts the various representatives will represent. Ideally, each state's share of the 435 representatives should equal its share of the population. In practice, however, compromises are unavoidable. What kind of compromises? Why are they "unavoidable"? To set the stage for answering these questions, here is an artificial example, deliberately created to exaggerate some of the compromises that our system makes in order to ensure that every state has at least one representative while keeping the total size of the House fixed at 435.

THE APPORTIONMENT CHALLENGE

Apportionment is the distribution or allocation of seats in the U.S. House of Representatives among the states based on population. Every 10 years in the United States, there is a reapportionment (i.e., reallocation) of these seats among the states reflecting changes in population counts as measured by the nationwide census.

To keep things simple to start with, imagine a miniature country, the Amalgamated States of America, with only three states, Arkebama, Minnessippi, and Washconsin. Suppose that instead of 435 seats in the House of Representatives, there are only six. Take a look at the year 2000 population figures in Table 1, and think about how many

State	Population (millions)	Percentage of Population	Number of Seats	Percentage of Seats
Arkebama	42	70%	4	66.67%
Minnessippi	15	25%	1	16.67%
Washconsin	3	5%	1	16.67%
Total	60	100%	6	100.00%

TABLE 1 Population (millions) and apportionment of six seats for the Amalgamated States of America

of the six seats you would award to each state. Then look at how many each state would get using the U.S. apportionment rules, and decide whether you think the result is reasonable.

In the year 2000, Arkebama had 70% of the country's population, Minnesippi had 25%, and Washconsin only 5%. According to the United States' apportionment system, Arkebama would get four seats, and the other two states would get one each. Arkebama's four seats is about 67% of the total, which is reasonably close to its 70% share of the population, but although Minnesippi has five times the population of Washconsin, it only gets one seat, the same as its tiny rival. The reason is that the rules (because of the Constitution) say that every state has to get at least one seat. If a tiny state gets overrepresented as a result, then some other state will be underrepresented.

To prepare yourself for a look at how the apportionment system works, try your hand at giving out seats based on your own sense of what is reasonable. Suppose that over the three decades from 2000 to 2030, the state populations change as shown in Figure 1.

Over the three decades from 2000 to 2030, although Arkebama remains the largest state, it steadily loses population to the two others. Washconsin increases its population by 6 million people between 2000 and 2010, and then adds another 6 million people between 2020 and 2030. Assume that the total number of seats remains fixed at six. At what points would you take a seat away from Arkebama? Which state, Minnesippi or Washconsin, should get the extra seat or seats?

In 2000, Minnesippi, with 25% of the population, has only $\frac{1}{6}$ ($\approx 16.67\%$) of the seats. The reason is Washconsin has to get a seat even though it has only 5% of the population. Because Washconsin's share of the seats is so much bigger than its share of the population, it can grow without gaining a new seat. It may seem odd, but between 2000 and 2010, Washconsin triples its share of the population, from 5% to 15%, while Minnesippi's population stays exactly the same, but *Minnesippi gains a seat and Washconsin does not* (Table 2). However, Washconsin's share of the seats is now approximately 16.67%, close to its 15% of the population. Minnesippi still has 25% of the population, but its share of the seats has gone up to 33%—much closer to its share of the population.

State	2000	2010	2020	2030
Arkebama	42	36	30	24
Minnessippi	15	15	21	21
Washconsin	3	9	9	15
Total	60	60	60	60

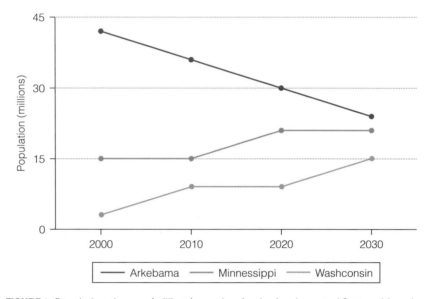

FIGURE 1 Population changes (millions) over time for the Amalgamated States of America

	2000		2010		2020		2030	
STATE	Percentage of Population	Number of Seats	Percentage of Population	Number of Seats	Percentage of Population	Number of Seats	Percentage of Population	Number of Seats
Arkebama	70%	4	60%	3	50%	3	40%	2
Minnessippi	25%	1	25%	2	35%	2	35%	2
Washconsin	5%	1	15%	1	15%	1	25%	2

TABLE 2 How the U.S. apportionment rule would change for the Amalgamated States of America

Over the next decade, from 2010 to 2020, Arkebama loses another 6 million people, and Minnessippi gains 6 million, but the distribution of seats doesn't change. In the third decade, from 2020 to 2030, Arkebama loses yet another 6 million people, Washconsin gains 6 million, and, as a result, Arkebama loses one seat to Washconsin (Table 2).

How are these numbers decided? What rule gives Minnessippi only one seat in 2000, and then increases its delegation to two seats in 2010, even though its share of the population hasn't changed? The rule, which has been used in every apportionment since 1940, is based on a particular way to measure how close an allocation of seats comes to the goal stated in our Constitution and the principle of one person, one vote. In what follows, we first give a quick overview of the history of the development of the U.S. form of representative democracy, then discuss a variety of strategies for allocating House seats, describe the rule actually used, and explain its relation to the principle of one person, one vote.

APPORTIONMENT AND THE U.S. CONSTITUTION: A BRIEF HISTORY

Just about everyone who has attended an American elementary or high school has participated in a democracy, either in his or her classroom's government, a school club, or through a community group. Most have voted for a classroom leader or voted for or against an issue in a club. In these votes, each person has equal authority, and the wishes of the majority determine the final actions. Such democracies are "pure" or "direct" because the people actually represent and rule themselves, as did the ancient Athenians around the 400s B.C. When groups become large and issues become complex, direct ("pure") democracies are not practical. In such cases, one alternative is for the people to select representatives who lead and represent the people. This representative form of government, sometimes called a republic, is the type put forth in the Constitution of the United States.

Resisting growing British control, abuse, and heavier tax burdens, the thirteen British colonies sought freedom and declared independence from England in 1776. To sustain the new U.S. society and help its people engage in a broad range of cooperative activities in a way that might be better than was the case with the Articles of Confederation (approved by Congress 1777, with ratification completed 1781), the Constitutional Convention produced the Constitution in 1787 by consent of the people in all states.

It is hard to think of a document more basic to the United States' democracy than the Constitution. Perhaps more than the ability of the people to elect the president and vice president (but only through the Electoral College) and senators, it is the election of the members of the U.S. House of Representatives that most makes us think of our representative form of democracy. The two-year terms of the members of the U.S. House of Representatives intentionally help to keep them responsive to the people in the congressional districts they represent.

Article 1, Section 2, Clause 3 of the Constitution makes it clear that the power in the U.S. House of Representatives granted to a state is reflected in that state's number of representatives, or seats, which is determined every 10 years by that state's population as determined by a census.

> Representatives and direct taxes shall be apportioned among the several states which may be included within this Union, according to their respective numbers, which shall be determined by adding to the whole number of free persons, including those bound to service for a term of years, and excluding Indians not taxed, three fifths of all other persons. The actual enumeration shall be made within three years after the first meeting of the Congress of the United States, and within every subsequent term of ten years, in such manner as they shall by law direct. The number of Representatives shall not exceed one for every thirty thousand, but each state shall have at least one Representative; and until such enumeration shall be made, the state of New Hampshire shall be entitled to choose three, Massachusetts eight, Rhode Island and Providence Plantations one, Connecticut five, New York six, New Jersey four, Pennsylvania eight, Delaware one, Maryland six, Virginia ten, North Carolina five, South Carolina five, and Georgia three.

The Constitution set the size of the first U.S. House of Representatives at 65 seats. Since that time, the size of the House tended to increase as new states were admitted to the United States and as the population grew, as recorded by each decennial census, until the size was fixed by Congress in 1911 at 435 seats. Table 3 gives some details on the changes in House size following each census.

Year of Census	Number of States	Number of Seats
By Constitution	13	65
1790	15	105
1800	16	141
1810	23	186
1020	24	213
1830	24	240
1840	26	223
1850	31	234
1860	33	241
1870	37	292
1880	38	332
1890	42	357
1900	45	391
1910	46	435
1920	*	*
1930	48	435
1940	48	435
1950	48	435
1960	50	435
1970	50	435
1980	50	435
1990	50	435
2000	50	435

TABLE 3 Number of seats in the U.S. House of Representatives

Source: Anderson (2000), p. 115–63.

* No reapportionment

House seats were not reapportioned among the states following the 1920 census because Congress did not pass an apportionment bill. The census results showed more people in urban areas than rural areas for the first time. There were also arguments over the fairness of the method of allocating the number of seats among the states (Anderson, 1988, 2000).

Like no other activity, the decennial census, a huge data collection effort that has been conducted every 10 years since 1790, is the fundamental ingredient that keeps the United States' form of democracy going.

AN OVERVIEW OF CENSUS 2000

The enumeration described in the Constitution is now known as the U.S. Census of Population and Housing (or the U.S. Census for short). The Census Bureau attempts to count all people in the United States and identify their homes as of April 1 of each census year. Counting people and associating them with their place of residence on a map as of a specific date is not easy. Some of the challenges in counting a finite set of people are illustrated by the following thought experiment (Wright, 1998):

> Think of 10 people asked to count the number of persons at a local high school basketball game during halftime. Assume all 10 are given the same instructions and are told to work independently from each other. During halftime, spectators come and go—some leave, some get refreshments, some switch seats—and the players and coaches go to the locker rooms. The ticket count will not do, because some are admitted without tickets, and some who bought tickets do not show. The dynamics of the population of persons in attendance at halftime suggest that some may be counted twice (those who change seats), and even more might be missed (those who were not in their seats when that area of the gym was being counted). If the 10 counters truly conduct their counting independently, the result will almost certainly be 10 different counts. The fact that almost surely no two counters would get the same count or the *true count* is an illustration of *measurement error*. Just as the estimates of attendance at the basketball game contain measurement error, censuses of the United States contain measurement error.

The thought experiment shows that errors in counting are likely, but doesn't tell us how big those errors are likely to be. These errors in counting include the omission of people as well as duplication (counting some more than once). When the level of omission is greater than the level of duplication, the result is called net under-counting. By using records and estimation of births, deaths, immigration, and emigration, demographers have documented estimates of net undercounting in the decennial census since 1940 (Robinson,

Ahmed, Das Gupta, and Woodrow, 1993): 1940 (5.4%), 1950 (4.1%), 1960 (3.1%), 1970 (2.7%), 1980 (1.2%), and 1990 (1.8%). This means that the count from the census was estimated to be too low in each of these years. Many worried that results would worsen in 2000 unless there were fundamental changes in counting methodology for Census 2000. A particular concern was "differential net undercounting" among different subpopulations. For example, people who rent tend to be harder to track down than people who own their homes, and as a result, the census is more likely to miss people living in rented housing. There has also been continuing evidence of undercounting minority groups to varying degrees. This tendency to count some groups less completely than others is what we mean by differential net undercounting. Any differential net undercounting can lead to unequal representation in Congress and inequality in the annual distribution of billions in federal funds, which depend on the counts from the decennial census.

The Census Bureau's initial plan for Census 2000 was to conduct a so-called one-number census. As Wright (1999) notes:

> In a one-number census, the best possible single set of results based on (conventional) *counting, assignment,* and *estimation* is used to produce the most accurate census. *Counting* means the full array of techniques by which direct contact is made with all respondents (using mail, personal visit, telephone, or other means), and it also means data obtained by proxy for another household, housing unit, or person. Historically people have been added to the census by obtaining information about their existence from administrative records. . . . *Assignment* is the use of indirect evidence from administrative records to add people to a specific geographic location without field (direct contact) verification. For Census 2000, research had convinced the Census Bureau that administrative records did not currently exist that would adequately or reliably account for people without field verification, and assignment was not planned. *Estimation* is the application of statistical techniques (such as sampling) to account for people or units not directly counted or assigned.

Not everyone agreed with this one-number census approach, especially the use of the statistical method called sampling. Two simple principles behind sampling are relevant here. First, sometimes you can get the information you need without tracking down every last detail ("You don't have to eat the whole ox to know that the meat is tough"). Second, when time and money are limited, you may actually get more accurate information by being really careful and thorough with a well-designed sample of people than by working with a much larger group and having to settle for less careful follow-up of the people who don't respond.

However, Congress had expressed concern about the constitutionality of sampling, the possibility that the use of sampling and estimation would allow the data to be manipulated for political advantage, and the magnitude of the sampling error in small geographical areas such as block levels. In fact, the U.S. House of Representatives sued to stop the one-number census. Eventually the case reached the U.S. Supreme Court. Although the Court did not address the question of whether the use of sampling was allowed by the Constitution, nevertheless the Court did rule (5–4) that one of the laws passed by Congress, Section 195 of the Census Act (Title 13, U.S. Code), "prohibits the proposed uses of statistical sampling in calculating the population for purposes of apportionment."

Thus Census 2000 was conducted using conventional counting methods complemented by an extensive partnership program with local groups across the nation and a paid advertising campaign. Relative to April 1, 2000, the conventional counting methods were applied to people living in households (with residential addresses) and people living in group quarters (e.g., college dormitories, prisons, health-care facilities, etc.). For counting people living in households, the following major steps were taken:

1. A nationwide list of all residential addresses was developed and updated periodically.
2. A questionnaire was delivered (mostly by mail) to each address on the list.
3. A partial count of the population was made from the returned questionnaires.
4. By knocking on doors, additional counts resulted from a follow-up of all those addresses for which a questionnaire was not returned.

5. Additional operations resulted in more people being counted in households.

For counting people living in group quarters, two major steps were undertaken:

1. A nationwide listing of group quarters was developed.
2. A count of persons at each group quarters was obtained.

A summary of the resulting counts by enumeration method is given in Table 4 (Treat, 2004; Abramson, 2004; and Buckley-Ess and

Enumeration Method	Number of Persons
1. IN HOUSEHOLDS	
Self-administered questionnaires	
Paper	197,418,790
Internet	173,291
Be Counted forms	347,410
Interviewer-administered follow-up questionnaires	
Nonresponse follow-up	57,346,012
Coverage improvement follow-up	4,394,067
Coverage edit follow-up	7,231,591
Telephone questionnaire assistance	155,695
Interviewer-administered enumerate questionnaires	
Nonresponse follow-up adds	662,284
Coverage improvement follow-up adds	191,478
Update/enumerate	1,627,023
List/enumerate	559,800
Remote Alaska	55,232
T-Night	35,320
Other	
Unlinked enumerator continuation forms	4,126
Whole households substitutions	3,441,154
Subtotal for households	*273,643,273*
2. IN GROUP QUARTERS	
Subtotal for group quarters	*7,778,633*
Total resident population	**281,421,906**

TABLE 4 Resident population counted in Census 2000 by enumeration method

Hovland, 2004). Explanations for the enumeration methods are given in Appendix A.

Table 5 shows the states (and District of Columbia) ordered by apportionment population and resident population from Census 2000. The apportionment population from Census 2000 includes the resident population (citizens and noncitizens) of the 50 states and U.S. Armed Forces personnel and federal civilian employees stationed outside the United States (and their dependents living with them) that can be allocated, based on administrative records, back to a home state (U.S. Census Bureau. [2000]. *What You Should Know about the Apportionment Counts.* December. Washington, D.C.). Departments and agencies of the federal government with overseas employees provided counts by home state mostly from their administrative records. For Census 2000, the overseas counts for the 50 states came to

State	Apportionment Population (April 1, 2000)	Resident Population (April 1, 2000)
California (CA)	33,930,798	33,871,648
Texas (TX)	20,903,994	20,851,820
New York (NY)	19,004,973	18,976,457
Florida (FL)	16,028,890	15,982,378
Illinois (IL)	12,439,042	12,419,293
Pennsylvania (PA)	12,300,670	12,281,054
Ohio (OH)	11,374,540	11,353,140
Michigan (MI)	9,955,829	9,938,444
New Jersey (NJ)	8,424,354	8,414,350
Georgia (GA)	8,206,975	8,186,453
North Carolina (NC)	8,067,673	8,049,313
Virginia (VA)	7,100,702	7,078,515
Massachusetts (MA)	6,355,568	6,349,097
Indiana (IN)	6,090,782	6,080,485
Washington (WA)	5,908,684	5,894,121
Tennessee (TN)	5,700,037	5,689,283
Missouri (MO)	5,606,260	5,595,211
Wisconsin (WI)	5,371,210	5,363,675
Maryland (MD)	5,307,886	5,296,486

TABLE 5 Apportionment and resident populations of the states from Census 2000

State	Apportionment Population (April 1, 2000)	Resident Population (April 1, 2000)
Arizona (AZ)	5,140,683	5,130,632
Minnesota (MN)	4,925,670	4,919,479
Louisiana (LA)	4,480,271	4,468,976
Alabama (AL)	4,461,130	4,447,100
Colorado (CO)	4,311,882	4,301,261
Kentucky (KY)	4,049,431	4,041,769
South Carolina (SC)	4,025,061	4,012,012
Oklahoma (OK)	3,458,819	3,450,654
Oregon (OR)	3,428,543	3,421,399
Connecticut (CT)	3,409,535	3,405,565
Iowa (IA)	2,931,923	2,926,324
Mississippi (MS)	2,852,927	2,844,658
Kansas (KS)	2,693,824	2,688,418
Arkansas (AR)	2,679,733	2,673,400
Utah (UT)	2,236,714	2,233,169
Nevada (NV)	2,002,032	1,998,257
New Mexico (NM)	1,823,821	1,819,046
West Virginia (WV)	1,813,077	1,808,344
Nebraska (NE)	1,715,369	1,711,263
Idaho (ID)	1,297,274	1,293,953
Maine (ME)	1,277,731	1,274,923
New Hampshire (NH)	1,238,415	1,235,786
Hawaii (HI)	1,216,642	1,211,537
Rhode Island (RI)	1,049,662	1,048,319
Montana (MT)	905,316	902,195
Delaware (DE)	785,068	783,600
South Dakota (SD)	756,874	754,844
North Dakota (ND)	643,756	642,200
Alaska (AK)	628,933	626,932
Vermont (VT)	609,890	608,827
Wyoming (WY)	495,304	493,782
District of Columbia (DC)	[a]	572,059
Total	281,424,177	281,421,906

TABLE 5 (Cont.)

[a] District of Columbia's population is not considered for the apportionment of the U.S. House of Representatives.

574,330 people (U.S. Census Bureau. [2001]. *Census 2000 Informational Memorandum No. 90.* January 11. Washington, D.C.). The apportionment population, not the resident population, is used to distribute seats.

Based on the apportionment population numbers in Table 5, the 435 seats in the U.S. House of Representatives were apportioned to the 50 states using the same rules that you saw in action in the miniature example (the Amalgamated States of America). The apportionment of seats, like the phrase "one person, one vote," has a goal of fairness. As nearly as possible, every person in the United States should have the same representation in the House. Although the goal of fairness is basic to our form of representative democracy, and although the principle of one person, one vote sounds straightforward enough, putting it into practice turns out to be harder than you might think.

APPORTIONMENT: IMPLEMENTATION

Could "As Nearly as Possible" Mean Minimizing Errors One State at a Time?

Ideally, we want each state's share of the House seats to be proportional to that state's population. In other words, we choose the number of seats a state gets so that the number of people in each congressional district is as close as possible to the "target size":

Target size = (Total U.S. population)/(Total number of House seats)

= (281,424,177)/435

= 646,952.13 people per district or representative

There are two side conditions: (1) every state must have at least one seat, and (2) the total number of seats must equal exactly 435.

Achieving the goal turns out to be quite a challenge, first because the straightforward-sounding phrase "as nearly as possible" is hard to pin down with a precise meaning, and second, because common-sense approaches to finding the right numbers of seats don't work.

To see a major difficulty with the phrase "as nearly as possible," suppose we take the states one at a time, and choose for each state the number of seats that makes the size of each congressional district as

Number of Seats	People per District (Target Size = 646,952.13)	Error
1	1,238,415.00	591,462.87
2	610,207.50	−27,744.63
3	412,805.00	−234,147.13
4	309,603.75	−337,348.38
5	247,683.00	−399,269.13

Note: Population = 1,238,145. The district size comes as close as possible to the target size when New Hampshire gets two seats.

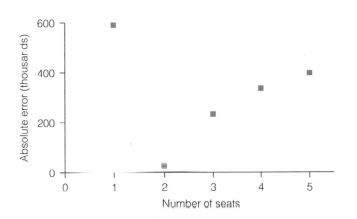

FIGURE 2 Choosing the number of representatives for New Hampshire, Census 2000

close as possible to the target size. Figure 2 shows how it would work for the state of New Hampshire.

According to the 2000 Census, New Hampshire had an apportionment population of about 1.2 million people, and the target size for a district was about 0.65 million people. Roughly, then, New Hampshire's population is about twice the target size, which suggests giving New Hampshire two seats. As you can see from Figure 2, two is in fact the number of seats that makes the (absolute) error as small as possible, at 27,745 people per district. Both of New Hampshire's seats have this error, so the total error for New Hampshire is about (2)(27,745) = 55,490.

Now consider how this approach works for the largest state, California (Figure 3).

Number of Seats	People per District (Target Size = 646,952.13)	Error
49	692,465.27	45,513.14
50	678,615.96	31,663.83
51	665,309.76	18,357.63
52	652,515.35	5,563.22
53	640,203.74	−6,748.39
54	628,348.11	−18,604.02
55	616,923.60	−30,028.53
56	605,907.11	−41,045.02

Note: Population = 33,930,798. The district size comes as close as possible to the target size when California gets 52 seats. The absolute error per district is only 5563.2, but all 52 districts have this error.

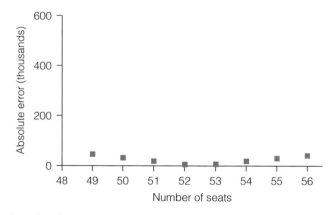

FIGURE 3 Choosing the number of representatives for California, Census 2000

California is the largest state, and the large population makes it possible to come fairly close to the target size. Notice that each time you increase the number of districts by one, the number of people per district goes down. With 50 districts, the size (people per district) is too high, at 678,616. The error (size minus target size) is 31,664. As the number of seats increases, the size decreases, until, at 52 seats, the size is just 5563.2 higher than the target size. Add one more seat, and the size is now below the target size, and the error is negative. Negative errors are allowed, but for 53 seats the absolute error is 6748.39, more than the error for 52 seats, so 52 seats gives the smallest possible absolute error. This error (5563.2) is the same for all 52 of California's seats, so the total error for California is approximately $(52)(5563.2) = 289,286.4$.

The tables for California and New Hampshire differ in a number of ways. First, because California is so much bigger, it gets many more seats, so the values along the horizontal axes are different for the two graphs. Second, having so many more seats makes it possible for California to come much closer to the target size than New Hampshire can, so the errors for New Hampshire (on the vertical axis) are much larger than for California, and the New Hampshire change in district size is much greater each time a seat is added than the change in district size when a seat is added for California. This difference in level of change reflects the fact that when there are few seats to begin with, adding just one more seat has a big effect on district size. One consequence is that the smallest achievable error of 27,745 for New Hampshire is almost five times as big as the smallest error for California. With a small state, you just don't have a lot of flexibility for getting close to the target size. However, because New Hampshire's error of 27,745 occurs for just two districts, the total error for the state is just $(2)(27,745) = 55,490$. For California, although the error for a single district is small, at 5563, that error gets repeated 52 times, making the total error for the state 289,286, or more than five times as big as for New Hampshire.

To conclude this section, take a look at Wyoming (see Figure 4), a state with only about half a million people. Wyoming's population is so small that giving the state just one district undershoots the target size by almost 151,648 people, and makes the error negative. For Wyoming, there is no way to obey the Constitution and make the error positive.

It would be straightforward to continue with the other 47 states in exactly the same way, choosing for each state the number of seats that gives the smallest possible absolute error. Although this might seem like a reasonable strategy—and for some purposes it would be—it has one shortcoming that prevents it from being used: we end up with too many seats. The Constitution says that even the smallest states must get a seat, and this means there aren't enough seats to ensure that all the other states come as close as possible to the target size. The Constitution also requires that the target size be at least 30,000.

Seats	People per District (Target Size = 646,952.13)	Error
1	495,304	−151,648.13
2	247,652	−399,300.13
3	165,101	−481,851.13

Note: Population = 495,304.

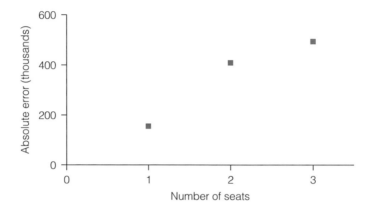

FIGURE 4 Choosing the number of representatives for Wyoming, Census 2000

The problem arises with this one-state-at-a-time method because each state is handled in isolation. The decision about California doesn't use any information about what will happen with New Hampshire or Wyoming. We need a method that deals with all the states at once.

"As Nearly as Possible" Means Minimizing the Sum of Squared Errors

If you've studied the standard deviation, you already know a way that statisticians use to deal with a collection of "errors" all at once: square them and add. The squaring turns negative errors, like the error for Wyoming, into positive values, and the adding gives a single number that you can use to measure how close an allocation comes to "as nearly as possible."

Returning to our simple example of the Amalgamated States of America, Table 6 shows how the sum of squared errors is computed

State	Population	Seats	Population/Seats	Error	(Error)²	(Seats)(Error)²
Arkebama	42	4	10.5	0.5	0.25	1
Minnessippi	15	1	15	5	25	25
Washconsin	3	1	3	−7	49	49
Total						75

TABLE 6 Computing the sum of squared errors for the Amalgamated States of America

APPORTIONMENT			
Arkebama	Minnessippi	Washconsin	Sum of Squared Errors
4	1	1	75.00
3	2	1	109.50
3	1	2	217.50
2	3	1	366.00
2	1	3	510.00
2	2	2	399.00
1	3	2	1243.50
1	2	3	1279.50
1	4	1	1229.25
1	1	4	1391.25

TABLE 7 Sum of squared errors for all possible allocations for the Amalgamated States of America

for the year 2000 for the allocation of four seats to Arkebama, and one seat each to Minnessippi and Washconsin.

The total population of 60 million and total of six seats gives a target size of 10 million people per district. For Arkebama, the four seats give a district size of 10.5 million people, with an error of 0.5 million people for each of the four seats. Each squared error is 0.25, and so Arkebama's four seats contribute $4(0.25) = 1$ to the sum of squared errors. For all six seats together, the sum of squared errors is 75.

Table 7 shows what you get if you apply the same computation to get the sums of squared errors for all possible apportionments for the three states; the one that minimizes the sum of squared errors assigns four seats to Arkebama, and one each to the other two states.

The sum of squared errors (SSE) is smallest for the apportionment that gives Arkebama four seats, and the other two states one each. The rule based on this meaning of "as nearly as possible" leads to the rule that has been used in the United States since 1940.

"As Nearly as Possible": Finding the Best Apportionment

In principle, the use of the sum of squared errors (SSE) solves the apportionment problem. All the Census Bureau has to do is make a complete list of all possible apportionments, as in Table 7, compute the SSE for each one, and pick the smallest. It sounds straightforward, but in practice, it turns out to be impossible, not because there is no smallest value but because it would take too long to list all possibilities in order to find the best one.

To see this, consider how the number of possible apportionments is related to the number of House seats. If the U.S. House of Representatives had exactly 50 seats, there would be only one apportionment: each state would get one seat. With a 51-seat House, there would be 50 possible apportionments: 49 states would each get one seat, one state would get two, and there would be 50 ways to choose the one state with two seats. Add just one more seat, to get a 52-seat House, and there are 1275 possible apportionments. With a 55-seat House, there are more than 3 million possibilities.

Suppose that you had a lightning-fast computer, faster than any now available, that could compute the SSE for a million different apportionments every second. With a 55-seat House it would take such a computer just three seconds to compute the SSEs for all possible apportionments. Increase the House size to 60 seats, and it would take the computer not three seconds but 17.45 hours. A 65-seat House would take five years, a 70-seat House more than 3000 years, and a 75-seat House more than a million years. If you could have started the computer 13.7 billion years ago, at the time of the Big Bang, by now it would only have finished enough computations for an 82-seat House.

How, then, does the apportionment for a 435-seat House get decided? Amazingly, it can be done on a hand calculator, although with a spreadsheet it would be a lot quicker and easier.

Here's a version of how it works, for the three-state example of the Amalgamated States of America. We have six seats to hand out, and each state must get at least one. Start by giving each state its one-seat minimum. That leaves three more seats to give out. Make a "priority table" as follows. There is one row for each state, and one column for each remaining House seat. Fill in the cells of the table with entries given by

Entry = (State population)/$\sqrt{[(\#seats)(\#seats + 1)]}$

The results are given in Table 8.

Because there are three remaining House seats to distribute, the easier version of the rule calls for locating the three largest entries (priority values) in the priority table and assigning an additional seat to a state each time one of these largest values is associated with that state. For this example, all three of the largest values are in Row 1, so all three remaining seats go to Arkebama.

Now consider the apportionment after the 2010 Census, using the population figures in Figure 1. Table 9 shows the new priority table.

This time, the three largest values are the two left-most values in Row 1 and the left-most value in Row 2: Arkebama gets two additional seats, and Minnessippi gets one.

	ADDITIONAL SEATS		
STATE	1	2	3
Arkebama	$42/\sqrt{(1)(2)} = 29.7$	$42/\sqrt{(2)(3)} = 17.1$	$42/\sqrt{(3)(4)} = 12.1$
Minnessippi	$15/\sqrt{(1)(2)} = 10.6$	$15/\sqrt{(2)(3)} = 6.1$	$15/\sqrt{(3)(4)} = 4.3$
Washconsin	$3/\sqrt{(1)(2)} = 2.1$	$3/\sqrt{(2)(3)} = 1.2$	$3/\sqrt{(3)(4)} = .87$

TABLE 8 Priority table for the Amalgamated States of America, Year 2000

Note: Each entry is a ratio whose numerator is the state's population, and the denominator is determined by the column number as $\sqrt{(\#seats)(\#seats + 1)}$. The three largest entries (priority values) are all in Row 1.

	ADDITIONAL SEATS		
STATE	1	2	3
Arkebama	$36/\sqrt{(1)(2)} = 25.5$	$36/\sqrt{(2)(3)} = 14.7$	$36/\sqrt{(3)(4)} = 10.4$
Minnessippi	$15/\sqrt{(1)(2)} = 10.6$	$15/\sqrt{(2)(3)} = 6.1$	$15/\sqrt{(3)(4)} = 4.3$
Washconsin	$9/\sqrt{(1)(2)} = 6.4$	$9/\sqrt{(2)(3)} = 3.7$	$9/\sqrt{(3)(4)} = 2.6$

TABLE 9 Priority table for the Amalgamated States of America, Year 2010

STATE	ADDITIONAL SEATS		
	1	2	3
Arkebama	$24/\sqrt{(1)(2)} = 17.0$	$24/\sqrt{(2)(3)} = 9.8$	$24/\sqrt{(3)(4)} = 6.9$
Minnessippi	$21/\sqrt{(1)(2)} = 14.8$	$21/\sqrt{(2)(3)} = 8.6$	$21/\sqrt{(3)(4)} = 6.1$
Washconsin	$15/\sqrt{(1)(2)} = 10.6$	$15/\sqrt{(2)(3)} = 6.1$	$15/\sqrt{(3)(4)} = 4.3$

TABLE 10 Priority table for the Amalgamated States of America, Year 2030

Finally, consider the 2030 Census, as shown in Table 10. For this census, each of the states gets one additional seat.

Our simple example of the Amalgamated States of America is small enough to do easily on a calculator, but even the real data for all 50 U.S. states can be done with just a calculator, although the population numbers are several digits long. The largest priority values will be those where the numerator is large (states with larger populations) and the denominator is small (that is, $\sqrt{(1)(2)}$, $\sqrt{(2)(3)}$, $\sqrt{(3)(4)}$, etc.).

Therefore, the official rule for the apportionment method used in the United States, known as the *method of equal proportions* (for some historical background and development dating back to 1911, see Wright, 2004), can be given in the following four steps:

Step 1: Assign one representative to each state.

Step 2: Order the states from largest to smallest according to apportionment population, and represent this ordering by $P_1, P_2, P_3, \ldots, P_{50}$.

Step 3: Compute the priority table.

$$P_1/\sqrt{(1)(2)} \quad P_1/\sqrt{(2)(3)} \quad P_1/\sqrt{(3)(4)} \ldots$$
$$P_2/\sqrt{(1)(2)} \quad P_2/\sqrt{(2)(3)} \quad P_2/\sqrt{(3)(4)} \ldots$$
$$\vdots \qquad \qquad \vdots \qquad \qquad \vdots$$
$$P_{50}/\sqrt{(1)(2)} \quad P_{50}/\sqrt{(2)(3)} \quad P_{50}/\sqrt{(3)(4)} \ldots$$

Step 4: Pick the 385 ($= 435 - 50$) largest priority values from the table in Step 3 along with the associated states. Each state gets an additional seat each time one of its priority values is among the 385 largest values.

It may be tedious, but it's not really hard, certainly not a problem to keep a supercomputer busy for billions of years. Table 11 shows the actual sequential assignment of the 51st through 435th seats in the U.S. House of Representatives based on Census 2000 along with priority values.

Seat	State	Priority Value	Seat	State	Priority Value	Seat	State	Priority Value
51	CA	23,992,697	78	CA	4,534,194	105	CO	3,048,961
52	TX	14,781,356	79	MA	4,494,065	106	CA	2,953,297
53	CA	13,852,190	80	IN	4,306,833	107	NY	2,932,531
54	NY	13,438,545	81	NY	4,249,641	108	FL	2,926,462
55	FL	11,334,137	82	WA	4,178,071	109	VA	2,898,849
56	CA	9,794,978	83	MI	4,064,450	110	MI	2,874,000
57	IL	8,795,731	84	TN	4,030,535	111	KY	2,863,380
58	PA	8,697,887	85	CA	3,998,783	112	SC	2,846,148
59	TX	8,534,020	86	MO	3,964,224	113	TX	2,793,414
60	OH	8,043,014	87	TX	3,816,530	114	IL	2,781,454
61	NY	7,758,748	88	WI	3,798,019	115	PA	2,750,513
62	CA	7,587,157	89	MD	3,753,242	116	CA	2,716,638
63	MI	7,039,834	90	AZ	3,635,012	117	MA	2,594,650
64	FL	6,543,767	91	IL	3,590,842	118	OH	2,543,424
65	CA	6,194,888	92	FL	3,584,169	119	NY	2,539,646
66	TX	6,034,463	93	CA	3,576,620	120	CA	2,515,118
67	NJ	5,956,918	94	PA	3,550,898	121	IN	2,486,551
68	GA	5,803,208	95	MN	3,482,975	122	FL	2,473,311
69	NC	5,704,706	96	NY	3,469,817	123	TX	2,463,559
70	NY	5,486,263	97	NJ	3,439,228	124	OK	2,445,754
71	CA	5,235,636	98	GA	3,350,484	125	NJ	2,431,902
72	IL	5,078,218	99	NC	3,293,614	126	OR	2,424,346
73	PA	5,021,727	100	OH	3,283,547	127	WA	2,412,210
74	VA	5,020,955	101	CA	3,235,175	128	CT	2,410,905
75	TX	4,674,275	102	TX	3,225,556	129	GA	2,369,150
76	OH	4,643,637	103	LA	3,168,030	130	CA	2,341,448
77	FL	4,627,142	104	AL	3,154,495	131	NC	2,328,937

TABLE 11 Ordered priority values and associated states from Census 2000

Seat	State	Priority Value	Seat	State	Priority Value	Seat	State	Priority Value
132	TN	2,327,030	167	NC	1,803,987	202	MN	1,421,918
133	MO	2,288,746	168	CO	1,760,318	203	MA	1,421,148
134	IL	2,271,048	169	IN	1,758,257	204	NV	1,415,650
135	PA	2,245,785	170	OH	1,755,130	205	OK	1,412,057
136	NY	2,239,758	171	CA	1,740,613	206	NY	1,408,742
137	MI	2,226,191	172	WA	1,705,690	207	OR	1,399,697
138	TX	2,203,474	173	FL	1,689,593	208	FL	1,395,136
139	WI	2,192,787	174	TX	1,673,659	209	CT	1,391,937
140	CA	2,190,224	175	IL	1,662,237	210	CA	1,385,219
141	MD	2,166,935	176	CA	1,655,653	211	IN	1,361,940
142	FL	2,141,951	177	NY	1,654,171	212	TX	1,349,347
143	AZ	2,098,675	178	KY	1,653,173	213	OH	1,340,502
144	OH	2,076,697	179	TN	1,645,459	214	CA	1,330,875
145	IA	2,073,183	180	PA	1,643,746	215	MI	1,330,404
146	CA	2,057,357	181	SC	1,643,224	216	WA	1,321,222
147	VA	2,049,796	182	MO	1,618,388	217	NY	1,311,468
148	MS	2,017,324	183	VA	1,587,765	218	IL	1,311,190
149	MN	2,010,896	184	UT	1,581,596	219	NJ	1,299,906
150	NY	2,003,300	185	CA	1,578,604	220	PA	1,296,604
151	TX	1,993,118	186	WI	1,550,535	221	VA	1,296,405
152	CA	1,939,695	187	TX	1,549,507	222	LA	1,293,343
153	IL	1,919,386	188	NJ	1,538,070	223	NM	1,289,636
154	KS	1,904,821	189	MI	1,536,218	224	AL	1,287,817
155	PA	1,898,035	190	MD	1,532,255	225	FL	1,283,338
156	AR	1,894,857	191	FL	1,528,295	226	WV	1,282,039
157	FL	1,889,023	192	NY	1,521,616	227	CA	1,280,635
158	NJ	1,883,743	193	OH	1,519,987	228	TN	1,274,567
159	GA	1,835,135	194	CA	1,508,408	229	TX	1,267,491
160	CA	1,834,767	195	GA	1,498,382	230	GA	1,266,364
161	MA	1,834,694	196	AZ	1,483,987	231	MO	1,253,598
162	LA	1,829,063	197	NC	1,472,949	232	NC	1,244,869
163	AL	1,821,249	198	IL	1,465,955	233	CO	1,244,733
164	TX	1,819,459	199	PA	1,449,648	234	CA	1,234,051
165	MI	1,817,677	200	CA	1,444,191	235	NY	1,226,766
166	NY	1,812,053	201	TX	1,442,513	236	NE	1,212,949

TABLE 11 (*Cont.*)

Seat	State	Priority Value	Seat	State	Priority Value	Seat	State	Priority Value
237	WI	1,201,039	272	MI	1,049,437	307	IL	922,043
238	OH	1,198,982	273	CA	1,044,148	308	ID	917,311
239	IA	1,196,953	274	TN	1,040,680	309	FL	916,311
240	TX	1,195,002	275	FL	1,034,660	310	UT	913,135
241	CA	1,190,738	276	NY	1,027,671	311	PA	911,786
242	FL	1,188,140	277	MO	1,023,558	312	WA	911,730
243	MD	1,186,879	278	TX	1,020,010	313	OH	910,692
244	IL	1,186,016	279	CA	1,012,972	314	KY	905,480
245	MI	1,173,306	280	LA	1,001,819	315	CA	904,902
246	PA	1,172,823	281	OK	998,475	316	ME	903,492
247	KY	1,160,070	282	AL	997,539	317	SC	900,031
248	MS	1,164,703	283	IL	995,920	318	MN	899,300
249	SC	1,161,935	284	NJ	992,820	319	TX	889,733
250	MA	1,160,363	285	OH	990,027	320	NJ	888,005
251	NY	1,152,346	286	OR	989,735	321	NY	884,191
252	CA	1,150,362	287	PA	984,842	322	CA	881,394
253	AZ	1,149,492	288	CT	984,248	323	TN	879,535
254	TX	1,130,359	289	CA	983,605	324	NH	875,692
255	NJ	1,125,752	290	MA	980,685	325	FL	866,743
256	CA	1,112,635	291	WI	980,644	326	MI	866,544
257	IN	1,112,020	292	NY	974,935	327	GA	865,091
258	FL	1,106,098	293	TX	972,542	328	MO	865,065
259	MN	1,101,413	294	FL	971,894	329	HI	860,296
260	KS	1,099,749	295	MD	969,083	330	CA	859,076
261	GA	1,096,703	296	GA	967,201	331	IL	858,375
262	VA	1,095,662	297	CO	964,166	332	TX	853,402
263	AR	1,093,996	298	CA	955,892	333	NC	850,407
264	NY	1,086,442	299	NC	950,784	334	MA	849,298
265	OH	1,084,520	300	MI	949,251	335	PA	848,827
266	IL	1,082,680	301	VA	948,871	336	IA	846,373
267	WA	1,078,773	302	IN	939,828	337	NY	844,874
268	NC	1,078,088	303	AZ	938,556	338	OH	843,137
269	CA	1,077,304	304	CA	929,698	339	CA	837,861
270	TX	1,072,352	305	TX	929,296	340	VA	836,826
271	PA	1,070,636	306	NY	927,348	341	WI	828,796

TABLE 11 (*Cont.*)

Seat	State	Priority Value	Seat	State	Priority Value	Seat	State	Priority Value
342	MS	823,569	374	TX	760,271	406	CA	699,641
343	FL	822,265	375	MN	760,047	407	WA	696,345
344	TX	819,922	376	IL	754,228	408	LA	691,321
345	MD	819,025	377	MO	749,168	409	NY	691,204
346	LA	817,982	378	MA	749,011	410	OH	689,683
347	CA	817,669	379	VA	748,480	411	AL	688,367
348	NV	817,326	380	PA	745,838	412	MI	687,017
349	AL	814,487	381	CA	745,777	413	AZ	686,953
350	IN	813,915	382	FL	745,731	414	CA	685,506
351	NY	808,905	383	NY	745,436	415	TX	685,469
352	NJ	803,231	384	NM	744,572	416	FL	682,235
353	IL	802,937	385	RI	742,223	417	VA	677,025
354	CA	798,427	386	WV	740,186	418	NJ	674,488
355	MI	797,104	387	KY	739,322	419	IL	672,626
356	PA	794,005	388	MI	737,975	420	CA	671,930
357	AZ	793,225	389	SC	734,872	421	TN	671,756
358	WA	789,581	390	OH	734,223	422	MA	669,936
359	TX	788,970	391	TX	733,586	423	NY	666,944
360	CO	787,238	392	NJ	733,246	424	CO	665,338
361	OH	784,918	393	CA	729,737	425	PA	665,144
362	GA	782,504	394	IN	717,806	426	TX	663,702
363	FL	782,130	395	WI	717,758	427	MO	660,704
364	CA	780,070	396	NY	717,296	428	CA	658,882
365	KS	777,640	397	CA	714,372	429	MN	658,220
366	NY	775,875	398	GA	714,325	430	GA	657,084
367	AR	773,572	399	FL	712,571	431	IA	655,598
368	OK	773,415	400	IL	711,093	432	FL	654,377
369	NC	769,222	401	MD	709,296	433	OH	650,239
370	OR	766,646	402	TX	708,712	434	CA	646,330
371	CA	762,538	403	PA	703,183	435	NC	645,931
372	CT	762,395	404	NC	702,201			
373	TN	761,699	405	NE	700,296			

TABLE 11 (*Cont.*)

THE SURPRISING FACT THAT MAKES IT ALL WORK

All this raises a pair of questions. We now appear to have two methods for finding the apportionment: minimizing the SSE and using the priority table. In the examples so far, the two methods give the same answer. Will this be true for every example? If so, why? The answers: (1) yes, it will always be true, and (2) explaining why involves a lot of algebra, although you can get the essential idea from an example. Take a look at Table 12, which comes from our previous examples (Tables 7 and 8).

Surprisingly, no matter which apportionment you look at, the sum of squared errors (SSE) plus the sum of the three associated squared priority values (SSP) is always the same. *A similar result is true for every apportionment problem.*

This result is the key to using priority values to decide how to allocate seats. Because the sum of squared errors plus the sum of associated squared priority values is the same for all possible apportionments, whichever apportionment maximizes the sum of squared priority values must also minimize the sum of squared errors. To find the best apportionment, then, all you have to do to distribute the remaining 385 seats among the 50 states is choose priority values one

APPORTIONMENT			Sum of Squared Errors (SSE)	Sum of Squared Priority Values (SSP)	Total
Arkebama	Minnessippi	Washconsin			
4	1	1	75.00	1323.00	1398.00
3	2	1	109.50	1288.50	1398.00
3	1	2	217.50	1180.50	1398.00
2	3	1	366.00	1032.00	1398.00
2	1	3	510.00	888.00	1398.00
2	2	2	399.00	999.00	1398.00
1	3	2	1243.50	154.50	1398.00
1	2	3	1279.50	118.50	1398.00
1	4	1	1229.25	168.75	1398.00
1	1	4	1391.25	6.75	1398.00

TABLE 12 SSEs and sums of the three associated squared priority values for all possible apportionments

at a time, starting with the largest, and working down one seat at a time, always choosing the largest remaining priority value. This method ensures that you end up with the largest possible sum of squared priority values and smallest possible sum of squared errors.

CONCLUSION

Taken together, the columns of Table 12 show that SSE + SSP has the same value for all possible apportionments. Whatever apportionment makes the sum of squared priority values as large as possible automatically makes the sum of squared errors as small as possible, and in that sense comes close "as near as possible" to representative democracy's goal of one person, one vote.

Now, at last, it is possible to explain why it was North Carolina, rather than some other state, that opposed Utah's lawsuit to get an extra seat in the U.S. House of Representatives following the 2000 Census. For the 2000 Census, North Carolina and Utah had priority values that ranked 385 (645,931 in Table 11 for seat #435) and 386 (645,684), respectively. If Utah's population figure had gone up (at least 857 people) as a result of its appeal to the U.S. Supreme Court, its priority value would have moved ahead of the 645,931 for North Carolina for the 435th seat, and Utah would have claimed the last seat to be allocated. As it turned out, however, the Court decided that Utah's challenge to the Census Bureau's population figures was not persuasive enough. North Carolina kept its seat (Wolfson and Lee, 2003; Cantwell, Hogan, and Styles, 2004).

ACKNOWLEDGMENTS

The views expressed in this essay are those of the authors and not necessarily those of the U.S. Census Bureau. The authors are grateful to Howard Hogan and James Treat for their careful and helpful reviews of earlier drafts.

REFERENCES

Abramson, F. (2004). "Special Place/Group Quarters Enumeration." *Census 2000 Testing, Experimentation, and Evaluation Program Topic Report No. 5, TR-5.* U.S. Census Bureau, Washington, D.C.

Anderson, M. J. (1988). *The American Census: A Social History*. New Haven, Conn.: Yale University Press.

Anderson, M. J. (Ed.) (2000). *Encyclopedia of the U.S. Census*. Washington, D.C.: Congressional Quarterly Press.

Balinski, M. L., and H. P. Young. (2001). *Fair Representation: Meeting the Ideal of One Man, One Vote*, 2nd ed. Washington, D.C.: Brookings Institution Press.

Buckley-Ess, J., and I. Hovland. (2004). "Puerto Rico." *Census 2000 Testing, Experimentation, and Evaluation Program Topic Report No. 14, TR-14*. U.S. Census Bureau, Washington, D.C.

Cantwell, P., H. Hogan, and K. Styles. (2004). "The Use of Statistical Methods in the U.S. Census: *Utah v. Evans*." *The American Statistician*, 58, no. 3:203–12.

Hallowell, J. H. (1977). "Democracy." In *The World Book Encyclopedia*, vol. 5, 104–8. Chicago: Field Enterprises Educational Corporation.

Holder, A. Roddey, and J. T. Roddey Holder. (1997). *The Meaning of the Constitution*, 3rd ed. Hauppauge, N.Y.: Barron's Educational Series.

Robinson, J. G., B. Ahmed, P. Das Gupta, and K. A. Woodrow. (1993). "Estimation of Population Coverage in the 1990 United States Census Based on Demographic Analysis." *Journal of the American Statistical Association* 88, no. 423:1061–71.

Treat, J. B. (2004). "Response Rates and Behavior Analysis." *Census 2000 Testing, Experimentation, and Evaluation Program Topic Report No. 11, TR-11*. U.S. Census Bureau, Washington, D.C.

Wolfson, L., and T. Lee. (2003). "Utah's Census Battles: The Inside Story." *Chance* 16, no. 3:21–29.

Wright, T. (2004). Equal Proportions: "As Near as May Be." Unpublished manuscript.

———. (1999). "A One-Number Census: Some Related History." *Science* 283, no. 5401:491–92.

———. (1998). "Sampling and Census 2000: The Concepts." *American Scientist* 86 (May–June): 245–53.

Wright, T., and H. Hogan. (1999). "Census 2000: Evolution of the Revised Plan." *Chance* 12, no. 4:11–19.

ADDITIONAL READING

This essay has focused on one small part of the work of the U.S. Census Bureau. The work of the Census Bureau involves statistics on the country's people and economy in a large number of ways that this essay hasn't considered in any detail. In particular, finding the population numbers used in the priority table is made challenging by the kind of issues illustrated by the thought experiment of counting people at a basketball game. What do you do about people who don't return the census form? Or people who move without leaving a forwarding address?

It is an important fact that some kinds of people tend to be easy to count, and others tend to be harder to track down. Those who are harder to count are more likely to be missed, and, as a result, certain groups of people get undercounted. You can read about the statistical challenge of getting accurate counts in Wright (1998).

For an accounting of selected events and related court actions leading up to the U.S. Supreme Court's ruling to not allow sampling to help produce the apportionment population for 2000, see Wright and Hogan (1999). Wolfson and Lee (2003) provide legal details on the legal ease involving Utah and Census 2000.

For some historical notes on the *method of equal proportions* and a more formal statistical derivation of the method outlined here, see Wright (2004) and Balinski and Young (2001).

QUESTIONS

1. Verify the entries in Table 12.
2. What are some advantages for having the size of the U.S. House of Representatives fixed at 435? Disadvantages?
3. The Census Bureau believes that it undercounted the population in each census from 1940 to 1990. However, in Census 2000, it believes that the population was overcounted by 0.49%. Thinking of Table 4 and Appendix A, how would you conduct Census 2010 so that there is no overcounting or undercounting?
4. Does the system of government in the United States achieve the "one person, one vote" principle? Explain.

5. If representatives were given a weighted vote (like 0.95 or 1.03 instead of 1), would it be possible for representative government to satisfy the "one person, one vote" principle?

APPENDIX A: EXPLANATIONS OF ENUMERATION METHODS IN TABLE 4 (*SOURCE:* TREAT, 2004)

A.1. In Households

Self-administered questionnaires are questionnaires completed by the respondents without assistance from an interviewer. The *paper* questionnaires were delivered mainly during March 2000 to the majority of households in the country and were completed by respondents and returned through the mail. Some respondents sent their information to the Census Bureau using the *Internet*. For the *Be Counted forms*, the Census Bureau distributed unaddressed paper questionnaires at targeted locations in the community. If a person felt he or she was not counted in the census, he or she could complete the form and return the questionnaire through the mail. The Be Counted forms were available from March 31 to April 17, 2000.

Interviewer-administered follow-up questionnaires are questionnaires that were completed with the direct assistance of Census Bureau interviewers and staff. For the *Nonresponse follow-up* operation, addresses for which a questionnaire had not been received on or before April 18, 2000 were visited by Census Bureau staff in order to collect information on the household members. This operation occurred from April 27 to June 26, 2000. The *Coverage improvement follow-up* operation occurred a few weeks after the completion of the Nonresponse follow-up operation. This operation was designed to enumerate housing units that were added late to the address list and thus could not be included in the Nonresponse follow-up operation. In addition, housing units classified as vacant or delete in Nonresponse follow-up were visited again during *Coverage improvement follow-up*. Also, if the enumeration located an address on the ground that was missing from their address register, they were able to add and enumerate the housing unit. This operation occurred from June 26 to August 23, 2000. For the *Coverage edit follow-up* program, the

Census Bureau reviewed the data from paper questionnaires returned through the mail and from Internet questionnaires. The review consisted of checks or edits to ensure that the respondent provided consistent data on the number of people in the household. This operation occurred from May 8 to August 13, 2000. For the *Telephone questionnaire assistance* program, the Census Bureau provided a toll-free 1-800 telephone number to assist respondents in completing the census questionnaire. The Telephone questionnaire assistance network was available to the public from March 3 to June 30, 2000.

Interviewer-administered enumerate questionnaires were questionnaires that were administered by Census Bureau interviewers to enumerate cases where the household was not provided the opportunity to self-respond to the census. *Nonresponse follow-up adds* and *Coverage improvement follow-up adds* are cases that were identified during the respective operations as new housing units. The field staff added the address and enumerated the household during the operations. From March to June 2000, Census Bureau staff canvassed the *Update/enumeration* blocks, updated both the address list and maps, and enumerated the housing units. The Update/enumerate blocks occurred primarily in communities across the country with low mail response rates in the 1990 census. *List/enumerate* areas contain mostly non-city-style addresses. These areas of the country are geographically remote with a low housing unit density. For these reasons the address list was developed during the enumeration of the housing units. From March to July 2000, Census Bureau staff canvassed the ground creating the address list, updating maps, and enumerating the housing units. These List/enumerate areas were in portions of 20 states mainly in the West and Northeast regions of the country. *Remote Alaska* areas contain mostly non-city-style addresses. These areas of Alaska are the most geographically remote with a low housing unit density. For these reasons, the address list was developed during the enumeration of the housing units. From late January to late April 2000, Census Bureau staff canvassed the ground, creating the address list, updating maps, and enumerating the housing units. The *T-Night* enumeration was designed to enumerate people at transient locations such as recreational vehicle (RV) parks, campgrounds, marinas, racetracks, fairs, and carnivals. People living or staying at these locations on Census Day were not likely to be at these locations year-round. T-Night enumeration occurred on March 31, 2000.

And finally, the enumerations in the *Other* category came from two operations. The *Whole household substitutions* represent households that were not enumerated by a data collection operation. Thus, the data for these cases were imputed. The remaining returns in the Other category are *Unlinked enumerator continuation forms*. These counts were generated during the *Interviewer-administered* follow up enumeration (all except the Telephone questionnaire assistance operation) and the Interviewer-administered enumeration operations. Continuation forms were used when there were more than five people in the household. The field staff was to transcribe the housing unit identification number (ID) onto the continuation form. The ID would permit the data from the continuation form to be linked with the parent form after the forms were data captured. A small number of continuation forms (2058) could not be linked with their parent form. These forms were included in the count of people and households, and they accounted for 4126 people.

A.2. In Group Quarters

The Census Bureau defines group quarters as places where people live or stay other than the usual house, apartment, or mobile home. Examples of group quarters include college and university dormitories, hospital/prison wards, and nursing homes. *Group quarters* enumeration was conducted April 1, 2000 until May 6, 2000. During the enumeration at each group quarters, enumerators obtained a list of residents, filled out a listing sheet, and (in most cases) distributed individual questionnaires to each resident. In some instances where residents could not fill out the forms themselves, enumerators were allowed to use administrative records to augment the process.

EVALUATING SCHOOL CHOICE PROGRAMS

JENNIFER HILL

Columbia University

—————— ⁂ ——————

Education reform has become an important public policy issue. The public focus on education reform is driven, at least in part, by statistics that suggest that our public school system may be failing. For instance, the U.S. Department of Education reports that currently only one-third of the nation's fourth graders read at a proficient level and only one-fourth of fourth and eighth graders perform math at a proficient level. The numbers are even more dire in urban areas. Whether these indicators reflect a genuine crisis or not, they have helped to bring the debate over education reform into the limelight. Two events that occurred in 2002 paved the way toward the possible adoption of one of the most controversial education reforms, publicly funded private-school vouchers.

In January 2002, President Bush's No Child Left Behind Act became law. One of the most important implications of this law is that any child attending a failing school (schools for which test scores for certain populations fail to improve over a two-year period) has to be provided with the opportunity to transfer to a school that is not failing. Critically, school districts cannot deny this right to a child simply due to capacity constraints in their schools. Given that overcrowding is already a problem in many of the affected school districts, this law may put pressure on these districts to turn to alternatives such as publicly funded private-school vouchers.

Also, in June 2002, the Supreme Court, in a 5–4 ruling, upheld the constitutionality of a publicly funded school voucher program in Cleveland that went into effect in 1996. The controversy arose because some (in fact, the vast majority) of the private-school scholarships were being used by participants to attend parochial schools. Opponents argued that this violated constitutional mandates regarding the separation of church and state. The Supreme Court decision removed this obstacle from states wanting to implement voucher programs.

Private-school vouchers are one manifestation within the more general rubric of "school choice." Other variations include magnet schools, charter schools, and various forms of intra– and inter–school district choice systems (that is, choice *within* the public school system). Private-school vouchers have been among the most contentious of these school choice initiatives. Critics maintain that publicly funded vouchers drain money away from already underfunded public schools, foster greater racial segregation, fail to provide better educational opportunities, steal the best students from the public schools, and create more instability for affected students. Opponents also argue that vouchers are not a realistic approach given the large number of children in public schools. Proponents of vouchers counter that private schools create greater racial integration, boost achievement, provide a safer and more orderly learning environment, allow parents to choose the most appropriate schools for their children, and can force public schools to compete and become more efficient.

Arguments about whether or not vouchers should be adopted in some form and for some populations (e.g., children in "failing" schools) traditionally have not been informed by strong empirical evidence. This essay reviews various approaches to measuring the effect of vouchers on children's educational achievement. Our first step is to consider why previous attempts at comparing public- and private-school achievement (a comparison that is relevant to our consideration of the likely effectiveness of voucher programs) have not provided convincing evidence. A second step is to consider the use of randomized experiments in evaluating educational programs. Randomized experiments, where a study population is randomly divided into two or more treatment groups, have been successful for determining the effects of treatments in fields such as medicine. This essay then examines a randomized experiment that was performed in the context of private-school vouchers. A private foundation's announcement that it planned to offer 1300 scholarships ($1400 each) to low-income children in New York City to use in private schools provided researchers with a naturally occurring randomized experiment. What this "natural" experiment can and cannot tell us is the subject of our story.

EARLY STUDIES AND PROBLEMS WITH OBSERVATIONAL DATA

It is instructive first to consider what was learned prior to the New York experiment (and other related experiments). In the absence of any programs implementing private-school vouchers it is quite difficult to examine the potential efficacy of vouchers. This is the conceptual problem that faced researchers in the 1970s and 1980s. A number of researchers attempted to make such an examination through comparisons of students attending public and private schools. One of the earliest and best known studies was carried out by James Coleman, Thomas Hoffer, and Sally Kilgore; the results were published in a series of articles and the 1982 book *High School Achievement: Public, Catholic and Private Schools Compared.* After conducting analyses on data from the High School and Beyond survey carried out by the National Center for Education Statistics, the authors concluded that private schools led to higher achievement than public schools. These findings were controversial and sparked a series of further studies and analyses. In general these studies did not fully support the strong findings of the original work; they showed nonexistent, much weaker, or mixed advantages to the private schools. Since then many more studies have been performed, with a range of findings. How is it possible that researchers were able to find so many different answers to the same question?

For one thing, there were a number of differences among the many studies. Not every study used the same survey data, or even the same type of student population (for example, age and race varied). The outcome used to measure student success also varied across studies; among the outcomes used are test scores, high school graduation, and college entry. Sample size also varied a great deal among studies, and it is important to remember that large data sets are more likely to yield statistically significant results than small data sets. These many differences are not the key issue, however. Even some studies that used the same data source, the same student population, the same outcome measure, and the same statistical methods reached substantially different conclusions. Why?

The major limitation of the early studies is that they rely on observational, rather than experimental, data. The students attending private schools were doing so because they (or their parents) chose to do so. The students attending public schools were doing so because they (or their parents) chose to do so (perhaps because of financial, logistical, or other constraints that ruled out private schools). This creates a situation where the students attending private schools are likely to be different than those attending public schools. They might come from families with more money (particularly in the case of those attending secular private school), be more motivated, or have higher educational expectations placed on them. Thus if there are differences in average test scores between these groups, it can be impossible to disentangle to what extent these differences were caused by variation in the types of students attending each (known as the effects of selection bias) and to what extent the differences were caused by the schools themselves (known as the causal effects of attending private school).

The early studies tried to adjust or control for the differences between the populations using statistical modeling. To illustrate how such an adjustment might work, consider the hypothetical data relating test scores, type of school, and family income in Figure 1. In the first panel (Figure 1a), it appears that private-school students, represented by the blue dots, outperformed public-school students, represented by the red dots, on the test.

Suppose, however, that we also have information regarding family income demonstrating that the children who go to private school come from families that have higher income on average. Perhaps income could explain this differential in test scores between private-school and public-school students. Figure 1b shows the same test scores with the corresponding incomes, along with the best-fitting line relating test scores to income. For these (hypothetical) data a single line does a good job for both types of students. This means that the

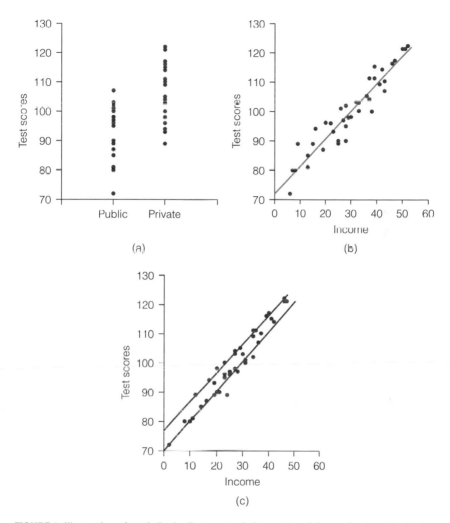

FIGURE 1 Illustration of statistical adjustment of observational data using hypothetical test score data. Panel A plots test score against school type indicating that private-school students have higher test scores on average. Panel B plots the same test scores against income, with blue dots for private-school students and red dots for public-school children. The best-fitting line is also shown; evidently the observed differences in test scores are all due to income. Panel C is similar to Panel B but for these hypothetical data there appears to be a private-school advantage even after controlling for income.

difference in test scores that we observed in the left panel is completely explained by differences in family income. If we make an adjustment we will find that students with comparable incomes perform similarly on the test regardless of which school they attend. Figure 1c shows a different hypothetical data set where even after adjusting for income it appears that private-school students have a test score advantage relative to public-school students (the best-fitting line for private-school students is higher than the best-fitting line for public-school students).

Clearly the many discrepant results described earlier suggest that this approach to addressing bias due to self-selection doesn't always work. One difficulty is that we can't be sure that we have properly adjusted for all relevant differences between groups. For instance, what if we have adjusted for age, race, and religion, but the most important variable to control for is income? Our statistical methods will generate an answer, and it will be wrong, but we won't have any way of knowing this from the result.

Another problem is that even if we somehow knew that we had included all relevant variables, the simple regression approach to adjustment relies on a number of difficult-to-verify assumptions. One key assumption is that the relationship between the outcome (e.g., test scores) and our background variables (e.g., income) has to be linear. If the relationship is not linear—for example, perhaps test scores increase with income but only to a point after which they plateau— then the regression adjustment will lead to a poor estimate of the difference between private- and public-school students.

There are, to be sure, more sophisticated approaches for trying to control for the differences between the private-school and public-school populations. Yet even among the subset of papers using more advanced techniques to address selection bias, the results still are mixed with some supporting private-school gains in general, some supporting gains for Catholic schools but not other private schools, and some finding no differences between educational sectors. Observational studies do not appear to be ideal for calculating the causal effect of being given a private-school education rather than a public-school education because people self-select into their school environment, yielding different types of students in the different types of schools.

THE ROLE OF RANDOMIZED EXPERIMENTS

The difficulties with observational studies are well known in many fields. For example, the same problem arises in medicine if we compare patients that select one particular form of treatment with a second group that prefers an alternative. In that context, and in many other fields, it has become common to rely on randomized experiments where the subjects are randomly assigned to the different treatments in order to assess the relative effectiveness of those treatments. The randomization process ensures that the groups being compared are similar on average. To study school choice, then, we might like to randomly assign students to either private or public schools. But, taken at face value, this seems an unethical and impractical plan. We can't easily manipulate the student population in this way. Would siblings be randomized into different schools? What about transportation issues? What if parents couldn't afford the school to which their child was randomly assigned? Sometimes there are situations, however, when the sought-after randomization occurs "naturally." Fortunately for school-choice research, several such opportunities arose in the 1990s.

One such "natural" randomized experiment occurred in New York City. In February 1997, a private foundation in New York called the School Choice Scholarships Foundation (SCSF) announced that it would offer 1300 scholarships (in the amount of $1400 each) to low-income children in New York City to use toward private-school tuition. Since the number of applicants was sure to exceed the number of available scholarships, a lottery was planned as the most equitable way of allocating the scholarships. This lottery performs the randomization needed for a fair comparison. During the time that the scholarship initiative was being developed, a Harvard professor, Paul Peterson, convinced the SCSF to take advantage of the randomization and allow the program to be evaluated by outside researchers. Mathematica Policy Research joined the project, with researcher David Myers as the lead principal investigator, joint with Peterson, and they helped to shape both the implementation and the evaluation of the SCSF program. The SCSF program targeted primarily low-income grade-school children from underperforming schools. Eligibility required the children to have family incomes low enough to qualify for school lunch programs and to be in public school in kindergarten through fourth grade when applying. Only 15% of the children awarded scholarships were allowed to be from schools whose average test scores were above the citywide median (high-performing schools) because the SCSF was more interested in vouchers as a potential policy initiative for students in troubled schools. The evaluation plan involved preliminary testing of students and then follow-up testing over the three years of the program for students receiving the scholarships and selected students among the unsuccessful applicants.

How does a lottery create "fair comparisons"? A lottery randomly divides the applicants into two groups: (1) a chosen group of scholarship recipients, and (2) a second group of unsuccessful applicants. Because of the randomization, the two groups should be similar in terms of their characteristics measured before the lottery took place (or characteristics not changed by the lottery, such as gender). Indeed, comparisons across groups for a host of characteristics ranging from child pretest scores and gender to parental income and education levels reveal that, as expected, the use of randomization creates groups that are similar but not identical. Any differences are caused by chance—they are not systematic. This means that if we were to perform the lottery again we would be just as likely to see differences in the opposite direction. Because our two groups are similar in terms of all their pretreatment characteristics, comparing the performance of the two randomized groups on achievement tests taken some time after the children have been randomized should allow for a fair evaluation of the effect of receiving a voucher/scholarship because now the only systematic difference between the two groups is that in one group people had access to scholarships to private schools and in the other group they did not.

Two large-scale randomized voucher evaluations similar in spirit to the New York City program followed in Washington, D.C., and Dayton, Ohio. There have also been randomized evaluations of voucher programs in other cities, but each had some problems with executing the randomization or in analyzing the data and thus are not discussed here. All of the randomized studies targeted low-income, primarily urban populations.

Major characteristics of the studies can be found in Table 1. A close examination of the table reveals two noteworthy features. First, the proportion of students awarded vouchers that actually used them is between 50% and 70%. This seems lower than one might expect. Second, a significant fraction of the children don't take the follow-up tests. Both of these occurrences, the failure of subjects to follow through with the assigned treatment and the failure to collect all

Location	Start Year	Years	Application Grades	Participation Rates	Follow-up Rates
New York City	1997–98	3	K–4	1st year: 64%	1st (1998): 78% (86% choice, 75% control)
				2nd year: 62%	2nd (1999): 65% (69% choice, 60% control)
				all 3 years: 53%	3rd (2000): 67% (69% choice, 65% control)
Washington, D.C.	1998–99	2	1–7	1st year: 53% both years: 38%	1st (1999): 63% 2nd(2000): 50%[a]
Dayton, Ohio	1998–99	2	K–12	1st year: 54%	1st (1999): 56% choice, 57% control
				2nd year: 50%	2nd (2000): 48% choice, 50% control

TABLE 1 Major features of the three randomized voucher experiments

[a]Wolf et al. (2001, p. 15) note that "approximately half the families in the treatment group actually used [a scholarship] during the program's second year."

Sources: Mayer et al., 2002; Howell et al., 2000.

needed data, are common in scientific studies involving humans, and they pose significant challenges to those evaluating the results. We later return to these issues.

For reasons that will become clear, the actual results are not the most important point of this statistical story. Table 2 presents results for each experiment from its final year as presented in the evaluation reports. This was year three for the New York experiment and year two for Dayton and Washington. The numbers in the table represent the average gain on standardized tests measured in percentile points for students that *attended* private school at least one year. A percentile score for a given student represents the percentage of other students taking the test nationwide who scored below him or her on the test; so, for example, a score of 40 for a given child indicates that 40% of

LOCATION	READING			MATH		
	Overall	AA	Other	Overall	AA	Other
NYC	0.4	5.5*	−2.7	2.3	9.7**	−0.1
DC		8.1*	−5.6		9.9**	5.8
Daytona		7.6	−0.4		5.3	0.0

TABLE 2 Effects of private school attendance on test scores (in national percentile rankings)

Sources: Mayer et al., 2002; Howell et al., 2000.

Note: Results are broken down by ethnicity and type of test (reading/math). "AA" stands for "mother is African American"; "Other" represents all other racial/ethnic groups.

*Significant at 0.05 level, 2-tailed test.

**Significant at 0.01 level, 2-tailed test.

the national sample scored below this child. In the New York City experiment math scores are about two percentile points higher on average for private-school attenders. Specifically, this "treatment effect" reflects the fact that the children who had attended private school scored 26.2 on average in terms of national percentile rankings and their comparison group from among the public-school children scored 23.9 on average in terms of national percentile rankings. The results are intriguing in that when the results are reported by ethnicity there are statistically significant improvements only for African Americans. "Statistically significant" means that the increases are larger than can be explained by good luck during the allocation of individuals via the lottery, and thus the increase can be attributed to attending private school rather than chance. On the other hand, the vouchers seem to hurt the performance of other ethnic groups in reading (though these results are not statistically significant) and do little for the other ethnic groups in math. It is not clear why there should be such a large difference among ethnic groups since the vouchers were not thought to be especially advantageous for a particular ethnic group. The purpose of the present essay is to demonstrate the role of statistical arguments in analyzing social experiments, so we won't treat these results as the end of our story but rather as a jumping-off point for discussing how they might have been affected by decisions made regarding how to analyze the data.

CHALLENGES IN DESIGNING AND ANALYZING SOCIAL EXPERIMENTS

Randomized experiments are supposed to be the gold standard in terms of identifying the true causal effects of an experimental program. So why do we downplay the results? Doesn't the randomized experiment address the question of private- versus public-school efficacy once and for all?

To be blunt the answer is no. Hints of the challenges faced by researchers were previously mentioned in the discussion of the study characteristics presented in Table 1, namely, a failure of voucher recipients to attend private schools and missing follow-up test scores. Even though these studies made use of randomized experiments, the participants were human subjects with free will who could not be forced to comply with study protocol. Of course such complications can and do occur in randomized medical studies but they are typically less dramatic in medicine, especially if participants are blinded so that they do not know which drug they have been randomized to receive. Here, those not selected to receive a voucher know that they have not been selected and there is no way to avoid that knowledge.

Noncompliance: People Don't Always Do What They Are Told

Not all of the students who were offered a voucher were able to actually use it to attend private school for a range of financial, logistical, or other reasons. The implications of this complication, often called noncompliance, for our analyses are a bit subtle. In general this means that a simple comparison between the randomized groups will not yield an estimate of the effect of *going to private school;* it will yield an estimate of the effect of *being offered a scholarship.* This is known as the "intent to treat" (ITT) estimate since it measures the effect of our intention to apply a treatment (offer a scholarship) without regard for whether that treatment was carried out or not. This is a common problem in randomized drug trials as well, particularly when the drugs cause unpleasant side effects.

In some situations it is argued that noncompliance is not a big problem because a simple comparison between the randomized groups—that is, the ITT analysis—is a conservative estimate of the true treatment effect (i.e., it should be closer to zero if anything). In other words, the ITT effect should be a "watered down" version of the true treatment effect; therefore, the results would only get stronger if everyone had followed through. Unfortunately this reasoning does not necessarily hold when we have missing data as well (as we will see more clearly in the next section). In this case our estimate could potentially overstate the benefits of the treatment—leading you to believe a treatment is effective when in fact it is not.

How can one get a more realistic estimate of the value of attending private school? Some obvious proposals don't work very well. For example, we can't just compare the scores of those who used the vouchers to the randomized comparison group because the types of people who took advantage of the voucher opportunity could be quite different from those who did not. This means that we lose the benefit of our randomized experiment and end up with a form of observational study.

It is, however, possible to generate a good estimate of the value of attending private school. It requires a fairly sophisticated statistical approach using what is known as an instrumental variable. A detailed discussion is beyond the scope of this essay, but the intuition in this scenario is that the approach tries to make comparisons between the group of people who *actually attended* and the group of public school attendees who we believe *would have attended* private school if they had instead won a scholarship. This subset of the control group should be more similar to the private-school attendees than the full control group. The results in Table 3 rely on such instrumental variable adjustments.

| SCHOOL | READING | | MATH | |
	ITT	Attended	ITT	Attended
Low	1.5	2.2	3.2*	4.7**
High	0.4	0.6	2.8	4.2

TABLE 3 Effect of one year's treatment on test scores (in national percentile rankings)

Source: Barnard et al. (2003).

Note: The table compares the intention-to-treat (ITT) effect and the effect of attending private schools (Attended). Results are broken down by low-scoring and high-scoring schools and type of test (reading/math).

* Significant at 0.05 level, 2-tailed test.

** Significant at 0.01 level, 2-tailed test.

For insight into the differences in conclusions that might be reached between analyses that estimate the ITT effect and analyses that focus on the treatment effect for those who actually attended a private school, consider the reading and math results in Table 3. The results in Table 3 are from the first year of the New York study for children in grades 1–4 (when they applied to the program) from families where only one child participated (see Barnard, Frangakis, Hill, and Rubin [2003]). The results are broken down by the type of school the child was attending at the time of application (low test score/high test score). As discussed earlier, fewer students from high-test-score schools were allowed to win scholarships (only 15%) because it was assumed that they were in less need of the potential benefits of private school. For each test (reading/math) and for each type of school (low/high) the estimated effects of actually attending private school (these are also adjusted for missing data) are larger in magnitude than the estimated ITT effects.

Some researchers argue that the ITT effect is the most appropriate for evaluating the potential average effect of a policy because it reflects the fact that not all eligible individuals will choose to participate in the program being investigated. However, for this logic to be correct, the effect when the "actual" policy is implemented would have to be just as "watered down" (that is, just as high a percentage of people would have to choose not to participate) as observed in the preliminary investigation. In addition, the same types of people would have to choose to participate. We prefer to report and consider both the ITT effect and the estimated actual effect of attendance. That way, if we can assume that the same types of people would choose to participate in the future, then we can use the estimated treatment effect for those who participated (here, the effect of actually attending private school) to calculate ITT effects corresponding to any potential compliance rate specified.

Missing Data: You Can't Force People to Stay in the Study

Missing outcome data occurred in these studies because not all children actually showed up to take follow-up tests. Although participants were offered a range of incentives (including money) in each study to encourage them to show up for the posttests, the follow-up rates achieved are well below perfect (though still quite good for studies of this kind), as can be seen in Table 1. This reduces the number of people in the study but wouldn't necessarily be a major problem—if we could rely on randomization to ensure that the students missing posttest scores were just like the students who took the posttests. Unfortunately randomization doesn't help us here because the follow-up tests happened after the lottery! We see in Table 1 that the follow-up rate is generally lower for the control group than for the scholarship group in most of the locations, which is troubling. The no-shows may be different on other traits as well; they may be less well motivated or they may have parents who work on weekends and hence they were not able to attend the weekend testing. If the people who failed to show up for their posttests are systematically different from those who did attend, and we have reason to believe this is true, then a simple analysis of the observed test scores can be biased.

There are a number of statistical approaches to addressing these missing follow-up test scores. The approach that was used in Table 2 uses *weighting* to adjust for the missing outcome data. The basic idea for this is as follows. Background information on each person (e.g., pretest scores, income, race/ethnicity) is used to develop a statistical model that can predict the probability that each person will have missing test scores. If this probability is low for a group, then the individuals in that group that do show up for the follow-up test are given increased weight in the analysis. So, for example, if students with lower pretest scores were less likely to show up for the posttest, the weights would adjust analysis results to reflect the fact that more students with low pretest scores should have had their posttest scores included in the analysis. The weights will give the test scores for students with low pretest scores who did show up more weight in the final analysis.

In order to explore the potential differences in results caused by missing data, consider the analyses of first-year New York data presented in Table 4 (again for children who applied in grades 1–4 from families where only one child participated). The table presents results computed in two different ways. The "complete case" analysis (denoted "CC" in the table) uses only data for New York City students who showed up for the posttest. The "all data" analysis uses data from

	READING		MATH	
SCHOOL	CC	All data	CC	All data
Low	0.6	1.5	2.2	3.2**
High	1.6*	0.4	0.5	2.8

TABLE 4 Effect of one year's treatment on test scores (in national percentile rankings)

Source: Barnard et al. (2003).

Note: The table compares estimates from complete-case analyses (CC) with estimates from more formal missing data models (All data). Results are broken down by low-scoring and high-scoring schools and type of test (reading/math).

* Significant at 0.05 level, 2-tailed test.

** Significant at 0.01 level, 2-tailed test.

all of the students, these are the same results reported as the ITT results in Table 3. The "all data" analysis uses an approach for the missing data (called "multiple imputation") that is more formal and complicated than the weighting strategy just described but relies on the same type of intuition.

The treatment effect estimates can change dramatically in either direction when the missing data are accounted for correctly. Here the reading scores for students from high-performing schools are significantly higher in the voucher group when only the complete cases are used but not in a more thorough analysis. The math scores increase for both high-performing and low-performing schools when the missing data are handled more formally. These comparisons highlight the need to appropriately handle missing data issues, as they could lead to different conclusions about the effectiveness of private-school education for these children.

"What If": Results May Be Sensitive to Data Analysts' Choices

Statistical analyses are often criticized with remarks like "you can get statistics to say anything" or "if you torture the data long enough it will confess to anything." Though unethical uses of statistics are possible, these criticisms also reflect another legitimate concern in a complex study such as the New York City school voucher program. Analyzing the data from such a study requires that a number of choices be made by the program evaluators. We have already seen that one must decide what exactly to estimate (the intent-to-treat effect or the actual effect of attending private school) and how to handle the missing data. In addition, a number of other decisions are required in carrying out the data analysis. In the New York City study, the initial analysis omitted kindergarten students (because they had no pretest scores and such scores were part of the preferred analysis) and defined children as African American based on the race only of their mother. Would different choices change the results? One subsequent study by economists Alan Krueger and Pei Zhu suggests that they could. They perform a variety of analyses making different choices regarding these issues (e.g., including kindergarteners, performing complete case analyses, defining race of child differently), and estimates from their analyses, while still positive for African Americans, are not all statistically significant.

CONCLUSIONS

Addressing social policy questions, such as whether there are educational benefits of school vouchers, presents unique statistical challenges. It can be difficult to carry out randomized experiments for such questions. And, even when randomized experiments can be implemented we must deal with missing data, noncompliance, and the sensitivity of our results to the choices made by the program evaluator. In fact, the challenges of evaluating social policies go beyond these technical issues. There are also a number of issues related to the representativeness of a given study. Are the people who volunteer to participate in programs like the School Choice Scholarships Foundation program in New York similar to those that government policies would ultimately target? Can our results with a relatively small number of students be generalized to forecast what would happen with a large program? For example, what would happen to a public school if a large proportion of its students left? What would happen to a private school flooded with applications?

Providing sound quantitative evidence to aid in the public debate concerning school reform is clearly a difficult problem. In the school voucher case, the natural randomized experiments that arise through the lotteries used to distribute the vouchers provide a unique opportunity. It is important to remember that these randomized experiments, though challenging to analyze, are much more likely to yield reliable conclusions than the observational studies available up to this point. The challenge to modern statistical science is to develop methods that allow program evaluators to reliably address important real-world complications such as missing data and noncompliance.

REFERENCES

Barnard, J., C. E. Frangakis, J. L. Hill, and D. B. Rubin. (2003). "Principal Stratification Approach to Broken Randomized Experiments: A Case Study of School Choice Vouchers in New York City." *Journal of the American Statistical Association* 98, no. 462: 299–310.

Howell, W. G., P. J. Wolf, P. E. Peterson, and E. D. Campbell. (2000). "Test-Score Effects of School Vouchers in Dayton, Ohio, New York City, and Washington D.C.: Evidence from Randomized Field Trials." Technical report. Program on Education and Policy Governance, Kennedy School of Government, Harvard University, Cambridge, Mass.

Mayer, D. P., P. E. Peterson, D. Myers, C. C. Tuttle, and W. G. Howell. (2002). "School Choice in New York City after Three Years: An Evaluation of the School Choice Scholarships Program." Technical report. Mathematica Policy Research, Cambridge, Mass., and Program on Education and Policy Governance, Kennedy School of Government, Harvard University, Cambridge, Mass.

Wolf, P. J., P. E. Peterson, and M. R. West. (2001). "Results of a School Voucher Experiment: The Case of Washington, D.C., after Two Years." Technical report. Program on Education Policy and Governance, Kennedy School of Government, Harvard University, Cambridge, Mass.

ADDITIONAL READING

Brandl, J. E. (1998). *Money and Good Intentions Are Not Enough, or Why A Liberal Democrat Thinks States Need Both Competition and Community*. Washington, D.C.: Brookings Institution Press.

Coleman, J. S., T. Hoffer, and S. Kilgore. (1981). "Public and Private Schools." Technical report. National Center for Education Statistics, Washington, D.C.

Cookson, P. W. (1994). *School Choice: The Struggle for the Soul of American Education.* New Haven, Conn.: Yale University Press.

Coulson, A. J. (1999). *Market Education: The Unknown History.* Bowling Green, Ohio: Social Philosophy and Policy Center and Transaction Publishers.

Fuller, B. and R. F. Elmore. (1996). *Who Chooses? Who Loses? Culture, Institutions, and the Unequal Effects of School Choice.* New York: Teachers College Press.

Krueger, A. B., and P. Zhu. (2003). "Another Look at the New York City School Voucher Experiment." Technical report. National Center for the Study of Privatization in Education, Teachers College, Columbia University, New York.

QUESTIONS

1. What is the main motivation for a study of the effect on children's educational achievement of the provision of publicly funded vouchers to attend private schools?

2. What is the key ingredient in the New York City study that was to make its findings more compelling than previous studies?

3. Explain why Figure 1 shows that a raw comparison between achievement in public and private schools is not, by itself, informative of the effectiveness of the two school systems?

4. The hypothetical study shown in Figure 1 adjusts test performance for family income. Explain why we can't be sure this yields a fair comparison.

5. The New York City study preselected children with low family incomes. How does this affect the conclusions that we can draw from this study?

6. The study was interested in the effect on student achievement of being offered a scholarship to a private school (ITT effect). But also of interest was whether the scholarship students that did actually attend a private school were benefited by this compared

with the similar students that stayed in the public system. How did the randomization influence the information available about these two study objectives?

7. Do you agree with the author that "these randomized experiments, though challenging to analyze, are much more likely to yield reliable conclusions than the observational studies available up to that point"? Discuss with reference to the essay.

DESIGNING NATIONAL HEALTH CARE SURVEYS TO INFORM HEALTH POLICY

STEVEN B. COHEN

Center for Financing, Access, and Cost Trends, Agency for Healthcare Research and Quality

TRENA M. EZZATI-RICE

Center for Financing, Access, and Cost Trends, Agency for Healthcare Research and Quality

——————— ⁑ : : ⁑ ———————

Health care expenditures represent nearly one-seventh of the United States gross domestic product. These expenditures grow at a rate faster than other sectors of the economy, and they constitute one of the largest components of federal and state budgets. To understand health care spending, policy makers need to assess the demand for health care services and the performance of the health care delivery system in meeting those demands. They must also determine which components of the health care delivery system and which segments of the population command the greatest resources and how this distribution of resources changes with time.

Policy makers need reliable information to help answer the following types of questions:

- How many uninsured people are there in a given year?
- What proportion of individuals have medical expenses, and how much did they pay out of their own pockets?
- How are medical expenses distributed by type of service and by source of payment?
- How do per capita expenses vary according to characteristics such as age, race, ethnicity, income, and health status?

Answering these and other questions requires data that are up-to-date and that provide an accurate picture of the health care utilization, expenditures, insurance coverage, sources of payment, and access to care for people in the United States. For example, for those without health insurance, we need to know how long they have gone without coverage and whether or not their status might change in the near future. As another example, when someone receives medical service, such as an eye examination or chemotherapy, we need to know how the service is paid for; if it is expensive, we need to understand how likely it is that the person will continue to require these services in the future. Furthermore, since the health care market is rapidly changing, it is critical to understand how a person's decision to use health care (e.g., to go to the doctor for a checkup or to schedule elective surgery) changes with different financial incentives and institutional arrangements and with different individual characteristics.

The demand for accurate and reliable information on health care served as the catalyst for the implementation of a comprehensive survey called the Medical Expenditure Panel Survey (MEPS). This survey was originally designed in 1996 by the Department of Health and Human Services to improve the analytic capacity of its programs, fill major data gaps, and establish a framework in which the agency's data activities are streamlined and prioritized. Today, MEPS is sponsored by the Agency for Healthcare Research and Quality. MEPS is unparalleled for the degree of detail in its data and its ability to link health care service use, medical expenditures, and health insurance data to the demographic, employment, economic, and health status of survey respondents. Recently, the MEPS data have been used to estimate the cost of providing a prescription drug benefit to Medicare beneficiaries, which culminated in the adoption of this new drug benefit provision as part of the recent Medicare Modernization Act. With its wealth of data on health conditions, prescribed medications, utilization, and expenditures and associated therapeutic drug classifications, the MEPS data have also served to identify potentially inappropriate medication use in the community.

CHALLENGES FACED BY THE MEDICAL EXPENDITURE PANEL SURVEY

The analytical and technical requirements for a national health care survey of this sort present a formidable set of challenges to survey designers. When we consider how to answer the questions just raised, several of these challenges immediately surface:

- A comprehensive health care survey needs to provide an *accurate picture* of the specific health services that Americans use, how frequently they use them, the cost of these services, and how they are paid.
- There is a need to effectively address critical health policy and health delivery issues for relevant subgroups of the population (e.g., minorities, the poor, the elderly, or other special populations, such as those with specific health conditions) because a number of studies have documented that these subgroups do not have the same access to and use of health care services.
- Timely estimates of the country's health insurance status are essential to health policy makers, federal and state governments, and the private sector to assess what percentage of the population has private coverage, or public-only coverage, and what percentage is uninsured. With continual changes in health care coverage and access, information can become out-of-date quickly. Consequently, the MEPS data are targeted for release to the public within 12 months after the completion of data collection.
- A snapshot in time of the state of health care and expenditures does not adequately characterize patterns in health care over

time. At the individual level, there are often dramatic transitions in health insurance coverage status over the course of a year, and the uninsured are the most vulnerable to financial risks.

- While a person's perceived health status, insurance coverage, and visits to the doctor and dentist can be obtained from the individual with reasonable accuracy, individuals may not be the most reliable source of information for the medical expenses incurred for a service or for the source of a payment for a service.

The questions asked on the survey, the size of the sample, the data collection strategies, and the task of making estimates for the population are factors to consider in addressing these challenges. Moreover, the analytic capacity afforded by the survey design needs to be counterbalanced by the complexity and cost of the survey. MEPS has several unique design features that were incorporated into its structure to meet these competing challenges. Three key features are described below.

Linking to the National Health Interview Survey

The target population for the survey consists of those individuals who are civilians, non-institutionalized (e.g., not in prison), and living in the 50 U.S. states or the District of Columbia. For practical reasons, the survey is administered to only a subset of this population, and the responses of the people in this subset are used to make estimates of national health-care parameters, such as the proportion of uninsured in the target population.

One of the main requirements for the MEPS specifies that the subset (i.e., sample) must yield estimates that have no systematic error and obtain a certain level of precision. To accomplish this, we use a chance mechanism to select people for the sample, where everyone has a known chance to be included in the sample. Such a sample will be representative of the population, and estimates for national health-care parameters can be made without bias and with quantifiable precision. A further requirement for the MEPS is that the estimates in each of the four census regions must also be without bias and with a pre-designated precision. To this end, the MEPS design requires the sample to be spread over 195 separate geographic areas.

In addition, the sample must meet a specified precision in estimates for certain subgroups of the population, for example, the poor (people who have a family income less than 200% of the poverty level), the elderly, and some racial and ethnic groups. In order to produce precise and unbiased estimates for a subgroup, such as the proportion of Asians who are uninsured, we need to further select a sufficient number of people from the particular subgroup, and this selection must also be accomplished in a representative and cost-efficient manner. Larger sample sizes translate to improved precision in resultant survey estimates. As a result of this additional sampling, individuals from specific subgroups are overrepresented in the total sample, and the national-level estimates must be adjusted to take this into account.

To help achieve acceptable precision of estimates for policy-relevant subdomains of the population, the selection of the MEPS sample is achieved by reusing a sample of households participating in another large, ongoing national health survey, the National Health Interview Survey (NHIS). The NHIS is also a survey of the U.S. civilian non-institutionalized population, and it includes a nationally representative sample plus a disproportionate sample (compared to the population) of blacks and Hispanics. Each year, the MEPS staff selects a random sample of households from the households that participated in the NHIS in the previous year. The selection of a subsample of respondents to the NHIS results in cost savings to the MEPS data collection because it eliminates the need to independently list and screen households to locate selected subgroups of the population. In addition to oversampling blacks and Hispanics, MEPS can take advantage of its unique linkage to the NHIS to oversample other subgroups. In particular, since 2002 the MEPS also oversamples Asians and low-income people.

Note that the NHIS takes a sample of households and subsequently obtains detailed information for a subset of individual members within the household, while the MEPS also samples households and acquires information on all the members within those households. The 2004 annual MEPS sample includes approximately 15,000 households and 37,000 individuals. This means that the MEPS supports both individual- and family-level analysis; however, the survey estimates must account for this added complexity.

Cross-sectional and Longitudinal Data

Data collected on the insurance status of the population at a specific time are cross-sectional, and so they provide a "snapshot in time." But, in order to examine the patterns of health insurance coverage over time, individuals selected for the sample need to be interviewed multiple times over an extended period. This longitudinal feature may yield substantial gains in analytical capacity, but it must be counterbalanced by the greater complexity in design and associated incremental costs.

In the MEPS, data are collected for each new sample (panel) of households to cover a two-year period, and a new panel is started each year. For example, the first two MEPS panels span 1996–97 and 1997–98, respectively. To produce health care estimates for a calendar year, data are pooled across the two distinct nationally representative MEPS samples. More specifically, annual cross-sectional estimates from the MEPS are based on combined data from the second year of one panel and the first year of the subsequent panel.

Each year a new nationally representative sample (panel) of households is selected from the NHIS for MEPS. Each new panel of households is interviewed a total of five times over a 30-month period to obtain health care information covering two calendar years (Figure 1). This number of contacts reflects a careful balance between managing the data collection budget and controlling the length of recall periods (time elapsed between the event and the interview) to minimize respondent reporting errors.

With the MEPS longitudinal design (five rounds of interviews spanning two calendar years), analysts can, for example, assess the persistence of high health-care expenditures by examining whether individuals in high expenditure percentiles of the health care expenditure distribution in a given year remain in upper percentiles in the following year or shift to another higher or lower percentile. The MEPS also has the capacity to assess the impact of survey attrition on the resultant survey estimates by comparing the national health-care estimates produced by the first year of a sample panel (with a higher response rate) in contrast to the estimates derived from the second year of a MEPS sample panel covering the same time period. In addition, with the linkage of MEPS and NHIS files, longitudinal analyses of transitions in health insurance coverage and health status

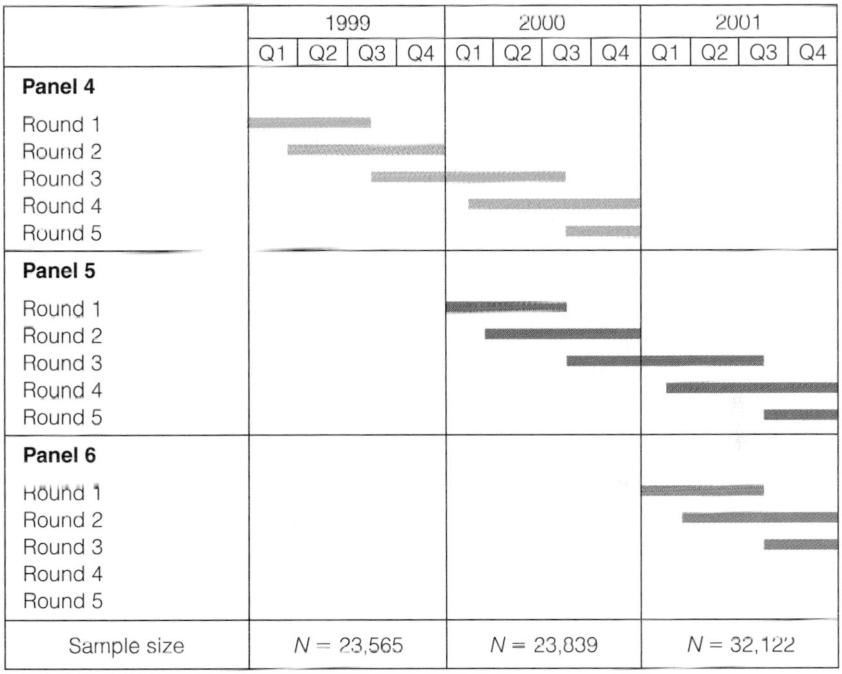

	1999				2000				2001			
	Q1	Q2	Q3	Q4	Q1	Q2	Q3	Q4	Q1	Q2	Q3	Q4
Panel 4												
Round 1	▬	▬										
Round 2		▬	▬									
Round 3			▬	▬	▬							
Round 4					▬	▬						
Round 5							▬					
Panel 5												
Round 1					▬	▬						
Round 2						▬	▬					
Round 3							▬	▬				
Round 4									▬	▬		
Round 5											▬	
Panel 6												
Round 1									▬	▬		
Round 2										▬	▬	
Round 3											▬	
Round 4												
Round 5												
Sample size	*N* = 23,565				*N* = 23,839				*N* = 32,122			

FIGURE 1 MEPS panel design: data reference periods

Source: Center for Financing, Access, and Cost Trends (CFACT), AHRQ, Medical Expenditure Panel Survey, 1999–2001.

characteristics can be examined over a three-year period. All the survey estimates derived from the MEPS account for this further complexity in the survey design and include adjustments for survey nonresponse.

Multiple Sources

The core component of the MEPS collects information from household respondents via computer-assisted personal interviewing. While household respondents reliably report their insurance coverage status and visits to doctors, dentists, and health care facilities, it is essential for the MEPS to go directly to providers of health care services for information on health care expenditures and sources of payment. Consequently, the MEPS includes two additional survey components. It includes a medical provider survey in order to supplement the household-reported data with expenditure reports from hospitals, physicians, home health-care providers, and pharmacies. Data collected in the medical-provider survey are used in conjunction with the household-component data to produce MEPS national medical-expenditure estimates.

In a parallel manner, in order to supplement the insurance information collected from household respondents, the MEPS insurance component collects detailed information from a sample of employers on costs and characteristics of health insurance plans held by and offered to respondents in the household component. Information on the amount, types, and cost of health insurance offered in the workplace is also collected from a nationally representative sample of employers. The insurance component provides a unique source of national, regional, and state data on the health insurance offerings available to Americans through their employers and the costs employers incur for providing such coverage. Starting in 2000, the Bureau of Economic Analysis in the Department of Commerce adopted the MEPS insurance component as the primary data source for estimating the cost to employers of providing their employees with health insurance coverage as a component of their estimates of the nation's gross domestic product.

The three components of the MEPS are administered under one umbrella to produce both national and regional estimates for the U.S. civilian non-institutionalized population and information on employer-sponsored health insurance. The collection of data from multiple sources to improve the accuracy of its survey estimates and enhance analytic utility is a third notable design feature of the MEPS.

SURVEY RESULTS

Two of the questions raised in the introduction are addressed in the following paragraphs, and they serve as examples of the MEPS's ability to inform health policy. All the estimates derived from the MEPS account for its complex survey design.

How Many Uninsured People Are There in a Given Year?

The possession of health insurance coverage helps ensure greater access to necessary medical care and offers greater financial protections against the risk of costly medical treatment and catastrophic health events. An understanding of the characteristics and prevalence of those individuals that experience long periods of being uninsured is critical to informing decisions targeted to those most vulnerable to health and financial risks. Comparable information on the characteristics and prevalence of the population ever experiencing an "uninsured spell" over a fixed time period, and the duration of the spell (or spells), is necessary to fully assess the dimensions of the population at risk.

The MEPS design structure permits estimates of health insurance for the first part of the calendar year in addition to the annual calendar year estimates in order to examine time-sensitive health-policy data, such as the proportion of the population not covered by health insurance. When estimates of the insurance-coverage-status distribution are derived from the MEPS, particular care is given to framing the time period used in calculating the national estimates. To illustrate the variation in the coverage estimates for the nation for a given year, the following distinct national estimates were derived from the MEPS: Data from the MEPS indicate that 25.9% of the population under age 65 were uninsured at some point during 2000, 18.2% were uninsured throughout the first half of 2000, and 13.3% were uninsured throughout the entire year (Figure 2). The data also serve to identify trends in coverage over time. For example, from 1996 to

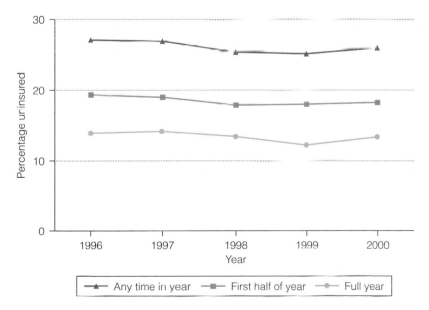

FIGURE 2 MEPS, 1996–2000, uninsured status, nonelderly

Source: CFACT, AHQR Medical Expenditure Panel Survey, 1996–2000.

1999, the percentage of the population uninsured at some point during the year declined from 27.0% to 25.1%, and the percentage uninsured all year declined from 13.8% to 12.2%. Following a similar pattern over time, the percentage of the population uninsured during the first half of the year declined from 19.2% to 17.8% from 1996 to 1998.

How Concentrated Are Health Care Expenditures?

No study of the medical expenditure patterns for the nation is complete without an examination of the concentration of health care spending. This analytical capacity was a core design requirement of the MEPS, which serves as the nation's primary source for these investigations. Based on analyses of the distribution of 1996 health care expenditures, it is clear that a relatively small proportion of the population (5%) accounts for the majority of health expenditures, and this pattern has remained quite stable over the prior two decades. The highly skewed nature of the health-care-expenditure distribution reveals that 1% of the population accounts for approximately 27% of total expenditures, while the bottom 50% of the population incurs only 3% of the overall total (Table 1). These results provide

Percentage of U.S. Population Ranked by Expenditures	1977	1980	1987	1996
Top 1 percent	27%	29%	28%	27%
Top 2 percent	38	39	39	38
Top 5 percent	55	55	56	55
Top 10 percent	70	70	70	69
Top 30 percent	90	90	90	90
Top 50 percent	97	95	97	97

TABLE 1   Distribution of health expenditures for the U.S. population, by magnitude of expenditures, selected years 1977–96

Sources: For 1977, the 1977 National Medical Care Expenditure Survey (NMCES); for 1980, the National Medical Care Utilization and Expenditure Survey (NMCUES); for 1987, the 1987 National Medical Expenditure Survey (NMES); and for 1996, the 1996 Medical Expenditure Panel Survey (MEPS). Center for Financing, Access, and Cost Trends (CFACT), Agency for Healthcare Research and Quality (AHRQ). Berk, M. L., and A. C. Monheit. (2001). "The Concentration of Health Care Expenditures, Revisited." Health Affairs 20, no. 2:9–18.

policy makers with evidence that there will be serious constraints to the application of cost containment strategies, financial incentives, and development of institutional arrangements that are limited to 90% of the population that account for roughly 30% of the health care expenditures.

SUMMARY

The Medical Expenditure Panel Survey collects data on the specific health services that Americans use, how frequently they use them, the cost of these services and how they are paid, as well as data on the cost, scope, and breadth of private health insurance held by and available to the U.S. population. The MEPS is unparalleled for the degree of detail in its data and its ability to link health service use, medical expenditures, and health insurance data to the demographic, employment, economic, health status, and other characteristics of survey respondents. Moreover, the longitudinal design of the MEPS provides a foundation for estimating the impact of changes affecting access to insurance or medical care on economic groups or special populations of interest, such as the poor, the elderly, veterans, the uninsured, and racial and ethnic minorities.

The MEPS data have recently figured prominently in investigations by the Institute of Medicine to estimate the costs of delivering uncompensated care, defined as care the uninsured receive but do not pay for fully themselves. The MEPS data have also been used pervasively in the first National Health Care Quality Report, completed by the Department of Health and Human Services in 2003 to provide an annual profile of the nation's quality of health care and to help measure improvements over time.

In summary, the MEPS design reflects a collaboration between researchers, survey designers, statisticians, policy makers, and data processing staff. The scope and depth of this data collection effort reflects the needs of government agencies, legislative bodies, and health professionals for comprehensive national estimates needed in the formulation and analysis of national health policies. The MEPS is presented as a model to guide related efforts. Few other surveys provide such comprehensive data.

ACKNOWLEDGMENTS

The views expressed in this essay are those of the authors and no official endorsement by the Department of Health and Human Services or the Agency for Healthcare Research and Quality is intended or should be inferred. The authors wish to thank Dr. Deborah Nolan for her careful review of this essay and her insightful recommendations regarding content, which guided the final revisions. In addition, the authors wish to thank Dr. Jessica Banthin for her helpful comments.

REFERENCES

Botman, S., T. Moore, C. Moriarity, and V. Parsons. (2000). "Design and Estimation for the National Health Interview Survey, 1995–2004." National Center for Health Statistics. Vital Health Stat 2(130). DHHS Publication no. (PHS) 2000-1330.

Cohen, J. W. (1997). "Design and Methods of the Medical Expenditure Panel Survey Household Component." Agency for Health Care Policy and Research. MEPS Methodology Report, No. 1. AHCPR Publication no. 97-0026.

Cohen, S. B. (2000). "Sample Design of the 1997 Medical Expenditure Panel Survey Household Component." MEPS Methodology Report no. 11. AHRQ Publication No. 01-0001.

Rhoades, J. A., J. P. Vistnes, and J. W. Cohen. (2002). *The Uninsured in America: 1996–2002.* Rockville, Md.: Agency for Healthcare Research and Quality, 2002. MEPS Chartbook, no. 9. AHRQ Publication No. 02-0027.

QUESTIONS

1. The Medical Expenditure Panel Survey (MEPS) provides information about changes in individual and family expenditures over time, and yet samples new households each year. How did the survey design provide for longitudinal information to be obtained from an annually replaced sample?

2. The essay states that there is a need for precise estimates for certain subsets. For example, policy makers may wish to estimate the percentage uninsured in the Hispanic population. Why is this an important constraint on the design of the sample survey?

3. Why was the ongoing National Health Interview Survey (NIIIS) a key resource for the MEPS?

4. The finding that only 10% of the population incurs 70% of the health-related expenditures is reported in Table 1. Does this suggest that disease management programs should focus on this small sector of the population to help mitigate the increase in overall future year health care expenditures?

PART TWO

SCIENCE AND TECHNOLOGY

———— ⁑::⁑ ————

Monitoring Tiger Prey Abundance in the Russian Far East

Ken Gerow, Dale Miquelle & V. V. Aramilev

Predicting the Africanized Bee Invasion

James H. Matis & Thomas R. Kiffe

Statistics and the War on Spam

David Madigan

Should You Measure the Radon Concentration in Your Home?

Phillip N. Price & Andrew Gelman

Statistical Weather Forecasting

Daniel S. Wilks

Space Debris: Yet Another Environmental Problem

David R. Brillinger

MONITORING TIGER PREY ABUNDANCE IN THE RUSSIAN FAR EAST

KEN GEROW

University of Wyoming

DALE MIQUELLE

Wildlife Conservation Society

V. V. ARAMILEV

*Institute for Sustainable Use of Natural Resources,
Vladivostok, Russia*

———————— ⁘ : : ⁙ ————————

The snow swirls in the bitter cold of a Khabarovsk winter morning, deep in the heart of Siberian tiger country. A pair of experienced hunters prowl side by side, searching the shifting landscape for sign of red deer, wild boar, or sika deer, favorite prey of one of the largest living cats in the world. One stops and looks down, his partner instantly motionless. He peers intently at the snowy ground, and then turns. "Sergei, do you think this might be the same deer track we saw one kilometer back?"

So goes another morning of fieldwork monitoring prey abundance in the range of the Amur tiger (Figure 1). Popularly called the Siberian tiger, the Amur tiger is a tiger subspecies named after the

FIGURE 1 Amur tiger and Russian taiga forest

FIGURE 2 The city of Khabarovok is in the southeastern corner of Russia. The Amur River is shown to the left (west) of the city, forming part of the border with China.

Source: www.lib.utexas.edu/maps

Amur River, which drains a major portion of the tiger's range in Northeast Asia (Figure 2). Scientists now question whether the original definition of seven distinct subspecies is valid, but the common naming of them as such continues.

As a group, tigers are in trouble: their population has declined by more than 90% over the last hundred years. It has been estimated that at the beginning of the 20th century 100,000 tigers flourished throughout Asia, from eastern Russia and Korea through eastern and southern China, Southeast Asia, the Indian subcontinent, and into Pakistan, with separate populations around the Caspian Sea and on the Indonesian islands of Bali, Java, and Sumatra. Even then, when tiger populations were flourishing (relative to the present) the Amur tiger represented the only population that lived in an extremely cold habitat, where snow covered the landscape for three to four months a year. By the start of the 21st century, tigers had already been eliminated from the majority of their range, with only approximately 6000 individuals remaining in the wild. Tigers are now extinct in Bali, Java, South Korea, and around the Caspian Sea. In China, a few remnant individuals hang on, and in North Korea their status is simply unknown.

More than 80% of today's tiger habitat is located outside of national parks or other protected areas, where they must share these forests and their resources with loggers, hunters, and grazing livestock. In addition, as human populations increase around the globe, those unprotected areas that are currently suitable tiger habitat could be lost to clear-cutting, agriculture, and urbanization.

The Amur tiger is the only tiger population whose habitat is comprised of temperate and boreal forests, with a climate characterized by long cold winters. The existent population of approximately 350 adult Amur tigers is distributed across approximately 180,000 km^2 (just over 70,000 m^2) in Primorskiy and Khabarovskiy Krais (provinces), with a few individuals wandering across the border into China.

Long cold winters and short summers limit the production of the vegetation that deer depend on, which in turn limits the abundance of prey for the tigers, relative to other populations across Asia. Consequently, while the dense ungulate populations ("ungulate" is a catch-all word biologists use for deer, moose, antelope, and the like) in productive habitat in India allow tigresses to patrol areas of only 20 km^2, Amur tigresses cover territories up to 600 km^2 in search of low-density prey. Red deer, sika deer, and wild boar are the primary prey species of Amur tigers, but roe deer, musk deer, moose, and even bears are eaten on occasion.

One of the preferred prey species of the Amur tiger in the southern part of its range is the sika deer (see Figure 3). Formerly found from the Russian Far East to Japan and southeastern China, it has disappeared from most of its historic range due to habitat loss. It is still common in the cool oak and coniferous forests of the southern Russian Far East, where it is a staple in the Amur tiger diet. In some parts of its range, it is also hunted for its tasty meat by local villagers. To be able to balance the needs of local people and at the same time ensure there are adequate numbers of prey for tigers, local government managers need to be able to accurately estimate sika deer numbers and monitor changes in their abundance. Despite the importance of prey abundance as the key limiting factor in tiger abundance in the Russian Far East, accurate estimates of prey populations are notoriously difficult to attain in forested environments. Russian biologists, in cooperation with scientists from around the world, have developed statistical methods for estimating the size

FIGURE 3 Male sika deer

of these prey populations to help ensure a future for tigers in the Russian Far East.

ESTIMATING ABUNDANCE

There are several well-established methods of estimating animal abundance; the problem faced by the local biologists and managers is that these methods don't work well in the forested environments of the Amur. Therefore, it behooved us to create our own approach, adapted to the specifics of this particular problem. We'll illustrate some of the difficulties by describing one of the more popular of the established methods and its poor fit to the Amur tiger prey species scenario.

One important method for estimating animal abundance is the so-called capture-recapture method. It really isn't a single estimation method but rather a diverse family of sophisticated methods with the same simple idea at their core. The idea is to capture some members of the target population and mark them so they can be identified (as having been captured once) in subsequent recaptures. Depending on the species, the capturing and marking needn't be literal: abundance estimation by capture-recapture has been done on tigers in India and elsewhere by photographing them using cameras set up at bait stations. The pattern of stripes on a tiger are unique, and depending on the quality of the photo (i.e., how well the animal posed!), an animal can be identified as having been previously seen when its photograph is subsequently "recaptured." In general, it isn't necessary to be able to identify individuals, only that an animal has been previously "captured." For example, colored plastic tags have been clipped onto fins of fish or ears of animals in order to mark them.

Some proportion of the animals captured in a subsequent capture event will be marked, and some will not. This proportion of marked to unmarked is the key to the method. To illustrate, suppose that 100 animals are marked in a first capture event. Then, suppose that 20 of the 50 animals captured in a second period are noted as having marks from the first event. We then estimate that the 100 marked animals represent $\frac{20}{50}$, or 40%, of the entire population, which we thus estimate to be 250.

There are variations on capture-recapture methodology that allow for animals to move in or out of the study areas, for animals to be born or die during the study period, and so on. To get back to prey of the Amur tiger, the coup de grâce against using capture-recapture methods in estimating their abundance is their low densities spread over incredibly large tracts of land. Simply, it is logistically infeasible to capture and mark enough animals (and then do a large enough recapture) to obtain estimates that have any sort of acceptable precision.

Other methods, such as line transect methods, rely on visual sightings of animals. Given the dense vegetation, low density of ungulates, and their capacity to bolt from approaching humans (all potential hunters to the deer) before being detected, methods that depend on seeing the animals will fare poorly. These difficulties have forced the biologists to develop their own methods that can be applied in their situation.

COUNTING METHODS

One method that has been commonly used in Russia to estimate the number of deer is an indirect counting method based on tracks in the snow. Plots are established in typical areas for a given habitat, and during winter the plot boundary is traversed by field-workers, recording tracks to estimate the number of animals that have walked into the plot versus the number that have walked out during the previous 24 hours. Plot size and shape are chosen (subjectively, given understanding by the biologists of sika deer movement patterns) to minimize the chance that deer within the plot would go undetected (i.e., not cross the plot boundary during the previous 24-hour period). A schematic of such a plot is shown in Figure 4.

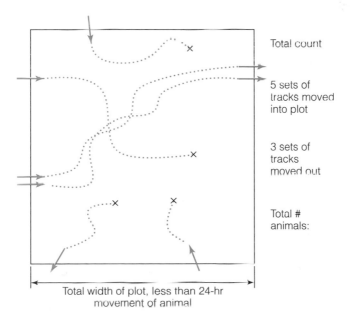

Total count

5 sets of tracks moved into plot

3 sets of tracks moved out

Total # animals:

Total width of plot, less than 24-hr movement of animal

FIGURE 4 Schematic of plot used to estimate deer numbers

The process of counting animals is much more complicated and subjective than simply "number of tracks in" minus "number of tracks out." First, it is often difficult to determine whether a given set of tracks is indeed fresh (less than 24 hours old), since estimating the age of tracks that have thawed during sunny days and then re-frozen at night becomes exceedingly difficult. In an alternative field method that has been developed to counter this problem, the perimeter of the plot can be swept of tracks one day, then surveyed the next, thus assuring that any tracks recorded are indeed fresh. Unfortunately, this method requires much more field-worker time. Further, there is a risk of a substantial bias: the activity on the first day often scares animals out of the area, leading to artificially lower numbers on the second day.

Another complication is related to group size and activity. The biologists assume that groups remain stable during a given 24-hour period. Thus if tracks of a group of two deer are recorded as entering a plot, and tracks of a group of three animals are recorded as leaving the plot, then the biologists assume that the group of two is still in the plot, and they are added to the total. Sightings of animals within the plot (even if no tracks are recorded for them) are included in the count of animals within the plot.

The biologists working on the problem believe that the method produces reasonably reliable counts, but they nevertheless have several concerns. One is the dependence on expert field-workers, who might not always be available. A second problem is that counting in plots is time-consuming and expensive because it requires three to five people to count each plot. Finally, while the counts appear to be fairly reliable, the result does not allow any statistical measures of precision. For these reasons, we sought a means to predict animal density with a simpler, quicker method that would provide statistical estimates of precision.

MODELING THE RELATIONSHIP BETWEEN COUNTS AND TRACKS

We sought to create a statistical approach that is more efficient, that is not dependent on expert interpretation, and that provides an estimate of statistical precision. We decided to explore the potential for "tracks counted" as the basis for a model to predict animal density.

This removes the issues around deciding which groups have entered the plot, then left again, and even whether the tracks are going in or out of the plot (not always easy to tell after the sun has thawed out the tracks). It still leaves the necessity of determining whether an observed track should be counted as "fresh" (i.e., less than 24 hours old) or not, but even so, the new approach, if successful, would constitute an improvement in the move toward a more objective methodology. Just as important, tracks can be counted along simple transects that can be randomly placed in tiger habitat, eliminating the cumbersome need to locate and count entire plots. These transects would not be tied conceptually to a specified plot (in the same way that traversing the perimeter is connected to a given plot). Nonetheless, the resulting density estimate could be scaled up into an abundance index by multiplying it by the area of land thought to be represented (by having similar habitat characteristics, for example) by the transect.

Because not all plots are exactly the same shape and size (in one study area, they range from 5 to 20 km², depending on restrictions imposed by the landscape), there isn't a stable relationship between perimeter length and area. Changing the count data to density (number of animals per plot divided by the size of the plot, yielding numbers per km²) and the track counts to a track rate (number of tracks per surveyed km) resolves some of the issues presented by the data in their original form.

Figure 5 shows a graph of animal density against track rate for deer species in Primorye Krai (a province), one of the two study areas

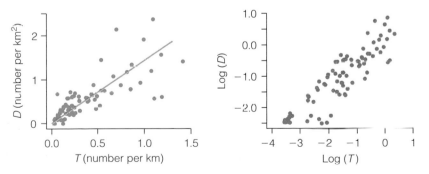

FIGURE 5 Observed densities (D) plotted against track rates (T) for deer in Primorye Krai (left panel). The right panel shows the plot for log-transformed D and T. The straight line in the left panel shows average D over the range of T.

(the other being Khabarovsk Krai). The data are combined for roe deer, red deer, and sika deer. These data were obtained by their historical methods (presumed to be valid, but very time-consuming and requiring expert field personnel). Our goal is to see if we can estimate the densities (on average, with decent precision) by using only the track-count-rate information (i.e., ignoring whether tracks are going in or out of the plot), thus making the process cheaper and less dependent on the availability of expert field personnel.

The data have a feature, illustrated by the line in the left panel, that motivates our modeling. If you fit a line to the data, it is natural to choose one that goes through the origin: the point (0,0) corresponds to a track rate of 0 and a density of 0. The equation of a line through the origin has the form *density* = $C \times$ (*track rate*). Biologists call C, which is the slope of the line, the correction factor (the number by which an observed track rate T should be multiplied to yield an estimated density D). In other words, the average density is proportional to the track sighting rate. This relationship appeals to biological intuition as well. If in one plot you see twice as many tracks as another plot, you might expect there to be twice as many animals in the plot (on average, there will of course be variation in the data caused by individual animals moving more or less and by where they were actually located with respect to the transects walked by the biologists).

There is, however, a problem with this data set. When the vertical scatter is about the same throughout the plot, we say the data are homoscedastic (literally, equal scatter). In the left panel of Figure 5 the variability among the densities is larger for high sighting rates (right side of the graph) than for low (left side of the graph). The sort of steadily patterned increasing variation (called heteroscedasticity) seen in the left panel of Figure 5 is common in biological data, particularly data based on counts such as we have here, and is problematic. If you try fitting a line segment to the points by eye, you can be quite confident where one end of the line should go, in the lower left part of the plot; but it is much less certain where the other end of the line should go, because the points toward the upper right are so much more spread out. In more quantitative terms, note, for example, that for a track rate of 0.25 tracks per km, the densities range from about 0.2 animals per km^2 to about 0.75, a range of about 0.55 animals per km^2. That same calculation done for a track rate of 1.0 tracks per km yields a range of about 1.7 animals per km^2: the range is more than three times as large.

Now that we've chosen a form for the relationship, we need to estimate the correction factor C. It is at this point that the unequal scatter creates a problem. Some data values are more variable and thus less informative than others. Fortunately, there is a solution ready at hand, one that requires only a little algebra and a few logarithms. Statisticians know that when the size of variability is roughly proportional to the size of what is being counted or measured, taking logs will "equalize the spreads": in the log scale, the size of the variability will be roughly constant.

The right panel of Figure 5 shows the same data as the left panel, plotted this time in the log scale. As you can see, the pattern is reassuring: the vertical spread does seem to be roughly constant. Such data have two advantages from a statistician's point of view. First, they make it reasonable to treat all points equally when you fit a line to the plot, and second, they make it possible to summarize the size of the variability using a single number, a feature that is important if you want to be able to specify a margin of error for any estimates based on the data.

Based on the results of these checks on the model, we can now go ahead with the estimation of the correction factor C. Our model, log-transformed, appears as $\log(D) = \log(C) + \log(T)$, which we can rearrange to $\log(C) = \log(D) - \log(T)$. In this form, $\log(C)$ can be easily estimated: for each data point, compute the difference $\log(D) - \log(T)$; the average of these will be our estimate of $\log(C)$.

For this data, $\log(C)$ is estimated to be 0.5596, with a standard deviation of 0.055. Approximately two standard deviations (0.11) is the usual margin of error for a 95% confidence interval. We can thus say that we are 95% confident that $\log(C)$ is between 0.45 and 0.67 (estimate minus margin of error, and estimate plus margin of error). Of course, we are really interested in the correction factor C, not its log-transformed version. Easily remedied: we simply exponentiate the relevant values. Thus we estimate C to be $e^{0.5596} = 1.75$ and are 95% confident that the true value is between $e^{0.45} = 1.57$ and $e^{0.67} = 1.95$.

Here is a brief illustration of how the correction factor might be used. An observed 0.25 tracks per kilometer translates into an estimated density of $1.75 \times 0.25 = 0.44$ animals per km², while an observed rate of 1.5 tracks per km (slightly higher than that observed in this data) yields an estimate of 2.63 animals per km². We can easily create a 95% confidence interval for these estimates if we take the observed track rate as a given. Simply multiply the rate by each of the two limits given for C. For instance, we can be 95% confident that, in the area that gave us 1.5 tracks per km, there are between $1.57 \times 1.5 = 2.36$ and $1.96 \times 1.5 = 2.94$ deer per square kilometer. As an aside, we note that if we treat the track rate as an estimate, with its own imprecision, the calculations are more involved than this.

MANAGEMENT IMPLICATIONS

The primary objective of this modeling exercise was to develop a means of estimating animal abundance without the cumbersome task of laying out and counting large plots. By developing the correction factor C, we are now able to estimate animal density (which can be easily scaled by area into an index of abundance) simply by counting tracks. Conducting track counts on transects is a quicker and more effective way to collect data, and it frees us from relying on expert interpretation of tracks to estimate animal abundance, as was the case for plot counts.

Second, we are able to attach some sense of precision to our estimates (which was not possible using the purely wisdom- and experience-based counting approach). The margin of error and resulting confidence statement meet that objective. In our case, the limits (1.57 and 1.95) are within (approximately) 10% of our estimate (1.75) for the correction factor C, which is a reasonable amount of precision.

Using track counts will provide a basis for estimating prey abundance across tiger habitat. Because the territory is vast, it would be simply impossible to establish a sufficient number of plots to reliably estimate animal density. However, by correctly placing transects in tiger habitat and counting ungulate tracks, we can use the correction factor to estimate ungulate density, and thus gain a better understanding of ungulate distribution and density across the entire range of tigers in the Russian Far East. Accurate estimates of ungulate abundance, and regular monitoring of changes in abundance, will allow managers to better control human harvest to ensure adequate food resources for both tigers and local people.

Because ungulate density is one of the key factors defining the quality of tiger habitat, we will be able to use data on ungulate densities to determine the importance of different forest patches across the region for tiger conservation. By identifying priority areas for protection, we can work with local officials to mitigate development activities and develop a zoning process that provides for the needs of local people but does not threaten tigers—or the prey base they depend upon. Information on prey density is thus a key piece of the puzzle needed for securing a future for tigers in the Russian Far East.

REFERENCES

Karanth, K. U. (1995). "Estimating Tiger *Panthera tigris* Population from Camera Trap Data Using Capture-Recapture Methods." *Biological Conservation* 71:331–38.

Krebs, C. J. (1999). *Ecological Methodology,* 2nd ed. Menlo Park, Calif.: Addison Wesley Longman.

Miquelle, D. G., et al. (1996). "Food Habits of Amur Tigers in Sikhote-Alin Zapovednik and the Russian Far East, and Implications for Conservation." *Journal of Wildlife Research* 1, no. 2:138–47.

———. (1999). "A Habitat Protection Plan for the Amur Tiger: Developing Political and Ecological Criteria for a Viable Land-Use Plan." In *Riding the Tiger: Meeting the Needs of People and Wildlife in Asia,* edited by J. Seidensticker, S. Christie, and P. Jackson, 273–95. Cambridge: Cambridge University Press.

———. (1999). "Hierarchical Spatial Analysis of Amur Tiger Relationships to Habitat and Prey." In *Riding the Tiger: Meeting the Needs of People and Wildlife in Asia,* edited by J. Seidensticker, S. Christie, and P. Jackson, 71–99. Cambridge: Cambridge University Press.

"Tiger." Microsoft® Encarta® Online Encyclopedia 2003. http://encarta.msn.com © 1997–2003 Microsoft Corporation. All Rights Reserved.

QUESTIONS

1. What changes in the tiger and deer populations over the last century have motivated this study?
2. How will the results of the study help to maintain a balance of human and animal populations in the study area?
3. Why is it difficult to estimate the abundance of sika deer in the study area?
4. What is the capture-recapture method of estimating animal abundance, and why doesn't it work in this study area?
5. Why would an instance of two animals moving into a plot and three animals moving out suggest that the two are still in the plot?
6. What motivated the statistical approach to animal density estimation based on counted tracks alone?
7. Why is the perimeter of the plot examined important in estimating animal abundance?
8. "So, all else being equal, one might expect to see a larger number of tracks (per animal inside the plot) for the smaller plots." Show why this is so by considering what might be observed in two square plots of 2 km per side and 5 km per side. (Hint: Is the number of tracks on the perimeter of a square plot approximately proportional to the number of animals in the plot?)
9. The relationship between the number of perimeter tracks counted and the number of animals depends on the plot size. How is this complication eliminated by the method proposed in the essay?
10. What is the technical simplification achieved by the logarithmic transformations proposed for Figure 5?
11. Track rates are easy to obtain; density is more useful, but harder to obtain. What is the relationship of log(track rate) and log(density), based on the data in Figure 5?
12. How can we assess the precision of density estimates using that relationship?

PREDICTING THE AFRICANIZED BEE INVASION

JAMES H. MATIS

Texas A&M University

THOMAS R. KIFFE

Texas A&M University

———— ❋ : : ❋ ————

The spread of Africanized honey bees (AHB) has been called one of the most remarkable biological events of the century. From the accidental release of 26 swarms of AHB in Brazil in 1956, the population had grown by 1989 to many millions of colonies spread over more than 20 million km^2 (roughly the size of the United States and Canada). The relentless range expansion, illustrated in Figure 1, is one of the best documented biological invasions ever recorded. It was clear in 1989 that the AHB would soon arrive in the United States.

The impending arrival of the AHB attracted great media attention at the time. Though the AHB is visually indistinguishable from the common, domesticated European honey bee (EHB), it differs in other ways that cause concern. First, the AHB has what might be called a "personality disorder," and such colonies are recognizable immediately by their extremely defensive nature. When a colony is disturbed, perhaps even unintentionally by such things as vibrations

FIGURE 1 Range expansion of Africanized honey bee populations, 1956–89

Source: Courtesy of John Thomas

from a lawn mower or weed eater, numerous bees (typically well over 100) from the colony attack the perpetrator. An AHB sting is no more toxic than an EHB sting, but the accumulated effect of so many AHB stings can be lethal to humans and domestic animals. The media focused on this public health concern, and the AHB became known as the "killer bee."

AHB colonies also differ in that they tend to swarm (i.e., multiply) and abscond (i.e., relocate) more frequently, which, combined with their increased defensiveness, make them more difficult to manage. These traits, besides just complicating honey production, impact the agricultural economy in another way. The yield of many crops, particularly certain fruits and nuts, may be increased by enhancing crop pollination. This is usually accomplished by importing large numbers of EHB beehives into the fields for a short period during the growing season. However, counties with known AHB presence are quarantined, which prohibits the importation of EHB colonies into the fields. The traits previously mentioned make the AHB much more difficult to use for crop pollination, which is estimated to have a multibillion-dollar impact on the agricultural economy.

The economic and public health concerns led to a flurry of research among biologists, economists, statisticians, and other scientists in southern Texas, the area in which the bees would enter the United States. The key questions were how long would it be before the bees reached the United States and how prevalent would they be once they arrived.

AHB ARRIVAL TIME PREDICTIONS

Early Studies of AHB Movement

Data were available in 1989 from east–west trap lines that were set up to detect the arrival of the AHB front in northern Guatemala and in the Atlantic and Pacific coastal areas of Mexico. These traps consisted of bait hives, each hive being a 25 I. cardboard box baited with a chemical attractant. The hives were hung from trees about 3 m above the ground, usually in clusters of two to three every 300 to 500 m along roadways. The hives were checked at least monthly to determine whether an AHB swarm had been captured. The distances between the trap lines ranged from 10 to 300 km, with a mean of 93 km. The data on the distances and time intervals between 45 consecutive first-capture dates along these trap lines are given in Table 1. Figure 2 illustrates the locations and capture dates for a subset of 19 trap lines.

We searched for suitable statistical models for the data on which we could base predictions of future movement. Previous predictions were based on calculating the speed (distance/time) at which, under various environmental conditions, the leading edge was proceeding northward. Some theoretical models suggest that under certain dispersal conditions the distance traveled by the leading edge in a fixed time interval would follow a normal distribution. We calculated the

Distance (km)	Time (mo.)	Transit Time (mo./100 km)	Distance (km)	Time (mo.)	Transit Time (mo./100 km)
75	4	5.3	85	7	8.2
65	3	4.6	140	2	1.4
40	1	2.5	120	1	0.8
60	5	8.3	105	2	1.9
50	1	2.0	90	3	3.3
55	1	1.8	80	1	1.3
130	1	0.8	65	1	1.5
45	1	2.2	70	1	1.4
105	5	4.8	50	1	2.0
140	4	2.9	115	4	3.5
120	5	4.2	100	4	4.0
80	5	6.3	95	3	3.2
75	5	6.7	85	3	3.5
60	12	20.0	30	2	6.7
60	11.5	19.2	10	2	20.0
35	2	5.7	60	3	5.0
70	2	2.9	35	3	8.6
35	2	5.7	40	7	17.5
55	2	3.6	55	3	5.5
90	1	1.1	250	4	1.6
260	10	3.8	300	5	1.7
175	1	0.6	90	2	2.2
70	7	10.0			

TABLE 1 Spread of Africanized honey bee through Mexico and Northern Guatemala

estimated speeds from the data in Table 1, but the distribution was not even close to being normal. We decided, therefore, to use a different approach based on "transit time," the reciprocal of speed. This approach would model directly the time required to travel some fixed distance, which after all was the desired end point of the study. One technical advantage of this new approach is that the observed time intervals in the data, which were usually available only to the nearest month and hence had large inherent lack of precision, would be in the numerator of our variable (i.e., time/distance). Placing the measurement with the greater uncertainty in the numerator, rather

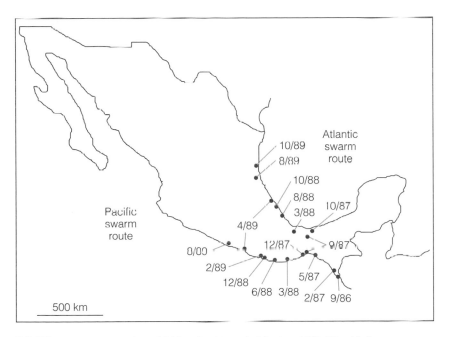

FIGURE 2 Range expansion of Africanized bees in Mexico, 1986–89, with first-capture dates at selected trap line locations

than in the denominator as in the speed (distance/time) variable, would give the subsequent statistical analysis more stability. Therefore, the data in Table 1 were transformed to give calculated consecutive transit times in standardized units of mo./100 km. A histogram of these 45 observations is illustrated in Figure 3. These data are not even close to being normally distributed either, but the distribution does have a recognizable shape.

One striking feature immediately apparent in the histogram is the large variability of the data, with the estimated times required to travel 100 km ranging from less than one month to 20 months. The data set also has four values that stand out as being unusual. We tried to identify any common characteristics shared by these four data points, such as a common location (for example, Atlantic vs. Pacific coast) or a common season, because if variables that might be associated with these unusual values could be identified, predictions of future movement could be improved by incorporating these variables into our analysis. Unfortunately, we were not able to identify any such variables; so, based on the histogram in Figure 3, it was

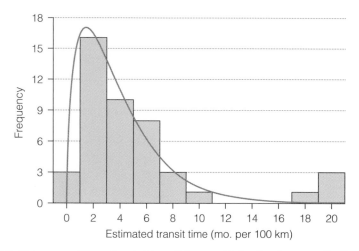

FIGURE 3 Histogram of the estimated transit times, with the fitted gamma distribution ($\alpha = 1.61, \beta = 2.37$)

decided to model the transit times using a single unimodal, skewed distribution. A gamma distribution was chosen for this purpose. Gamma distributions are flexible, ranging in shape from very skewed to symmetric, and are often used as a model for variables that, like transit time, can only take on positive values. A gamma distribution that provides a good fit to the data is shown by the smooth curve that is superimposed over the histogram in Figure 3.

Predicting the U.S. Arrival

When we began our work, the northernmost capture of the AHB occurred in September 1989 in Mexico's Tamaulipas Province, a distance of 215 km from the U.S. border. By thinking of the time to travel 215 km as

time to travel 100 km + time to travel 100 km + 0.15(time to travel 100 km)

we were able to use the gamma distribution in Figure 3 (which describes the distribution of time to travel 100 km) to develop a new model to describe transit time for a distance of 215 km. This new model is also a gamma distribution, and is shown in Figure 4.

The mean transit time for the gamma distribution in Figure 4 is 8.2 mos., with a standard deviation of 4.4 mo. Our point estimate for the most likely date for the leading edge to be observed in the United States was therefore 8 mos. after the last sighting in

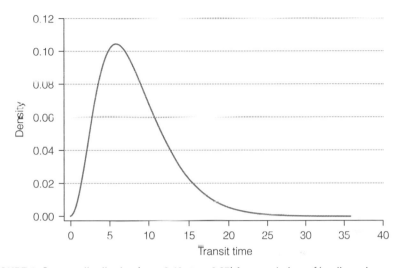

FIGURE 4 Gamma distribution ($\alpha = 3.46$, $\beta = 2.37$) for transit time of leading edge to travel 215 km

September 1989, giving a predicted arrival in May 1990. A number of other, mostly biological investigators had also given point predictions. As statisticians, we added an interval estimate for the transit time of 2.0 to 18.8 mos. This 95% prediction interval extended from November 1989 to March 1991. The wide interval provided a quantitative manifestation for scientists that predicting the movement of the leading edge from the existing data was very imprecise.

The first capture in the United States actually occurred on October 15, 1990, in Hildalgo County, Texas. The capture was confirmed and released to the media on October 18. It was a huge media event. Our prediction turned out to be the most accurate one, and we gained instant recognition in the beekeeping community and in some biological modeling circles.

We also made a second prediction in September 1989 that was published separately. This prediction concerned the expected transit time from the northernmost AHB invasion along the Pacific coast to the expected second U.S. entry point near Nogales, Arizona. The northernmost Pacific coast capture at that time was in Guerrero Province, a distance of 1856 km from Nogales. Our best estimate for the transit time was 58.2 mos., which gave a July 1994 arrival date prediction. The corresponding 95% prediction interval for the transit time was 39.1 to 80.6 mos., yielding an interval from December 1992 to May 1996 for the capture date. The actual data of first capture in Arkansas was July 1993. Our predictions again turned out well, but the event was largely ignored in the national media as it occurred nearly three years after the Texas event.

A Local Prediction

After the initial capture in Texas, there was a long, mystifying pause until the next AHB swarm was captured six months later on April 15, 1991, at a nearby location in the Rio Grande Valley in Texas. By the end of April, there were another 24 captures, and the number of captures started to grow exponentially, reaching 120 by September 1991. We were invited at that time to write a news report in which we made a third prediction. In order to call attention to our hometown, which at the time had a statistician (Larry Ringer) as its elected mayor, we predicted that the AHB would be captured in College Station, Texas, a distance of 620 km from the Rio Grande Valley, in April 1993. Our 95% prediction interval was from June 1992 to July 1994.

There was no capture in College Station, or in its surrounding county, prior to April 1993 or even by July 1994. In fact, the AHB leading edge did not arrive until seven years later, in July 2001. This was a bit humbling because our previous successes were well-known and because we had touted the practical utility of the interval estimate.

This failure could have led to one of two possible general reactions on our part. One would be to dismiss it as just bad luck—the method we used has a 5% error rate, and this could just be one of the times when the method failed. Another would be to investigate why the prediction could have failed so completely. Predictions, of course, are conditional upon constant environmental conditions. It was apparent through the mid-1990s that the leading edge was slowing down, but there were no obvious environmental changes to account for this phenomenon. Soon afterward, however, the reason for the slowdown was discovered as beekeepers began to observe a new invader, the varroa mite, in their hives.

A Second Biological Invasion

Figure 5 shows several bees with varroa mites. The varroa mite is a bee parasite that was well-known to beekeepers but had not

Lilia de Guzman, USDA/ARS Honey Bee Breeding, Genetics and Physiology Lab

FIGURE 5   Bees with varroa mites (appearing as red dots on the bees)

previously been discovered in Texas. The mite was first discovered in Java in 1904, where it has little effect on the local bee species, the Indian bee. The mite had migrated to the USSR by the '60s and to western Europe by the '80s, with disastrous consequences on EHB colonies. The EHB is totally susceptible to the mite, and the presence of even a single mite in a colony is an indicator of its likely eventual demise if the colony is left untreated. The first reported sighting of the mite in the United States was in 1987. Thus unbeknown to us at the time, the two pest invasions, the AHB from the south and the mite from the east, met in Texas in the mid-1990s. The mite was decimating the feral (i.e., wild) EHB colonies, and drastically slowing down the AHB invasion.

We are still confident, of course, in the transit time prediction methodology based on the gamma distribution model. However, we are now more cautious and will make future predictions only when new baseline data under the new environmental conditions become available.

AHB DENSITY PREDICTIONS

Local Data

A key question for the subsequent management of the AHB is its predicted density in an area after its arrival. Data for such predictions were gathered at the Welder Wildlife Refuge (WWR) from 1993 to 2000. WWR is a 7800-acre ranch located 50 km north of Corpus Christi, Texas, roughly half way between the Rio Grande Valley and College Station. The first AHB colony was captured in the county (San Patricio) in 1992. Table 2 gives the observed numbers of AHB and EHB colonies for each year at the 6.25 km^2 study area at WWR.

The trends are clear. The number of AHB colonies started increasing rapidly from 1993 to 1995. The first varroa mite was observed at WWR in 1995, with resulting immediate consequences. The number of EHB colonies declined drastically, with a 76% reduction in 1996, and never recovered. The AHB is known to be more resistant to the mite. Indeed, the number of AHB colonies never decreased, though its population growth was damped somewhat for

Year	EHB Colonies	AHB Colonies
1993	24	1
1994	59	3
1995	67	14
1996	16	16
1997	13	22
1998	10	28
1999	12	49
2000	15	61

TABLE 2 Bee density at Welder Wildlife Refuge, 1993–2000

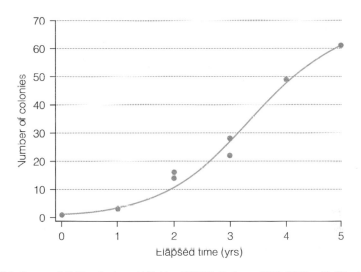

FIGURE 6 Count of AHB colonies at Welder Wildlife Refuge, 1993–2000, with fitted logistic growth curve

four years as the AHB population adapted to the presence of the mite.

The logistic growth curve is a well-known model in ecology for natural population growth. The curve fits the growth data well, as illustrated in Figure 6. The time scale in Figure 6 condenses the four-year period of slow growth into two normal years to account for the temporary lull in natural population growth. The asymptotic value of the logistic curve is called the "carrying capacity" of the environment. This value is 68.5 for the fitted curve, which in the present context would be interpreted as the number of AHB colonies that are sustainable by the natural environment in the study area at the WWR.

A Stochastic Model

A model that takes random birth and death events into account, developed by the British statistician M. S. Bartlett and his colleagues in the early '60s, provides much additional insight. This model assumes that the population size N is an integer random variable. The count N may either increase by one (for a "birth"), decrease by one (for a "death"), or remain unchanged in any future time interval. The probabilities of these birth and death events depend upon the size of the population at the time. Because N is a random variable, the solution to this model at any given time is not a single value, as in Figure 6, but is instead a probability distribution for the population size. Bartlett and his colleagues describe a numerical procedure for finding this distribution when the process is in long-term balance, that is, in "equilibrium." This distribution is called the equilibrium size distribution, and it is the stochastic analog of the carrying capacity.

As an illustration, we propose, based on the fitted logistic curve in Figure 6, a stochastic logistic growth model with birth and death rate functions given in Figure 7. The rate functions, when scaled to a suitably small time interval, give the probabilities of a birth or a death event in the time interval for a population of any given size N. For example, the birth rate from Figure 7 for $N = 50$ is 79.25/yr., or $79.25/365 = 0.217$/day; hence, the probability of a birth in a one-day interval for population size 50 is 0.217, whereas the probability of a death for a population of the same size in the same interval is $62.25/365 = 0.171$. Such probabilities are sufficient to simulate this process from any initial size in daily time intervals, though in practice other simulation procedures are also available. The two quadratic curves in Figure 7 intersect at 68.5, which is the aforementioned carrying capacity. Therefore, the probability of a birth exceeds that of a

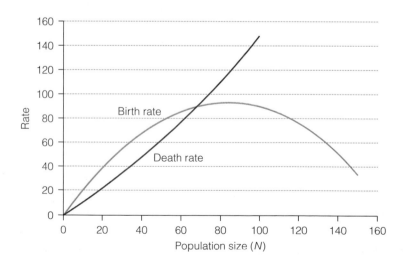

FIGURE 7 Assumed annualized birth and death rate functions at WWR. The intersection at $N = 68.5$ gives the estimated carrying capacity.

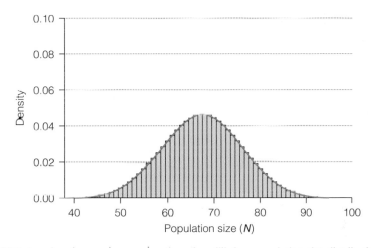

FIGURE 8 Exact and normal approximation of equilibrium population size distribution for AHB colonies at WWR from the assumed stochastic logistic model

death for any $N \leq 68$, whereas the opposite is true for $N \geq 69$. These population size-dependent, or in ecological terms "density depen-dent," probabilities create a natural equilibrium size distribution in which population sizes much smaller than 68 become increasingly unlikely, as do population sizes much larger than 69. The exact solu-tion for the equilibrium size distribution for this assumed model, using the Bartlett et al. (1960) procedure, is given in Figure 8.

Bartlett et al. (1960) also discuss simple methodology for approxi-mating the equilibrium distribution. For example, the approximate mean and standard deviation of 67.4 and 8.6, respectively, for the equi-librium distribution may be calculated by substituting the parameters of the curves in Figure 7 into algebraic expressions. These results imply that in the long run the number of AHB colonies, instead of reaching the fixed capacity of 68.5 as given in Figure 6, will fluctuate with a mean of 67.4 colonies and a standard deviation of 8.6. It can also be shown that the distribution is approximately symmetric, and so a normal distribution can be used to approximate the equilibrium distribution, as illustrated in Figure 8. It is apparent that the normal approximation is indeed accurate. Hence based on this birth-death model, one may conclude that a 95% prediction interval for the number of AHB colonies in this area at WWR, in the long run, is roughly 50 to 85.

Implication of Prediction

Clearly the natural variability in population size inherent in this model is considerable. This model takes into account only the variability resulting from assumed random single-birth and single-death events, and more sophisticated models are also available. Nevertheless, the present simple interval is a useful management tool, as it quantifies the inherent variability that one would expect in the absence of any new environmental impact, such as some new invading interacting species. As an illustration, based on these findings, managers know that there is no statistical evidence of a change in the assumed conditions unless the AHB colony count exceeds 85 or falls below 50. Such findings are obviously not available from the fitted logistic growth curve alone.

To give some perspective to the results, we note that the predicted mean of 67.4 colonies gives a predicted density of 10.8 AHB colonies/km^2 in this area at WWR. This density is roughly 50% higher than any previously recorded density in the scientific literature for feral bees in an area, including both suitable and unsuitable habitat. Moreover, the stochastic model implies that even higher densities will certainly occur randomly over time under present conditions. In short, Texas has been hospitable to both recent pest invaders, the AHB and the varroa mite, and further study of their interactions is important for the agricultural economy and is fascinating scientifically.

CONCLUSIONS

These predictions of future arrival dates and densities relating to the AHB invasion illustrate some strengths and limitations of the underlying statistical methodology as applied to the biological world in general. One obvious strength of the predictions is that they incorporate the inherent variability of the processes, as reflected by the prediction intervals. On the other hand, one limitation is the assumption of constant conditions, which may be unrealistic in the long run in a biological world in constant flux. For example, in addition to the sudden changes in the projected AHB arrival dates and AHB density counts due to the unexpected mite invasion, there is growing evidence that the EHB is also gradually becoming more mite resistant. Moreover, though commercial beekeepers are committed to maintaining pure EHB colonies, the feral EHB colonies are being replaced by African hybrid colonies (which unfortunately retain the extreme defensiveness). An overall conclusion is that the present statistical tools, though effective under stated assumptions, should be applied only with wisdom and good judgment to the ever-changing biological world.

REFERENCES

Bartlett, M. S., J. C. Gower, and P. H. Leslie. (1960). "A Comparison of Theoretical and Empirical Results for Some Stochastic Population Models." *Biometrika* 47, no. 1:1–11.

Baum, Kristen A. (2003). "Feral Africanized Honey Bee Ecology in a Coastal Prairie Landscape." PhD diss. Texas A&M University, College Station, Tex.

Baum, K. A., et al. (2004). Spatial and Temporal Distribution and Nest Site Characteristics of Feral Honey Bee Colonies in South Texas. Unpublished manuscript.

Kaplan, J. K. (2004). "What's Buzzing with Africanized Honey Bees." *Agricultural Research* 52 (March): 4–8, http://www.ars.usda.gov/is/AR/archive/mar04/.

Matis, J. H., and T. R. Kiffe. (2000). *Stochastic Population Models.* New York: Springer-Verlag.

Matis, J. H., W. L. Rubink, and M. Makela. (1992). "Use of the Gamma Distribution for Predicting Arrival Times of Invading Insect Populations." *Environmental Entomology* 21, no. 3:431–40.

Matis, J. H., et al. (2003). "A Simple Saddlepoint Approximation for the Equilibrium Distribution of the Stochastic Logistic Model of Population Growth." *Ecological Modelling* 161:239–48.

Renshaw, E. (1991). *Modelling Biological Populations in Space and Time.* Cambridge: Cambridge University Press.

QUESTIONS

1. Why is the AHB disruptive to the EHB population?
2. Why is the AHB disruptive to human populations?
3. What do Figures 1 and 2 show, overall?
4. Figure 3 shows how the data compare with the best-fitting gamma model. The data show longer transit times per 100 km than the model. If the gamma model is used to predict the speed of advance of the AHB from northern Mexico northward, how would the model estimate differ from an estimate that used the actual data distribution?
5. What unexpected natural factor limited the advance of the AHB?
6. Figure 6 shows the logistic growth curve for the AHB population. Why is it natural for a growth curve to have this sigmoid shape?
7. In Figure 7, when the population is small, the birth rate exceeds the death rate and the population grows. Can the population grow past 68 colonies? (Suppose the intersection is at 68 colonies.)
8. Why do you think the 95% interval was used rather than, for example, a 50% or 99% interval?
9. If a predictive model provides erroneous predictions, such as the model that predicted the AHB capture at College Station in 1993, is there any justification for continuing to use the model?

STATISTICS AND THE WAR ON SPAM

DAVID MADIGAN

Rutgers University

────────── ❋ : : ❋ ──────────

"Are you an Inventor? Have a Great Idea? We can help"
"Now available—cash maker"
"Find out if your mortgage rate is too high, NOW. Free Search"
"Instantaneously Attract Women"
"Confirming your FREE PDA"

These are the subject lines from unsolicited e-mail messages I received in a one-hour period. Once a mild annoyance, unsolicited bulk e-mail—also known as spam—constitutes close to half of all message traffic on the Internet. Spam is a big problem for several reasons:

- Spam imposes a significant burden on Internet service providers; the resulting costs ultimately trickle down to individual subscribers.
- Spam supports pyramid schemes, dubious health products and remedies, and other fraudulent activities.
- Spam exposes minors to inappropriate material.
- Spam results in a loss of productivity; the cumulative costs add up quickly when all e-mail users spend a few minutes a day dealing with and disposing of spam.

Defenders of spam draw comparisons with print catalogs. Catalog companies use regular mail to send catalogs to potential new customers.

Why is it OK to send unsolicited catalogs, and not OK to send unsolicited e-mails? One key difference is cost—spammers can send millions of messages for little or no cost. Paul Graham, an antispam activist, proposed the following thought experiment (see http://www.paulgraham.com for further detail):

> Suppose instead of getting a couple print catalogs a day, you got a hundred. Suppose that they were handed to you in person by couriers who arrived at random times all through the day. And finally suppose they were all just ugly photocopied sheets advertising pornography, diet drugs, and mortgages.

Given the central role of e-mail in our professional lives, the analogy with spam is not so far-fetched.

The problem of spam is attracting increasing media attention and a sort of War on Spam has begun. In what follows I describe the critical role statistics plays in this battle.

SPAM FILTERS

Systemic progress against spam may require complex worldwide legislation. Individual users, however, can protect themselves with so-called spam filters. A spam filter is a computer program that scans incoming e-mail and sends likely spam to a special spam folder. Spam filters can make two kinds of mistakes. A filter might fail to detect that a particular message is spam and incorrectly let it through to the inbox (a false negative). Equally, a filter might incorrectly route a perfectly good message to the spam folder (a false positive). There is a trade-off between controlling these two kinds of mistakes: aggressively labeling e-mail as spam will keep spam out of your inbox but will misclassify your good messages (too many false positives). On the other hand, being lax routes more good messages to your inbox but lots of spam get past the filter, too (too many false negatives). An effective spam filter will have a low false negative rate and, perhaps more critical, a low false positive rate.

Anyone can build his or her own spam filter. In fact, early spam filters used handcrafted rules to identify spam. Here are the antecedents for some typical spam rules:

- <Subject> contains "FREE" in CAPS
- <Body> contains "University Diploma"
- <Body> contains an entire line in CAPS
- <From:> starts with numbers

This approach can produce effective spam filters. However, the handcrafting approach suffers from some significant drawbacks. First, creating rules is a tedious, expensive, and error-prone process. Second, humans are unlikely to spot more obscure indicators of spam such as the presence of particular HTML tags or the use of certain font colors. Plus, the ever-evolving nature of spam requires continual tweaking of the rules, otherwise they might get outdated. In the last couple of years, the statistical approach to constructing spam filters has emerged as a method of choice and provides the core technology for several leading commercial spam filters. The statistical approach starts with a collection of e mail messages that have been hand-labeled as spam or not-spam; creating these "training data" is the time-consuming part. Next, we build a statistical algorithm for detecting spam by scanning the collection of e-mail messages and identifying discriminating features of the spam. Finally, the trained algorithm scans new e-mail messages as they arrive and automatically labels each one as spam or not-spam. In what follows, we describe "Naive Bayes," a straightforward and popular approach to building spam filters that has achieved excellent results on some standard test collections.

SIMPLE ONE- AND TWO-WORD SPAM FILTERS

We begin with a simple example that introduces some of the basic ideas behind a statistical approach to spam filtering. Suppose you build a filter that merely looks for the word "free" in an e-mail message. This approach catches many of the spam in my collection because "free" occurs frequently in the spam messages and almost

IS "FREE" IN MESSAGE?	IS MESSAGE SPAM?		
	No	Yes	Total
No	395	240	635
Yes	5	360	365
Total	400	600	1000

TABLE 1 Occurrence of the word "free" for a hypothetical set of 1000 messages

never in the other messages. To see how this might work, imagine that the training data contain 1000 messages, 600 of which are spam. You count occurrences and create Table 1.

In this example, "free" occurred in 60% of the spam messages, and only five times out of the 400 other messages. Now suppose you declare a message to be spam if the word "free" is present. How well does your filter work? There are 365 messages that contain the word "free," and 360 of these are spam, or about 99%. Also, the false alarm rate is about 1%: not bad. Unfortunately, the miss rate—the rate of failing to detect spam and declaring it to be real—is not anywhere near as good. There are 635 messages without the word "free," and 240 of these, about 38%, are spam. Our single-word filter would let all these messages pass through. Clearly, we need something more effective.

The logic we just used is easy to extend to create filters based on more than one word. To illustrate how this might work, suppose you add the word "mortgage" to your table. For each message in the training set, you now record whether "free" is present, whether "mortgage" is present, and whether or not the message was spam. Table 2 presents the additional detail in the 1000 hypothetical messages.

By reading down the columns of numbers, you can see a "numerical profile" that tells how spam and regular messages differ statistically. Most of the genuine messages contain neither word, whereas 47% of the spam messages contain "free" only, 17% contain "mortgage" only, and 13% contain both words. By reading across the rows, you can see what fraction of messages of a given type is spam. If neither word is present, the data show that $\frac{140}{485}$, or about 29%, of the

IS "FREE" IN MESSAGE?	IS "MORTGAGE" IN MESSAGE?	IS MESSAGE SPAM?		
		No	Yes	Total
No	No	345	140	485
No	Yes	50	100	150
Yes	No	4	280	284
Yes	Yes	1	80	81
Total		400	600	1000

TABLE 2 Occurrence of the word "free" and/or the word "mortgage" for a hypothetical set of 1000 messages

messages are spam. If "mortgage" is present but "free" is not, $\frac{100}{150}$, or about 67%, of the messages are spam. If "free" only is present, 99% of the messages are spam, and if both words are present, it is also 99%.

By using two words instead of just one, you can create a filter with error rates that are better than what we got with the single-word filter. The improvement is a change in the right direction, but some of the miss and false-alarm rates are unacceptably high. Fortunately, there is nothing that forces you to stop with just two words. Why not try three, or 10, or all of the words in the training data? First you will need an efficient way of tracking the different words in the e-mail messages.

REPRESENTING E-MAIL FOR STATISTICAL ALGORITHMS

All statistical algorithms for spam filtering begin with a quantitative representation of individual e-mail messages. Researchers have studied many different representations but most applications use the so-called bag of words, which is a complete list of all distinct words taken from all the e-mail messages in the training database. For each e-mail message in the training set, we count the number of occurrences of each word in the bag of words. The string of counts gives the quantitative representation of the message.

If the training data comprise thousands of e-mail messages, the number of distinct words often exceeds 10,000. Two simple strategies

Order	Message	Spam
1	the quick brown fox	No
2	the quick rabbit ran and ran	Yes
3	rabbit run run run	No
4	rabbit at rest	Yes

TABLE 3 Flow chart of the process of data reduction for training data comprising four labeled e-mail messages

to reduce the size of the list somewhat are to remove uninteresting words, called "stop words" (such as *and, of, the,* etc.), and to reduce words to their root form, a process known as stemming (so, for example, "ran" and "run" reduce to "run"). The bag-of-words representation has some limitations. In particular, it contains no information about the order in which the words appear in the e-mail messages!

Table 3 presents a simple set of training data comprising four e-mail messages. The complete vocabulary list contains 10 distinct words: *and, at, brown, fox, quick, rabbit, ran, rest, run,* and *the.* After eliminating stop words and stemming, the reduced vocabulary list contains six distinct words: *brown, fox, quick, rabbit, rest,* and *run.* The word counts for the second e-mail message, "the quick rabbit ran and ran," are (0, 0, 1, 1, 0, 2) because the message does not contain the words *brown, fox,* and *rest,* but it does contain one occurrence each of *quick* and *rabbit* and two occurrences of the word *run* (in its "ran" form). The words *and* and *the* are not included in the word count. We simplify the word counts to a binary form (0, 0, 1, 1, 0, 1), which tracks whether the word appears at all in the message (1) or not (0).

NAIVE BAYES FOR SPAM

The Naive Bayes approach seeks to build a model for the probability that a message is spam given its binary representation. In principle, such a filter could give error rates that were quite tiny. In practice, unfortunately, a direct extension of the approach that the previous section described does not work. The reason is that there are just too many possible combinations of words in comparison to the number of messages in the training set. How many is that? It would not be unusual for the reduced vocabulary list to contain at least 1000 different words. With 1000 words, we would need to build a table with rows that correspond to a set of 1000 yes/no answers. With just one word, we needed two rows (Table 1). With two words, we needed $2 \times 2 = 2^2$ rows (Table 2). With 1000 words, we'd need 2^{1000} rows. That number is almost as big as the number of grains of sand in the world—squared! It looks hopeless, but it turns out, somewhat surprisingly, that making an unrealistic assumption about the content of e-mail messages leads to a simplification of the table and a filter that works extremely well—better, in fact, than the rules used by some of the commercial spam filters.

This "naive" but effective simplification assumes that for each word in the reduced vocabulary list, whether that word is present in a spam message does not depend on any of the other words in the message. Similarly, whether that word is present in a non-spam message does not depend on any of the other words in the message. Even though this assumption is unlikely to reflect reality (e.g., in many spam messages the word "mortgage" is strongly associated with the word "interest"), the simplification often performs well in practice. The extreme nature of the simplifying assumption earns this statistical approach its "naive" name; some earlier literature refers to the model as "Idiot's Bayes."

Such a simplification means that for a bag of words made of a 1000 words, we do not need to compute the table with 2^{1000} rows. Instead, we make 1000 two-row tables like Table 1—that is, one for each word in the reduced vocabulary list. To show how we would use these minitables in place of the big table, consider our simple two-word example with "free" and "mortgage" again (Table 2). If we wish to estimate the proportion of spam messages that, say, contain "free" and not "mortgage," we notice that $\frac{360}{600}$, or 60%, of the spam messages contain "free" and $\frac{420}{600}$, or 70%, of the spam messages do not contain "mortgage." The naive approach says that because 60% of the spam contains "free," then about 60% of the spam without the word "mortgage" should also contain "free." So we estimate the proportion of spam with "free" and without "mortgage" to be 0.6 × 0.7 = 0.42, or 42%. Actually, $\frac{280}{600}$, or 47%, of the spam contain "free" and not "mortgage." This may seem like a poor way to go about estimating these proportions, but remarkably, it proves quite effective!

Notice that we have found a simple way to approximate the chance that a spam message contains "free" and not "mortgage," but what really interests us is the chance that a message with "free" and not "mortgage" is a spam message. It appears that we have approximated the wrong probability! Bayes' rule comes to our rescue here, which is where the "Bayes" part of the name of this approach comes from. When we look across the third row in Table 2 we see the counts that we need to compute the proportion of messages with "free" and not "mortgage," $\frac{280}{284}$. Bayes rule shows us how to reexpress this probability using proportions taken from the columns of the table.

To do this, it turns out to be easier for us to work with the ratio of the proportion that we are after (the proportion of "free" and not "mortgage" messages that are spam) to the proportion of these messages that are not spam, $\frac{4}{284}$. The ratio of these two chances, $\frac{280}{4}$, is called the odds ratio. The odds ratio will be near 0 when it is unlikely that the message is spam, and it will get very large when the message is highly likely to be spam. In our simple example, the odds ratio is very large, indicating that a message with the words "free" but not "mortgage" is probably spam. An alternative computation for this odds ratio follows.

$$\frac{280}{4} = \frac{600/1000 \times 280/600}{400/1000 \times 4/400} = \frac{600/1000}{400/1000} \times \frac{280/600}{4/400}.$$

We see from this alternative expression that the odds ratio is the product of two other ratios—namely, the ratio of the proportion of spam to good e-mail in the training set and the ratio of the proportion of spam that contains "free" and not "mortgage" to the proportion of good e-mail that contains "free" and not "mortgage." Now we are back in business with column proportions that we can approximate simply with our naive approach:

$$\frac{600/1000}{400/1000} \times \frac{360/600 \times 420/600}{5/400 \times 350/400} = \frac{600/1000}{400/1000} \times \frac{360/600}{5/400} \times \frac{420/600}{350/400} = 58.$$

Our computation is the product of simple ratios taken from our two-row tables. For the full bag of words, the odds ratio will be computed from the product of thousands of these types of ratios.

How does it work in practice? From the training set, we create a bag of words. Then we convert the e-mail messages into 0-1 representations, and prepare two-row tables for each word in the bag. Now the filter is ready. A new e-mail message arrives. Does the Naive Bayes filter predict that this message is spam? After stemming and stop-word removal on the message, the filter creates the binary representation of the message that tells which words in the bag of words are present in the message. To find the odds ratio for whether or not the message is spam, we go through the binary representation of the message one word at a time and compute the product of the appropriate ratios. If the resulting odds ratio is large, the filter routes the message to the spam folder. Otherwise the filter lets the message through to the inbox. How large does it need to be? The spam-filter builder chooses a threshold. When the probability exceeds the threshold it is labeled spam. The choice of the threshold must reflect the builder's relative concerns about the two kinds of errors mentioned earlier. A low threshold may result in a higher false positive rate—more legitimate messages ending up in the spam folder. A high threshold may result in a higher false negative rate—more spam messages ending up in the inbox. Most statistical spam filters allow the user to select the threshold.

NAIVE BAYES EFFECTIVENESS

The actual implementation of the Naive Bayes algorithm has a few more complications to it. For example, because there are so many words in the training data, some of the proportions may be 0. This happens when a particular word only occurs in good messages in the training set and not in any spam, or vice versa. Division by 0 would create problems for our product of ratios. The standard solution to this problem is to "smooth" the estimates by adding a small positive constant to each cell in all of our two-row tables. This adjustment typically leads to more accurate estimates of the probabilities. Also, in practice, it is not the odds ratio that is computed but the logarithm of the odds ratio. The logarithm of a product of values is the sum of the logarithm of each value. Computing the sum of the log ratios rather than the product of the ratios is more stable.

All in all, the Naive Bayes approach is simple to implement and scales well to large training data. The literature on spam filtering reports a number of experiments where Naive Bayes delivers competitive false positive and false negative rates. Sahami et al. (1998) conducted experiments with a corpus of 1789 e-mail messages, almost 90% of which were spam. They randomly split this corpus into 1538 training messages and 251 testing messages. A Naive Bayes classifier, as previously described, achieved a false negative rate of 12% and a false positive rate of 3%. Extending the quantitative representation to include handcrafted phrases and some domain-specific features reduced the false negative rate to 4% and the false positive rate to 0%. Androutsopoulos et al. (2000) report experiments using the "Ling-spam" corpus of 2893 e-mail messages, 481 of which are spam. They reported a false positive rate of 1% computed via 10-fold cross-validation. Their Naive Bayes algorithm outperformed a set of rules built into a widely used commercial e-mail client. Carreras and Marquez (2001) on yet another corpus (PU1) report a Naive Bayes false positive rate of 5%. Neither Androutsopoulos et al. nor Carreras and Marquez report false negative rates.

Further, the binary representation of the messages ignores the number of occurrences of each word in a message, yet these word counts may provide useful clues for identifying spam. A number of researchers have reported success with nonbinary versions of the Naive Bayes approach. Newer statistical algorithms such as regularized logistic regression, support vector machines, and boosted decision trees can achieve higher effectiveness. Researchers have reported experiments in which boosted decision trees outperformed Naive Bayes by a few percentage points on the Ling-spam corpus.

DISCUSSION: BEYOND SPAM FILTERING

Spam filtering has provided the primary focus for this essay and we have highlighted the central role of statistics. In fact spam filtering is one particular example of the broader topic of "text categorization." Text categorization algorithms assign texts to predefined categories. The study of such algorithms has a rich history dating back at least 40 years. In the last decade or so, the statistical approach has dominated the literature. The essential idea, as with statistical spam filtering, is to infer a text categorization algorithm from a set of labeled documents (i.e., documents with known category assignments), where features of the documents are represented quantitatively.

Applications of statistical text categorization include:

- Building web directories such as the dmoz open directory and Yahoo!
- Routing incoming customer-service e-mail
- Identifying "interesting" new stories for intelligence analysts
- Classifying financial news stories as "significant" or "insignificant"

While Naive Bayes is rarely the top-performing algorithm in any of these applications, it invariably provides reasonable effectiveness with minimal computational effort. More important, Naive Bayes illustrates the concepts that allow one to understand how more sophisticated spam filters of text categorization programs work.

REFERENCES

Androutsopoulos, I., et al. (2000). "An Experimental Comparison of Naive Bayesian and Keyword-Based Anti-Spam Filtering with Personal E-mail Messages." In *Proceedings of the 23rd ACM SIGIR Annual Conference*, 160–67. N.p.

Carreras, X., and L. Marquez. (2001). "Boosting Trees for Anti-Spam Email Filtering." In *Proceedings of RANLP-01, 4th International Conference on Recent Advances in Natural Language Processing*, 58–64, http://citeseer.nj.nec.com/article/carreras01boosting.html.

Cox, D. R. (1970). *The Analysis of Binary Data*. London: Chapman and Hall.

Eyheramendy, S., D. D. Lewis, and D. Madigan. (2003). "On the Naive Bayes model for Text Classification." In *Proceedings of the Ninth International Workshop on Artificial Intelligence and Statistics*, edited by C. M. Bishop and B. J. Frey, 332–39. N.p.

Lewis, D. D. (1998). "Naive (Bayes) at Forty: The Independence Assumption in Information Retrieval." In *ECML'98, the Tenth*

European Conference on Machine Learning, edited by C Nedellec and
C. Rouveirol, 4–15. N.p.

McCallum, A., and K. Nigam. (1998). "A Comparison of Event Models
for Naive Bayes Text Classification." In *Proceedings of the AAAI-98
Workshop on Learning for Text Categorization.* Menlo Park, Calif.:
AAAI Press.

Mosteller, F., and D. L.Wallace. (1984). *Applied Bayesian and Classical
Inference,* 2nd ed. New York: Springer-Verlag.

Sahami, M., S. Dumais, D. Heckerman, and E. Horvitz. (1998). "A
Bayesian Approach to Filtering Junk E-Mail." In *Learning for Text
Categorization: Papers from the 1998 Workshop.* AAAI Technical Report
WS-98-05, 55–62.

Sebastiani, F. (2002). "Machine Learning in Automated Text
Categorization." *ACM Computing Surveys* 34, no. 1:1–47.

Spiegelhalter, D. J., and R. P. Knill-Jones. (1984). "Statistical and
Knowledge Based Approaches to Clinical Decision Support Systems,
with an Application to Gastroenterology (with Discussion)." *Journal
of the Royal Statistical Society (Series A)* 147, no. 1:35–77.

QUESTIONS

1. Spam filters use keywords and other tests to identify spam
   e-mail. It is possible to instruct your mail program to
   automatically delete messages identified as spam. Does this solve
   the problem for the potential recipient of spam?

2. Each individual might have a different tolerance for the errors
   made by a spam filter in identifying messages the individual
   considers to be spam. Is it necessary to modify the criteria used
   by the spam filter to match each individual's needs, or is there a
   way to adapt a single spam filter to meet many individuals'
   needs?

3. Suppose you wanted to choose between two, one-word filters,
   where one filter used the presence of the word "free" and the
   other used the presence of the word "amazing" as indicators
   of spam. Of 1000 messages, of which 600 were spam, 360 of
   the spam messages used the word "free" while only five of the

non-spam messages used this word. On the other hand, in a different sample of 1000 messages to the same e-mail address, of which 400 were spam, 300 of the spam messages used the word "amazing" while 10 of the non-spam messages used this word. Which one-word filter would you prefer? Explain your preference using the appropriate error rates.

4. How successful would you be in constructing a spam filter based on the use of both "free" and "amazing," based on the data in Question 3? Or is it impossible to say based on the data provided? Explain.

5. The odds ratio based on data in Table 2 is found to be $280/4 = 70$. However, the odds ratio calculation using the Naive Bayes method is 58. Why are these results different? Which approach is more logical? Which approach is more useful in practice? Explain.

6. What is the nonbinary use of the word list in a training set, and why might it provide a better spam filter than the binary use of the word list?

7. Spam filtering is an example of a method for automated categorization of text excerpts based on a training data set of such text messages that have been categorized by experts. What applications, other than spam filtering, might use this technology?

SHOULD YOU MEASURE
THE RADON
CONCENTRATION IN
YOUR HOME?

PHILLIP N. PRICE

*Energy and Environment Division, Lawrence Berkeley
National Laboratory*

ANDREW GELMAN

*Department of Statistics and Department of Political Science,
Columbia University*

———— ☀ : : ☀ ————

In this essay, we discuss how statistical methods can potentially save lives, dollars, or both when applied to improve a public health decision, even when the estimates of various key inputs to the decision are very uncertain. The same principle can apply to many problems besides the one we discuss here, which is the health risk from home exposure to radon, a naturally occurring radioactive element.

Radon was discovered in the year 1900, and high lung cancer rates among miners were recognized hundreds of years before that, but it wasn't until the 1950s that researchers accepted that exposure to radon-decay products was one of the major causes of the high lung cancer rate among miners. Risks to the general population weren't recognized until the early 1980s, when it was discovered that even homes can have high concentrations of radon—indeed, the airborne radioactivity in some (very rare) homes is higher than is allowed in uranium mines!

Once a risk from radon was recognized, the U.S. Environmental Protection Agency and other organizations including the Department of Energy launched research efforts to help assess risks and remediation options: what is the statistical and spatial distribution of indoor radon, what methods can be used to reduce radon concentrations in homes, what is the risk as a function of exposure, and so on. In this essay, we summarize some of the data on radon distributions and some of the risk estimates, and then we discuss how the available information can be used in a decision analysis to make recommendations on who should perform radon measurements or remediation. This essay focuses on the decision analysis itself (originally discussed in Lin et al. [1999]), and not on the statistical model that fed into the decision analysis. For more information about the statistical methodology (hierarchical regression), see Gelman et al. (2003, section 22.4).

THE HOME RADON PROBLEM

Radon is a naturally occurring radioactive gas produced by decay of the element radium, which is present in small quantities in rocks and soil. Since radon is a gas, it can flow through the soil and into homes. Radon's decay products, which are themselves radioactive, are known to cause lung cancer if high concentrations are inhaled for a long time. When we speak of risk from radon, we really mean the risk from radon-decay products.

What is the risk from radon? Studies have been carried out for uranium miners, many of whom had extended exposures at a high concentration of 20 picoCuries per liter, abbreviated 20 pCi/L, or more. Those studies, and others, suggest that lifelong exposures of this magnitude increase the risk of lung cancer to about 0.06 for male nonsmokers (from a risk of less than 0.006 for those not exposed to radon) and 0.32 for male smokers (from an unexposed risk of 0.07); estimates for women are comparable but slightly lower (National Research Council, 1998).

Figure 1 shows the estimated additional lifetime risk of lung cancer death for male and female smokers and nonsmokers, as a function

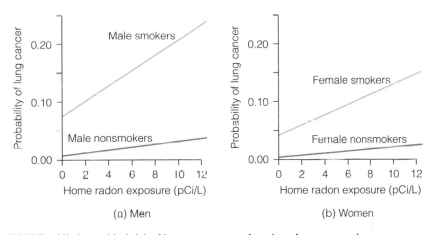

FIGURE 1 Lifetime added risk of lung cancer, as a function of average radon exposure in picoCuries per liter (pCi/L). The median and mean radon levels in ground-contact houses in the U.S. are 0.7 and 1.5 pCi/L, respectively (see Figure 3), and more than 50,000 homes have levels above 20 pCi/L.

Source: Gelman et al. (2003).

of average radon exposure over 30 years, under the assumption that observed risks at the concentrations faced by uranium miners can be extrapolated linearly to lower concentrations that occur in homes (typically below 5 pCi/L). The linear extrapolation assumes that half the exposure means half the risk; we return to this assumption later in the essay. The uranium miners were exposed to much higher radon concentrations than typically found in homes, but more than 50,000 of the United States' 70 million houses have levels above 20 pCi/L; people who live in these houses receive a radiation dose that exceeds the occupational safety limit for uranium miners. For almost everyone else, the risk of lung cancer is much lower than for uranium miners, but there are millions of people facing this risk. Assuming the linear dose-response assumption is accurate, between 5,000 and 25,000 people die annually in the United States due to lung cancer caused by inhaling radon-decay products in their homes, with most of these deaths being among smokers. The estimated number of annual deaths from radon far surpasses that from other chemical or radiological exposures. However, these estimates of radon deaths (including the uncertainties) are based on linear extrapolation, and there is no guarantee that the risk from radon is actually linear in exposure; the actual risk could be lower (or somewhat higher) than these estimates suggest.

THE EPA'S RECOMMENDATIONS

The Environmental Protection Agency (EPA) recommends that all homeowners in the United States test for radon and if the radon level exceeds a 4 picoCuries per liter (4 pCi/L) "action level," remediate the house to reduce the radon exposure to an acceptable level. Remediation costs about $2000, including maintenance and energy costs for operating the system. This quick summary of the EPA's recommendation doesn't address the important question of how the action level was set and what kind of tests should be used by homeowners.

The action level was originally set based on the "annual living-area average" concentration in U.S. homes, which is the mean concentration in the living areas of a home, averaged over a year. However, the EPA's recommended testing protocol doesn't measure the average concentration of the living areas of a home: the EPA recommends testing only on the lowest level of the home that is used as living space. Some state health departments, and many radon testing companies, recommend testing on the lowest level of the home that *can be* used as living space rather than the lowest level that *is* used as living space, so in practice most homes are monitored in basements, whether or not the basement is frequently occupied. Testing on the lowest level of the home that is (or can be) used as living space almost always leads to a measurement that is higher than the living-area average radon concentration, because concentrations are typically highest on the lowest level of the home. The concentration in the basement is often two to three times higher than the concentration on the upper floors.

There are two types of inexpensive radon measuring devices: one measuring the average concentration over just a few days, the other averaging over several or many months. Unfortunately, a short-term measurement gives little information about a home's long-term average concentration, which is believed to be the important parameter for radon risk; indoor radon concentrations vary substantially from day to day and from season to season, primarily due to changes in weather that affect the amount of soil gas that is drawn into the house. The EPA recommends either a long-term test or a short-term test plan that has two tiers: make a short-term test, and if the concentration exceeds 4 pCi/L, make another short-term test, taking the average of the two tests to see if remediation is recommended. In practice, relatively few people make long-term tests, and many people fail to follow the short-term protocol, especially when testing is performed as part of a real estate transaction and a rapid test is desired; the de facto testing plan for many homeowners who test for radon is to make a single short-term test on the lowest level of the home, and use that test as the basis for a decision. We will refer to this one-measurement decision plan as the "standard" plan even though it differs somewhat from the EPA's recommendation. Analyzing the EPA's actual recommended short-term testing plan would complicate our work without changing the results much.

The standard decision plan has two notable features: first, the measurements on which it is based are biased (because they are taken on a low floor) and highly variable (because they are taken over only a few days) if interpreted as estimates of the average annual living area concentration; and second, the same testing procedure is recommended for all homes, even though some areas of the country and some types of homes tend to have much higher radon concentrations than others do.

Performing a single short-term test in every house in the country would cost nearly $2 billion just for the measurements, and if those measurements are used to make remediation decisions based on the 4 pCi/L action level, billions more will be spent on remediations, most of them in homes that do not in fact have substantially elevated radon concentrations. We estimate that use of this decision plan in every house would cost $21 billion ($25 for each measurement and $2000 for each remediation) and save 55,000 lives over a 30-year period. It is natural to wonder if we can do better using area- and house-specific information to develop an alternative decision plan.

TOWARD A BETTER DECISION PLAN

Collecting and analyzing area- and house-specific information about radon risk is key to finding optimal homeowner-specific recommendations about what action to take. This can be done through a formal decision analysis that considers the possible homeowner decisions (test, remediate, do nothing) as part of a statistical analysis. The decision analysis requires that we determine the predicted outcomes for each possible decision, namely how much money would be spent and how many people would die, including some measure of the uncertainty in these predictions. Including the uncertainty is crucial so that decisions take into account the full range of possible outcomes. One must also develop a quantitative measure of the desirability of each outcome. In this decision problem, that amounts to attaching a dollar value to each life saved (or lost). We return to this important but hard question later. Based on the predicted outcomes (dollars spent and lives saved) and the relative desirability of each outcome, the decisions are analyzed to choose the one that has the best average outcome.

Data and Estimation of Radon Exposure

The first step in building a decision plan is to obtain information about the distribution of radon exposure across different homes in different regions of the country. There are several sources of residential radon data. In terms of the number of measurements, by far the largest data sets simply summarize all of the radon measurements performed by radon testing companies; often these are available by state, county, or even by zip code. Unfortunately, there are some serious problems with these data, most important that they overrepresent homes with high radon levels and often include multiple measurements from such homes. High-radon homes are more likely to be included because people who suspect their houses have high radon levels—perhaps because a neighbor has obtained a high measurement—are more likely to take measurements, and those who obtain a high measurement are more likely to measure again, and this measurement will also be included in the database. Because of this large and unpredictable bias in "volunteer" data, we instead work with data collected by the government from random samples of houses. This is an illustration of the importance of using chance in the collection of data to avoid having systematic biases enter the data.

The EPA worked with most of the state health departments to perform radon testing in representative random samples of homes in each state. Taken together, these EPA/State Residential Radon surveys contain about 55,000 radon measurements (Wirth, 1992). These are

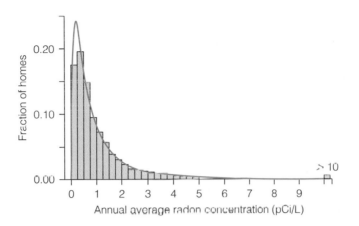

FIGURE 2 Distribution of average annual living area radon concentrations in U.S. residences. Bin heights show fraction of homes in each bin, and the last bin shows the fraction of homes with concentrations exceeding 10 pCi/L. A superimposed curve shows a lognormal distribution with geometric mean (GM) of 0.67 pCi/L and geometric standard deviation (GSD) of 3.1; that is, the logarithms of the radon levels have mean log(0.67) and standard deviation log(3.1). The nationwide distribution is very close to lognormal.

short-term measurements on the lowest level of the home, but at least they use a standard protocol and are not subject to the selection bias (favoring high radon homes) that affects the commercial data.

Additionally, the EPA conducted a National Residential Radon Survey, which measured annual living-area average concentrations (using long-term detectors on every floor of each home) in about 5700 homes across the country (Marcinowski, Lucas, and Yeager, 1994). Figure 2 shows the distribution of annual living-area average radon concentrations as determined by that survey. It is common when working with quantities such as concentrations, which are greater than or equal to zero, to work with the logarithm of the measurement. The logarithms tend to more readily allow the use of traditional statistical tools. For example, the logarithms of the radon levels follow a normal distribution with mean log(0.67) and standard deviation log(3.1); in standard terminology, the distribution is "lognormal" with a "geometric mean" (GM) of 0.67 pCi/L and a "geometric standard deviation" (GSD) of 3.1. Among other things, this implies that 6% of the homes have radon levels exceeding the EPA's

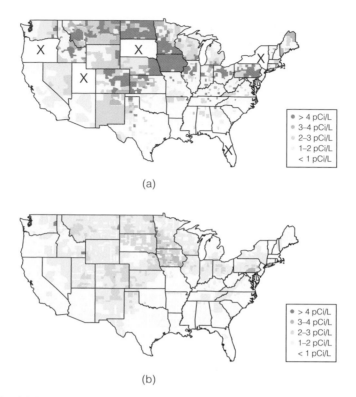

(a)

(b)

FIGURE 3 (a) Observed geometric mean indoor radon concentration by county, from the EPA/State Residential Radon surveys. Measurements in these surveys were taken in basements in the winter and thus are strongly upwardly biased estimates of average living-area concentrations. Some states (marked with X on the plot) did not participate in these surveys but did collect data in different surveys; those data, using different measurement protocols, are not displayed here but were used in our full analysis.
(b) Estimated county geometric mean indoor radon concentration, from a hierarchical Bayes regression model, which reduces small-sample variation, corrects for systematic measurement biases present in map (a), and also uses geologic and other information to predict long-term living-area concentrations.

threshold level of 4 pCi/L. To see this, note that a radon level of 4 pCi/L is $(\log[4] - \log[0.67])/\log(3.1) = 1.58$ standard deviations from the mean, and the probability that a sample from a standard normal distribution exceeds 1.58 is 6%.

The state radon surveys contain measurements in about 55,000 homes. The geometric mean measurements by county are shown in Figure 3a. There are about 3000 counties in the United States. Some of the counties were heavily sampled by the surveys, while others were missed altogether or included only a few houses. Radon concentrations vary a lot within a county, so a small number of measurements isn't enough to estimate accurately the distribution of radon concentrations within any given county. Also, as mentioned above, the measurement protocol for these surveys leads to measurements that are biased (because they are from the lowest level of the house only) and that are subject to a lot of variation (because they are short-term measurements).

The national and EPA/state survey data can be used to form a predictive model that allows us to draw inferences about the radon concentration in a house. For the moment we focus on the state survey data and ignore the known weaknesses (bias, variability) of the data. Our approach to learning about radon concentrations from the state survey data is to provide a statistical model linking the survey measurements to house and county factors (Price, Gelman, and Nero, 1996; Price, 1997). This is a traditional survey inference approach, learning about the population from a limited number of sampled units.

We analyzed the radon data with a technique known as Bayesian hierarchical modeling (sometimes called "multilevel modeling"). The key idea underlying this method is that information about one data sampling unit (for example, a county) also tells you something about other units (other counties). For example, suppose we were to make a lot of radon measurements in 86 out of the 87 counties in Minnesota using the state survey protocol, and that we find that almost all of the counties have geometric mean concentrations in the range 2–5 pCi/L. Even with no samples at all from the missing county, we would think that it's likely that its geometric mean would also be somewhere between 2 and 5 pCi/L. As this example illustrates, knowing the distribution of county geometric means can help in estimating the geometric mean for a specific county, even if that county is sparsely sampled or isn't sampled at all.

This concept can be applied even in more complicated cases that involve predictor variables. In our case, the state survey measurements are found to correlate with a number of home and county characteristics: the county's average surface radium concentration as measured with airborne detectors by the National Uranium Resource Evaluation; whether the home has a basement; the "geologic province" containing the county; and the county's heating-degree days (a measure of winter climate severity that is related to the convective forces that draw soil gas into homes). Even when all of these variables are taken into account, however, there remains considerable variation in radon concentrations, both within and between counties. Homes in some counties tend to have higher radon levels than expected given the county and home characteristics, while homes in other counties tend to have lower levels than expected. These tendencies are probably related to county characteristics that we have not measured or included in the study. We use a regression modeling approach to estimate the effects of all of the variables listed above on radon measurements, and we also include parameters to allow for "county effects" that quantify the extent to which radon levels in a county are higher or lower than would be predicted from the other variables. We estimated the regression model separately in each of 10 regions of the United States, yielding predictions (and uncertainties) of what every county's measured radon distribution would have been if they had all made large numbers of measurements in the (biased) state surveys.

By comparing the state survey data to the national radon survey data it is possible to estimate the degree to which the state survey data are biased. This can be thought of as a calibration step, relating the biased state survey data to the more accurate annual living-area average concentration from the national survey data. One way to think about this step is that we take the data from the national survey and compare each home's measurement with the prediction that would be made for that home using the regression model that was developed with the state survey data. If the national data differ systematically from the state survey prediction, we can figure out how to adjust the prediction to remove the systematic error. (In practice, the adjustment is done through a more sophisticated statistical model that looks at both data sources simultaneously [Price and Nero, 1996].) Once the bias is removed, we have a predicted radon distribution for almost every county in the United States; in some counties the distribution can be estimated rather precisely, but in the ones that were sparsely sampled the uncertainties are large. The estimated geometric mean average living-area radon concentration in each county is displayed in Figure 3b.

We shall use these estimated distributions to make decision recommendations for homeowners living in each of the 3000 counties in the United States, but first we must lay out the decision options and their costs and benefits.

Decision Analysis: Balancing Dollars and Lives

As with many public health decisions, radon measurement and remediation involves a balance between dollars and lives—or, more precisely, between resources spent on reducing radon risks and resources spent otherwise. Mathematically, the balance of dollars and lives is expressed as a loss function measuring the combined cost of measurements, remediation, and lives lost. For this problem a natural way to express the loss is to start with the cost of measurements and remediation (which are measured in dollars) and add to that a constant times the expected number of lives lost in the next 30 years (a reasonable guess of the time that a remediation system will be effective). The constant assigns a dollar value to saving a life. An approach for choosing a value for that parameter is described next.

By the way, when we discuss "expected" costs or "expected" lives lost, this is statistical terminology that means "If this identical decision were faced infinitely many times, what would be the average costs and average lives lost?" So, why not just say "average" instead of "expected"? The answer is rather subtle (and perhaps not very important): statisticians usually use the term "average" to refer to the actual average (the sum of a set of values divided by number of values), whereas the term "expected value" refers to the statistical expectation for what that number should be. For instance, if you flip a coin 100 times, the "expected value" of the number of heads you'll get is 50, but the average number of heads that you get if you perform an experiment in which you perform several repeats of flipping a coin 100 times might not be exactly 50. We'll use the "expected" terminology throughout this essay, but you can think of it as an average if that is easier to picture.

Consider a home with an average living-area radon concentration of 4 pCi/L, which is the EPA's recommended "action level." Suppose the home has the average U.S. household mix of 0.30 male and 0.27 female smokers and 1.07 male and 1.16 female nonsmokers. (It may seem odd to consider a household containing fractional people, but this allows us to discuss an average case that addresses all of the types of people of interest.) For such a household, a 4 pCi/L concentration represents an expected 0.02 radon-induced lung cancer deaths over a 30-year period. Let C represent the dollar value of a life. If this household does nothing about radon, then the expected loss is $0.02C$. Assume remediation costs about $2000 and reduces the average radon level to 2 pCi/L for 30 years (thus cutting the radon risk in half). For this household, the expected loss if remediation is performed is $2000 + 0.01C$. The fact that 4 pCi/L is the EPA's recommended action level means that this is the radon level for which the EPA thinks the preceding two expected losses are about the same. If true, this would tell us that the value of saving a life is $C = \$200,000/\text{life}$. Though $200,000 can buy a lot, it is actually low compared to the value of life assumed in most public health decisions in rich countries; if the linear dose response is true, then remediating your 4 pCi/L home for $2000 is something we should consider a good investment! Setting the threshold at other levels yields different implicit values of a life. For example, a threshold of 20 pCi/L, as used by Canada, corresponds to only $22,000 per life saved under our assumptions. The fact that such a low value is implied for a human life suggests that if the dose-response relationship is linear, then the Canadian threshold is way too high and puts too many people at risk.

Using the same action level in every house will actually lead to a wide variation in the dollars spent per life saved: for a household with several inhabitants who smoke, reducing radon would save more lives than would the same reduction in a home with one nonsmoking inhabitant. Conversely, if the trade-off between lives and dollars is the same for every person, then it is the action level that should vary among households. We take this latter view and assume that each household assigns the same dollar value ($200,000, as implied by the EPA recommendation) to a life. Then we compute the number of lives that would be saved by remediation for each house based on the predicted or measured radon level and the composition of the household. This allows us to calculate the expected loss with and without remediation and then select the homeowner's strategy that minimizes the expected loss.

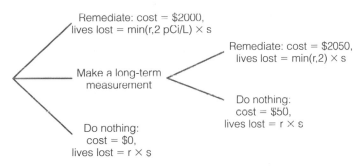

FIGURE 4 Decision tree for radon measurement and remediation. The first decision (whether to remediate, do nothing, or take a [long-term] measurement) depends on estimates from the county-level model displayed in Figure 3b. If a measurement is taken, the second decision depends on the value of the measurement. The expected loss of life depends on the radon concentration r, the post-remediation radon concentration (which we assume to be 2 pCi/L or r, whichever is lower), and on the risk (in dollars) per unit of radon concentration for the household, which we call s. The risk s depends on the number and type (smoker/nonsmoker, male/female) of people in the household. We assume a $50 cost for a long-term measurement covering all living areas of the home.

Advice for Homeowners

We perform two separate decision analyses, one of them geared toward individuals (should you measure the radon in your home?) and one geared toward public-policy makers such as the Environmental Protection Agency (should we recommend that everyone in a particular county perform a measurement?).

The homeowner's decision tree is shown in Figure 4. At the first stage, he or she must choose between doing nothing, remediating immediately, or making a long-term measurement and then deciding whether to remediate or not. (A short-term measurement is also possible but turns out not to be the optimal decision for almost everyone.) The second stage of the decision tree covers the case where the homeowner measures first and then decides. Remediating has the highest cost in dollars but the lowest expected loss of life; doing nothing costs zero dollars but has the highest expected loss of life, and measurement potentially followed by remediation is somewhere in between in terms of dollars spent and lives saved. Our analysis chooses the decision option with the lowest expected loss (combining dollars spent and lives saved as described earlier). Remember that each household will make decisions based on the number of males/females and smokers/nonsmokers in the house, whereas we discussed only the average case earlier.

The decision tree must be evaluated from right to left, starting with the final decision; that is, we first figure out what each home-owner should do if they do take a measurement. This is the only way to compare the measurement option to the other possible first-stage decisions. In detail, then, we proceed as follows:

- We must decide whether the homeowner should remediate the house, given a measurement of the average living-area airborne radon concentration. We assume the homeowner would remediate if the home's indoor radon concentration exceeds a household-specific action level, where the action level is set by equating the expected loss (given the measured radon concentration and the makeup of the household) under remediation and under no action using the value of $200,000/life that is implicit in the EPA's current recommendation. There is a slight difference between remediating if the home's long-term measurement exceeds the action level and remediating if the home's long-term average concentration exceeds the action level, because even long-term measurements do not exactly match the personal exposure of the people in the home, but this is not crucial for the discussion here.

- Given the rule for the inner decision, the homeowner must decide whether to remediate immediately, do nothing, or make a measurement in the initial (outer) decision. The value of the measurement depends on the probability that it will affect the decision of whether to remediate, which is why we first had to evaluate the measurement–remediation decision.

The inputs to the decision problem for any household include the predicted radon concentration from our statistical model, which is based on the county and the house type; the number of male and fe-male nonsmokers; the costs of monitoring and remediation; and the value of C, the "dollar value of a life" (which mathematically deter-mines, or is determined by, the radon concentration at which the own-ers would remediate). The inputs allow us to compute the expected cost of each action including "do nothing," which costs nothing and

County, State	House	Hhold	GM (pCi/L)	GSD	Pr(remed)	Expected Cost ($)	Expected Lives Saved
Lebanon,	NB	avg	3.7	2.2	0.47	990	0.013
PA	BL	avg	6.4	2.2	0.74	1530	0.030
Montgomery,	BNL	FS,MS	0.8	2.1	0.02	80	<0.001
MD	BL	FS,MS	1.4	2.1	0.08	210	0.003
	BL	avg	1.4	2.1	0.08	210	0.001
Sacramento,	NB	avg	0.4	2.1	0.00	52	0.000
CA	BNL	avg	0.6	2.1	0.01	62	0.000
	BL	avg	0.8	2.1	0.02	84	0.000

TABLE 1 Costs and benefits of measurement–remediation decision

Note: House type can be BL = basement is a living area, BNL = basement is not a living area, or NB = no basement. The variable Hhold represents household makeup; in the table we consider the average household (0.27 female smokers, 0.30 male smokers, 1.16 female nonsmokers, and 1.07 male nonsmokers) and a household with 1 female smoker (FS) and 1 male smoker (MS). GM and GSD are the geometric mean and geometric standard deviation of the predicted radon distribution for houses of the specified type in the county. The column labeled Pr(remed) shows the probability that remediation will be recommended if the "action level" is set at 4 pCi/L, independent of household composition. Expected lives saved are over a 30-year period.

saves no lives. Table 1 shows the expected costs and benefits for one possible action: carry out a strategy of making a long-term living-area measurement and then deciding whether or not to perform remediation. The table shows the costs and benefits for different types of homes and households in several different counties.

For example, the first line of the table shows that for a home in Lebanon County, Pennsylvania—a county with one of the highest radon levels in the country—following this approach in a nonbasement house will lead to a 47% chance of performing remediation. If remediation costs $2000 (including energy and maintenance costs for the system), then the expected total cost of following this plan is about $990, including the cost of measurements, and for a home with the national-average mix of female and male smokers and nonsmokers this plan is expected to save about 0.013 lives. Both the cost and the life savings presented in the table are for the entire decision plan: there's some chance that remediation will not be performed (because the radon measurement will not be high) and thus no expected lives will be saved, and some chance that remediation will be performed and there will be some expected life savings. The values of expected dollar cost and of expected lives saved include both of these possibilities.

For a home in Lebanon County that has a basement that is used as living space, the expected cost is higher than for a nonbasement house because there's even higher likelihood of remediating due to the higher expected radon concentration, but the number of lives saved is also much higher because of the larger radon reduction that would be provided by remediation. For such a home, using this decision plan has an expected dollar cost of $1530 but is expected to save almost 0.03 lives. That's quite a bargain: there aren't many other opportunities to have a 3% chance of saving a life for under $1600.

For a nonbasement home in Sacramento, California—a low-radon county—the expected lives saved would be of the order of 0.0003 if this measurement plan is followed in an average house (rounded to 0.000 in the table). There's a 99% chance that the long-term living-area measurement will fall below the 4 pCi/L action level, and even if it does exceed the action level it is likely to do so by only a small amount (in contrast to the situation in Lebanon County, where many houses will be far above the action level). In Sacramento County, if the "value of a life" is set to $200,000 as implied by the 4 pCi/L action level under a linear dose-response relationship, making measurements in a nonbasement house isn't worth the cost.

Montgomery County, Maryland, has a radon distribution that is fairly typical for the northern half of the continental United States, with radon concentrations generally higher than in Sacramento but lower than in Lebanon County. Table 1 also shows the difference in risk for different households: even though an "average" household contains about three people, remediation would actually save more lives on average in a household consisting of two smokers because of their higher risk for a given exposure.

The analysis just described is the analysis that a homeowner might carry out, given only some basic information about a home's location and the makeup of the household. Additional information can be used as well, as long as the relationship to radon concentrations can be estimated. For example, suppose you have knowledge of a neighbor's measurement. Such information can be included to improve the predictive distribution for the radon concentration in your house. Using information concerning the spatial correlation of radon—that is, the degree to which nearby homes have similar radon concentrations—we modified the decision analysis to incorporate information on measurements in nearby houses if they are available. You can try out the process at the website http://www.stat.columbia.edu/radon/ (Gelman and Price, 1999).

A public health official has a different decision problem to work on. Such a person would need to see the effect of different public policies and/or recommendations on all of the homeowners they serve. Individual decisions can be combined to determine nationwide costs and benefits for various public policies that could be implemented at the county level. For example, the maps in Figure 5 display, for each county, the fraction of houses that would measure and the estimated fraction of houses that would remediate if the recommended decision strategy were followed everywhere with an action level of 4 pCi/L (and if we treat all households as if they have the country-average number of male and female smokers and nonsmokers). About 26% of the 70 million ground-contact houses in the United States would monitor. The total monetary cost is estimated at $7.3 billion: $1 billion for measurement and $6.3 billion for remediation, and would be expected to save the lives of 49,000 smokers and 35,000 nonsmokers over a 30-year period. This compares quite favorably with the current strategy of having everyone make a remediation decision based on a short-term measurement: that plan would cost almost three times as much and save fewer lives.

DISCUSSION

The standard dose-response model for radon—which we have used—is that risk is linear with dose even at low concentrations, but there's no strong evidence that this is true. The Canadian government assumes that the risk per dose is less at low doses and, consequently, they recommend remediation at 20 pCi/L rather than 4 pCi/L. Several studies have tried to estimate radon risk from concentrations below 10 pCi/L, but the studies aren't conclusive about even the presence of risk at, for example, 4 pCi/L, much less the exact magnitude. Our analysis could, of course, be redone using different risk–dose assumptions. In the paper on which this essay is based

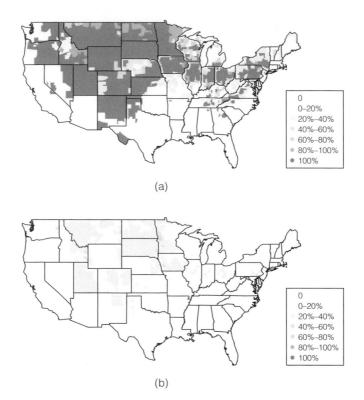

(a)

(b)

FIGURE 5 Maps showing (a) fraction of houses in each county for which measurement is recommended, given the perfect-information action level of Raction = 4 pCi/L. (b) Expected fraction of houses in each county for which remediation will be recommended, once the measurement y has been taken. In some counties no measurements would be recommended; these counties do have some high-radon homes, but for any individual home the odds of an elevated concentration are so low that it's not worth the paying for a measurement. Apparent discontinuities across the boundaries of Utah and South Carolina arise from irregularities in the radon measurements from the radon surveys conducted by those states, an issue we ignore here. We estimate that following the recommendations summarized in these maps would, over a 30-year period, save 84,000 lives and cost $7.3 billion. This is a great improvement over the current standard plan of measurement and remediation, which is estimated to save 55,000 lives at a cost of $21 billion.

Source: Lin et al. (1999).

(Lin et al., 1999), we considered how the costs and benefits would change if there is a threshold below which there is no health effect. Of course, we could also keep the same dose-response assumptions but try different "action levels" (or different quantities for the "value of a life saved").

Many people are uncomfortable with setting a dollar value for saving a life, and may even think it's immoral to do so, but actually, there's no avoiding it—if an expenditure of D is considered worthwhile because it will save N lives, then that implies a value per life. It is impossible to spend an infinite amount of money to reduce risks, so D/N will always be finite. Nobody should be ashamed at choosing a dollar value, although there are plenty of other opportunities for shameful choices in risk-related decisions (a common one is failing to distinguish between who bears the costs and who reaps the benefits, but that's not an issue here, because the risks of radon exposure and the costs of remediation are borne by the same people). In our analysis we have set the "dollar value of saving a life" at a low level, but this is the level that is implied by current radon policy and standard dose-response assumptions. Another value could be used instead, which would lead to different recommendations for which homes should be monitored for radon. The key is that whatever "value of a life saved" you select, you want to choose the optimal monitoring and remediation strategy for that value; otherwise, you're wasting lives by saving fewer people than you could for the same money.

Setting a radon policy involves many considerations other than those discussed here, including issues of practicality and politics that are difficult to formalize. (To give just one example, there is political pressure against creating lists of high-radon counties or high-radon areas because some residents of those areas are afraid that such a designation will harm property values.) We didn't try to take those issues into account. Instead, what we have illustrated is how statistical methods can be used to help inform a difficult public health decision in which exposures are both spatially variable and have variable uncertainty. We have used a statistical model to estimate the distribution of home radon levels within U.S. counties using data from national and state radon surveys. The estimated radon distributions can be used for individual level decisions and can be aggregated to test the costs and radon reductions if various radon policies are implemented by national policy makers. The individual decisions can be modified to take into account additional information such as available measurements on neighboring houses. Our calculations suggest that simple modifications to the EPA's monitoring recommendations could save more lives for less money than the current policy, assuming, of course, that people would actually follow the recommendations.

ACKNOWLEDGMENTS

We thank the editors and reviewers for helpful comments, Chiayu Lin and David Krantz for collaboration in the research presented here, and the National Science Foundation and the Department of Energy for financial support.

REFERENCES

Carlin, B. P., and T. A. Louis. (2000). *Bayes and Empirical Bayes Methods for Data Analysis,* 2nd ed. London: CRC Press.

Clemen, R. T. (1996). *Making Hard Decisions,* 2nd ed. Belmont, Calif.: Duxbury Press.

Gelman, A., J. B. Carlin, H. S. Stern, and D. B. Rubin. (2003). *Bayesian Data Analysis,* 2nd ed. London: CRC Press.

Gelman, A., and P. N. Price. (1999). "Radon Risk." http://www.stat.columbia.edu/radon/.

Lin, C., A. Gelman, P. N. Price, and D. H. Krantz. (1999). "Analysis of Local Decisions Using Hierarchical Modeling, Applied to Home Radon Measurement and Remediation (with Discussion)." *Statistical Science* 14:305–37.

Marcinowski, F., R. M. Lucas, and W. M. Yeager. (1994). "National and Regional Distributions of Airborne Radon Concentrations in U.S. Homes." *Health Physics* 66:699–706.

National Research Council, Committee on the Biological Effects of Ionizing Radiation (BEIR VI). (1998). *Health Risks of Radon and Other Internally Deposited Alpha-Emitters.* Washington, D.C.: National Academy Press.

Price, P. N. (1997). "Predictions and Maps of County Mean Indoor Radon Concentrations in the Mid-Atlantic States." *Health Physics* 72:893–906.

Price, P. N., A. Gelman, and A. V. Nero. (1996). "Bayesian Prediction of Mean Indoor Radon Concentrations for Minnesota Counties." *Health Physics* 71:922–36.

Price, P. N., and A. V. Nero. (1996). "Joint Analysis of Long- and Short-Term Radon Monitoring Data from the Northern U.S." *Environment International* 22:S699–S714.

Wirth, S. (1992). *National Radon Database Documentation: The EPA/State Residential Radon Surveys.* Washington, D.C.: U.S. Environmental Protection Agency.

QUESTIONS

1. If every home in the U.S. were tested, and those with over 4 pCi/L were remediated, what would be the cost per life saved? How does this compare with the cost based on a decision tree such as the one recommended in the essay?

2. Comparing smokers and nonsmokers, who will benefit most, on average, from reduction of radon levels?

3. Why does the decision to test for radon depend on whether it is the homeowner or the government who is considering the question?

4. If you were considering buying a home, and you were moving into an area with moderately high radon, with average home radon levels of 2 pCi/L, what would you do?

5. How do you think the discovery of the prevalence of high radon levels should affect the prices of homes?

6. Do you think it is right for state governments to pass laws requiring (a) that new houses be built with radon remediation, or (b) that houses be tested for radon before any sale? What are the costs and benefits of such laws?

7. What other public health problems do you think could be studied using decision analysis? How might such analyses be carried out?

STATISTICAL WEATHER FORECASTING

DANIEL S. WILKS

Cornell University

———— ⁑ : : ⁑ ————

Many persons find it quite natural to pray for rain or shine when they would think it ridiculous to pray for an eclipse. Poincaré

IS THE WEATHER RANDOM?

Does the atmosphere, and the evolution of weather within it, behave like a big machine whose future movements are determined by physical laws? Or are weather events random occurrences with obscure and unknowable origins?

In principle the atmosphere is a deterministic physical system whose future behavior is fully described by a set of coupled differential equations expressing conservation of momentum, conservation of mass, the first law of thermodynamics, and the ideal gas law—mathematical expressions of the physical laws that govern the behavior of the atmosphere. The apparent implication is that given a sufficiently precise set of measurements of the state of the atmosphere—globally comprehensive observations of pressures, temperatures, humidities, and so on, at a particular time—solutions to these governing equations beginning from that initial condition should be able to provide weather forecasts far into the future, in much the same way that it is possible to calculate tide tables a year or more in advance. The physicist and meteorologist Vilhelm Bjerknes was a prominent and influential proponent of this view early in the 20th century. The development of this perspective on weather by Bjerknes and others elevated the scientific legitimacy and respectability of meteorology, although the lack of adequate computing capabilities delayed practical efforts to realize their vision of deterministic atmospheric prediction for another 50 years or so.

By the latter half of the 20th century, electronic computers had advanced to the point that simplified versions of the deterministic equations describing atmospheric behavior could be used to make weather forecasts, both for research and operational purposes. Steadily increasing computing capacity allowed increasingly accurate and detailed representations of the atmosphere in these computer models, called dynamical models. But purely dynamical prediction did not work as well as had been expected: while computer forecasts might resemble actual atmospheric behavior for a few days into the future, at some point—a couple of weeks at most—these forecasts are no more accurate on average than a past state of the atmosphere chosen randomly from the historical record. The apparent inability to make good extended forecasts of the behavior of an important physical system (the atmosphere) using the equations known to govern that behavior was initially puzzling. For many years the problem was assumed to derive from computational limitations, although it remained as faster computers, allowing more accurate approximations to the physical laws, were successively developed through time.

A more fundamental cause for the loss of atmospheric predictability at increasingly distant future times was discovered in 1959 by Edward Lorenz, a meteorologist at the Massachusetts Institute of Technology. While working with a highly simplified version of the equations governing atmospheric behavior, he noticed that small round-off errors would amplify drastically over the course of a simulated forecast. That is, the forecast (i.e., the computer's solution to the governing equations) exhibited sensitive dependence on the initial conditions—what is now usually called "chaos." This mathematical phenomenon is important beyond just weather forecasting, but it explains why the atmosphere is not amenable to deterministic prediction: small differences in initial conditions lead rather soon to big differences in the forecasted weather, and the true initial state of the atmosphere is not (and for practical purposes cannot be) known exactly. A readable and nontechnical account of "chaos" and the development of our understanding of it has been written by Lorenz himself (see the reference list at the end of this essay).

Since the atmosphere is and will always be incompletely observed, the best characterization we can make of its state at a given time is in terms of a probability distribution, possibly centered at some "best estimate" that would constitute the initial condition for a conventional deterministic forecast. Imagine a random sample from this distribution of initial-condition uncertainty, which would consist of a set of specific, plausible initial states. There will be small differences among the members of this random sample, and none of them are likely to match exactly the true, unknown, initial state. Sensitive dependence on initial conditions implies that forecasts originating from our sample of plausible initial states will diverge (i.e., yield forecasts that are quite different), both from each other and from the course taken by the actual weather, even if there could be no errors or approximations in the solution of the physical governing equations. A relatively new and still developing approach to weather forecasting called "ensemble forecasting" is structured in exactly this way in order to account for the effects of initial-condition uncertainty on forecast uncertainty. It looks at the dispersion of forecast states that is produced as a consequence of uncertainty about the initial state, interpreting that dispersion as indicative of the probability distribution over future atmospheric states. An introduction to ensemble forecasting can be found at the website http://www.hpc.ncep.noaa.gov/ensembletraining/.

Because the atmosphere is "chaotic" the uncertainty associated with the state of the atmosphere at the time a dynamical forecast is initiated does not disappear but rather amplifies until physically based forecasts are no more accurate than climatological guesswork. Even though atmospheric behavior is determined by physical laws, from a practical standpoint its behavior is also random. Accordingly, complete approaches to describing the weather must involve probability, and statistical approaches to weather forecasting are useful at all lead times.

USING STATISTICS TO ENHANCE PHYSICALLY BASED WEATHER FORECASTS

Even though perfect deterministic weather forecasts are doomed to failure by chaos, these dynamical forecasts still contain quite useful information during the time that the amplification of initial-condition errors is still relatively small—currently up to about a week into the future. Within this forecast range, dynamical weather forecasts can be, and are operationally, improved upon through statistical post-processing. Broadly, these improvements are of two kinds: (1) correcting systematic errors, and (2) providing forecasts for quantities that are not explicitly represented by the dynamical model. The physical models are usually implemented on computers using a grid, with each point representing the average conditions within a layer of the atmosphere above a horizontal grid point. Forecast quantities of interest that are not explicitly included in the dynamical model include such things as near-surface temperature at particular locations not corresponding to grid points and probability estimates for events of interest (e.g., at least 0.01″ of rainfall).

Corrections and enhancements of this sort to raw dynamical weather forecasts are known by the acronym MOS, or model output statistics. MOS forecasting is used to predict a desired value, for example, daily maximum temperature (T_{max}), using values of other variables that are known to be correlated with the desired value, including especially related variables that are predicted by the dynamical model being enhanced. For example, a variable in the dynamical models that tends to have a relatively high positive correlation with T_{max} is the predicted temperature at 850 mb (about 1500 m above sea level).

Overwhelmingly, the most common approach to MOS forecasting is multiple linear least-squares regression, which quantifies the nature of the relationship of a variable of interest with one or more predictor variables. To develop a MOS equation one needs an archive of dynamical forecasts (generally spanning two years or more) representative of those to be post-processed in the future, together with corresponding observations of the quantity to be forecast by the MOS equation. A simple example of the result is the following equation, which was formerly used operationally to make forecasts of daily maximum temperatures at

Binghamton, New York, at 60 hours lead time, during the months of September, October, and November:

$$T_{max} = -363.2 + 1.541\, T_{850} - 0.1332\, RH_{0-490} - 10.3 \cos\left(\frac{2\pi\, DOY}{365}\right)$$

This equation contains two predictor variables (T_{850} and RH_{0-490}) from the dynamical model to which it was fit that would have been predicted to occur 60 hours after forecast initialization at the model grid point nearest Binghamton. Here T_{850} is the temperature at 850 mb, and RH_{0-490} represents the forecasted average relative humidity between sea level and 490 mb (about 5500 m above sea level). This relative humidity is an indicator of the cloud cover, so that higher values tend to be associated with lower maximum temperatures. The final term, involving the cosine of the day of the year, represents the tendency for the fall season to cool progressively between September 1 (DOY = 244) and November 30 (DOY = 334). Different MOS equations are fit and used for different seasons of the year, reflecting the possibility that the physical relationships between the forecast quantities, and possibly also the error characteristics of the dynamical forecast models, could change through the year. Different equations are also needed for different locations, and for different dynamical models being post-processed. Importantly, MOS equations relating forecasts and weather observations at different lead times are different, with regression coefficients for the dynamical model predictors (e.g., T_{850} and RH_{0-490} in preceding the equation) tending to shrink toward zero as lead time is extended. These shrinking regression coefficients yield predictions progressively nearer the climatological average, reflecting the progressive loss of predictability at longer lead times.

MOS products are useful to weather forecasters, although the general public rarely sees them. However, they are regularly posted on the Internet, for example on the website of the National Weather Service, at http://www.nws.noaa.gov/mdl/synop/products.shtml. Figure 1 shows an example set of MOS forecasts taken from this website, pertaining again to Binghamton, New York. The first line indicates that the forecast location is Binghamton (BGM), and that the dynamical model being post-processed is the "Eta" model, which had been initialized at 1200 UTC (i.e., Greenwich time, or 8 A.M. local

KBGM   ETA MOS GUIDANCE   5/07/2003   1200   UTC

DT/MAY   7/MAY   8   /MAY   9   /

HR	18	21	00	03	06	09	12	15	18	21	00	03	06	09	12	15	18	21	00
N/X							54				66				48				**69**
TMP	66	65	61	58	57	55	56	60	63	63	59	56	53	50	52	60	66	67	61
DPT	51	51	51	50	51	51	52	53	52	51	51	47	44	43	44	43	41	40	41
CLD	BK	OV	OV	OV	OV	OV	OV	OV	OV	BK	BK	BK	SC	CL	CL	CL	SC	CL	CL
WDR	29	32	36	02	03	01	36	33	33	32	33	34	30	27	29	29	29	30	35
WSP	08	06	04	04	03	03	03	06	06	07	05	04	03	02	04	06	08	07	04
P06	6		30		53		48		28		15		9		7		9		
P12							61				61				20				10
Q06	0		0		1		1		0		0		0		0		0		
Q12							1				1				0				0
T06	6/10		24/ 0		18/ 0		18/ 6		23/11		3/ 0		3/ 2		1/ 5		2/11		
T12							24/10				31/ 6				23/11				3/ 5

FIGURE 1 Example MOS forecast panel for Binghamton, New York, including forecasts for six through 60 hours after the initial time of 1200 UTC. The quantities forecast on lines 4–15 are, respectively, minimum and maximum temperature, 3-hour temperature, dew point, cloud cover, wind direction, wind speed, 3-hour probability of precipitation, 12-hour probability of precipitation, 6-hour precipitation amount, 12-hour precipitation amount (amount forecasts coded into discrete categories), 6-hour thunderstorm probabilities, and 12-hour thunderstorm probabilities.

daylight time) on May 7, 2003. The different entries in the table are produced by different regression equations, as would corresponding forecasts for different locations and for forecast models other than the Eta. The second and third lines indicate the days, and times of day, to which the forecasts on the subsequent lines pertain. For example, line 4 (N/X) shows forecasts of miNimum and maXimum temperatures during the forecast period. The 60-hour forecast of maximum temperature (for the 12-hour period ending 00 UTC on May 10), predicted by the currently operational counterpart of the preceding equation, was 69°F (bold entry on the fourth line). The actual maximum on that date was 64°F, and the preceding minimum temperature (predicted to be 48°F) was 49°F. The other forecast elements provided in this panel are 3-hour temperatures (TMP) and corresponding dew points (DPT); cloud cover (CLD), characterized as clear (CL), scattered (SC), broken (BK), or overcast (OV); wind direction (WDR) in tens of degrees; wind speed (WSP) in knots; probabilities (×100) of precipitation during 6-hour (P06) and

12-hour (P12) periods; nonprobabilistic forecasts of precipitation amounts during the same 6-hour (Q06) and 12-hour (Q12) periods (0 = zero precipitation, and 1 = 0.01 to 0.09 inches; other codes are used for larger amounts); and probabilities (again, $\times 100$) for thunderstorms (to the left of the slashes), and conditional (on occurrence of any kind of thunderstorm) probabilities of severe thunderstorms (to the right of the slashes), during 6-hour (T06) and 12-hour (T12) periods.

An interesting feature of Figure 1 is that some of the predictands are probabilities, for example for rainfall (P06 and P12). Of course rainfall probabilities are neither observed at rain gauges nor forecast by individual deterministic model calculations, but forecast probabilities can nevertheless be produced using regression equations. The operational experience with this kind of regression is that the forecasts derived from them behave correctly as probabilities. For example, the relative frequency of rain on those days when regressions of this kind produce a value near 0.9 tends to be quite near 90%.

PURELY STATISTICAL WEATHER AND CLIMATE FORECASTING

When the interest is in predicting weather in the range of six hours to perhaps 10 days into the future, the best results with current technology are obtained when physically based dynamical forecasts provide the initial component of the forecast process, for example as described in the previous section. However, outside this range of forecast projections, purely statistical forecast methods are competitive with, and are often more accurate than, forecasts based partly or wholly on dynamical forecasts. Purely statistical methods are those that forecast future atmospheric behavior from relationships extracted only from historical records.

For short lead times, purely statistical methods are used because dynamical forecasts are not practical. The collection and preparation of global weather observations, their summarization in a form that is suitable for defining initial conditions for dynamical forecast models, the actual running of those models, and the statistical post-processing of the results together impose a substantial lag after the data cutoff time. For example, the forecasts in Figure 1 were prepared using weather observations available through 1200 UTC, but they would not have been available until several hours later. R. G. Miller, in the essay "Very Short Range Weather Forecasting Using Automated Observations," which appeared in the previous edition of this volume, describes the use of statistical forecasts for these short lead times. These forecasts are usually also produced using methods such as linear regression (as in the previous section), with the difference that the predictors are all observed (not model-predicted) quantities that are known by the time a forecast needs to be made.

As the effects of sensitive dependence on initial conditions become more severe at longer lead times, dynamical models lose the ability to produce useful forecasts of specific weather events. One setting where this poses a problem is the production of forecasts for 90-day averages of temperature and precipitation, at lead times that range from weeks to about a year. These are called seasonal, or "climate" (i.e., time averages of daily weather), forecasts. Several groups issue such forecasts (see, for example, the websites at http://www.cpc.ncep.noaa.gov/products/predictions/90day/ and http://iri.columbia.edu/climate/forecast/index.html). At such long lead times, forecasting relies on consistent responses of the atmosphere to slowly varying (and thus more predictable) influences such as sea-surface temperatures, particularly those associated with El Niño and La Niña.

The format of these seasonal forecasts is itself interesting from a statistical perspective. At very long lead times it is clearly not credible to forecast a single temperature outcome with no expression of uncertainty. Accordingly, most seasonal forecasts are expressed in terms of adjustments to the climatological, or long-run frequency, distributions of the quantity being predicted. For example, Figure 2 shows a forecast for average temperatures over the United States for December 2003–February 2004, which was made 7.5 months earlier, in April 2003. The actual map was drawn by a human forecaster who subjectively combined information from several forecast methods. The shaded regions of the map indicate locations where the forecaster is calling for shifts in the probabilities of temperature outcomes for the lower, middle, and upper $\frac{1}{3}$ of local climatological distributions. For example, the frequency distribution for average

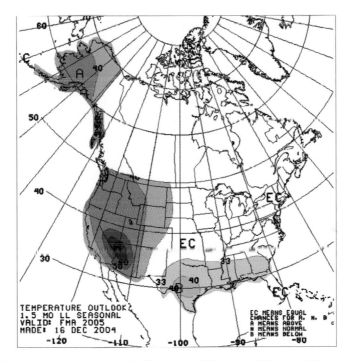

FIGURE 2 Seasonal forecast map for December 2003 through February 2004 average temperatures across the United States, made in mid-April 2003. The predicted quantities are "probability anomalies," with brown (or blue) shaded regions indicating greater (or lesser) chances of temperatures in the upper $\frac{1}{3}$ of the local winter temperature distributions, with correspondingly smaller (or larger) probabilities for temperatures in the lower $\frac{1}{3}$.
Source: Courtesy of NOAA.

December through February temperatures at Ithaca, New York (located near the center of New York state), is very nearly Gaussian (normal), with mean 24.7°F and standard deviation 2.8°F; so, this distribution is divided into thirds at 24.7 ± (0.43)(2.8°F), or 23.5 and 25.9°F. The forecast map in Figure 2 shows an above normal "A" (probability anomaly) of about 5% for Ithaca, implying a probability of 0.33 + 0.05 = 0.38 for average temperature above 25.9°F, and a probability of 0.33 − 0.05 = 0.28 for seasonal temperature below 23.5°F. Appropriately for the high level of uncertainty intrinsic to forecasts of this kind, the probability shifts in seasonal forecasts are typically modest. Experience has shown that these forecast probabilities correspond reasonably well to the subsequent event relative frequencies.

CONCLUSION

Uncertainty is intrinsic to atmospheric behavior because the physical laws governing atmospheric behavior are "chaotic"; that is, small initial differences lead to huge differences in future behavior, and the state of the atmosphere at any given time is never known comprehensively. While weather forecasts based purely on physical principles are useful for perhaps 10 days into the future, they are inherently uncertain at all lead times. The result is that a probabilistic perspective is useful in thinking about weather and climate, and that statistical methods are important in practical weather forecasting.

REFERENCES

Friedman, R. M. (1989). *Appropriating the Weather: Vilhelm Bjerknes and the Construction of a Modern Meteorology.* Ithaca, N.Y.: Cornell University Press.

Lorenz, E. N. (1993). *The Essence of Chaos.* Seattle: University of Washington Press.

Miller, R. G. (1989). "Very Short Range Weather Forecasting Using Automated Observations." In *Statistics: A Guide to the Unknown,* 3rd ed., edited by J. M. Tanur, F. Mosteller, W. H. Kruskal, E. L. Lehmann, R. F. Link, R. S. Pieters, and G. R. Rising, 261–67. Belmont, Calif.: Duxbury Press.

QUESTIONS

1. How does the Poincaré quotation at the beginning of the essay relate to the theme of the essay itself?
2. Is it reasonable to believe that the physical laws that govern the weather are deterministic, even though all the data-based evidence we have about the weather suggest that it behaves like a random process? Explain.
3. If tomorrow's weather cannot be predicted perfectly today, how do forecasters state their limited knowledge about tomorrow's weather?

4. "Ensemble forecasting" is described briefly as a relatively new technique in weather forecasting. What source of variation does it try to allow for?

5. "Corrections and enhancements of this sort to raw dynamical weather forecasts are known by the acronym MOS, or model output statistics." Give an example of a "correction" and one of an "enhancement" that demonstrate what this quotation means.

6. In the example of the MOS equation to predict the maximum temperature 60 hours ahead in Binghamton, New York, there is a trigonometric term. What is the reason for this?

7. When a forecast is attempted over a long time period, the forecast tends to be the long-term average. Why does this have limited usefulness?

8. Some of the MOS elements in Figure 1 are probabilities. Is it useful to give a forecast as a probability? Explain.

9. In what sense is a "purely statistical forecast" different from a "physically based dynamical forecast"? Why are both types needed?

10. For which time durations are statistical forecasts most effective? Explain why.

11. Figure 2 shows regions having different probabilities for higher or lower than average temperatures 90 days hence. Use the textual description in the essay to explain more exactly what property the "high" region has.

SPACE DEBRIS

Yet Another Environmental Problem

DAVID R. BRILLINGER

University of California, Berkeley

———————— ⁑ : : ⁑ ————————

The National Aeronautics and Space Administration (NASA) estimates there to be more than 200,000 pieces of space debris exceeding 1 cm in size in orbit around the Earth. In fact, there is so much debris that the space near Earth has been described as a garbage can. The debris is the result of more than 3000 space flights that have taken place since 1957; it includes spacecraft and rocket bodies, springs, clamp bands, bolts, lens caps, spacecraft carriers, breakup debris, solid rocket motor slag, and paint flakes. In order to make decisions, NASA needs to be able to quantify the risk of collision of a satellite or a space vehicle or an astronaut with a piece of orbiting space debris. A single paint chip can wreck a space mission. With all of this clutter in space, there is great concern.

What can be done to gain some handle on this situation? The managers of space programs place requirements on the risk of collision during a space mission. For example, according to NASA Flight Rule A 4.13-6, "an accepted residual collision risk for trackable objects is 1 in 200,000 during a 10 day mission." The NASA publication *Protecting the Space Shuttle from Meteoroids and Orbital Debris* states that the shuttle is "allowed to experience a 1-in-200

— *183* —

NASA

FIGURE 1 The International Space Station

probability per mission of critical failure from the impact of mete-oroids or orbital debris," and for astronauts, the probability of "pen-etration of 6 hour EVA (extravehicular activity) for 2 astronauts at the end of the mechanical arm" can be at most 1 in 4800.

Figure 1 shows a prime candidate for damage by space debris, the International Space Station and its appendages. It is the largest space-craft yet built, having a cross section bigger than a football field, and it has a planned lifetime of 15 years. Statisticians working with engi-neers, physical scientists, social scientists, and computer scientists seek answers to questions such as, What is the chance of some piece of space debris passing through the International Space Station in the next 15 years?

BACKGROUND

Space debris has been defined as man-made objects or parts thereof in space that do not serve any useful purpose.

The 1999 NASA report *Orbital Debris: A Chronology* reviews much of the history of space missions and resulting orbital debris.

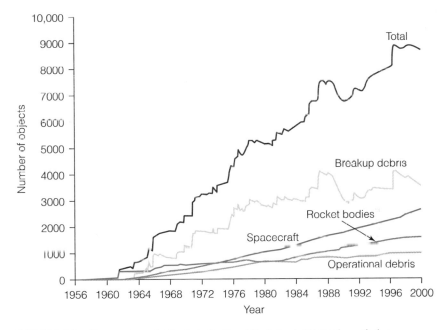

FIGURE 2 Monthly number of objects in Earth orbit of size ≥10 cm through the year 2000

Source: Orbital Debris News 6, no. 1:10

The first known breakup of an artificial object occurred in June 1961, when the American Transit 4-A satellite's upper stage exploded two hours after separation. The first known collision between an operational spacecraft and a cataloged space object took place in July 1996. Debris from a booster rocket used in 1986 to launch a French satellite severed the stabilizing boom of a British-built, French-owned microsatellite from the CERISE (Characterisation de l'Environment Radio-électrique par un Instrument Spatial Embarqué) program. Luckily there was no international incident because both items were French. Recently, in 2001, three upper stages and two upper-stage components of rockets exploded.

The problem is becoming steadily more serious, as Figure 2 highlights. The series plotted are counts of objects in low earth orbit, the region 200–2000 km above the Earth. The counts are taken from a catalog maintained by the U.S. Air Force.

The numbers fluctuate from year to year but increase steadily. The upward trend results from the regular launching of new satellites and the explosions of existing rocket upper stages. The small dips come from the Earth's gravity pulling objects down and from the pressure of particles flowing from the sun pushing them down. The dips in the growth of the "Total" series are found to correspond to periods of low solar activity.

Space debris risk assessment for the International Space Station might be defined as *the estimation of the probability that damage occurs to the International Space Station within a specified time period, as a result of a collision with some piece of space debris.*

This essay focuses particularly on the problem of how to estimate the probability of a collision, focusing on the case of an object at a given location. In order to make these estimates, we use both the physical laws of orbiting objects and experimental data. The tools of mathematics, probability, and statistics are employed.

LARGER OBJECTS

To study the population of orbiting objects, data are collected both from the ground and in space. Collection methods include radar, liquid mirror telescopes, optical telescopes, and impact panels. The objects in low earth orbit that are greater than 10 cm across are tracked individually by the air forces of Russia and the United States (Figure 2). Also, the European Space Agency maintains its Database and Information System Characterising Objects in Space (DISCOS) catalog. For a particular orbiting object, one would like to know if it passes through some given patch (or local surface) of sky, say where the International Space Station (ISS) is. One might also like to know the times at which passage occurs. If a passage looks likely, the ISS will be moved to another location. Figure 3 shows one such patch in the sky. Also graphed are three orbital paths, with one passing through the patch.

The mathematics and physics of orbits have been studied for centuries by renowned scientists such as Kepler, Newton, and Lagrange. They were concerned with the orbits of the moon and the planets. Following Newton's laws of dynamics and his law of gravity, equations of motion have been developed. According to these equations, an orbiting object's trajectory orbit can be found deterministically.

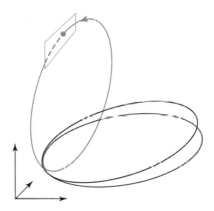

FIGURE 3 Realizations of random debris orbits illustrating possible passages through a patch

The orbit is shaped like an ellipse, and the motion may be described by six parameters, the so-called orbital elements. Three define the kind, shape, and inclination of the orbit. The others describe the orientation of the orbit on the orbital plane. These parameters have been estimated for many years by a form of least squares (see for example *Satellite Orbits* by Oliver Montenbruck and Eberhard Gill [Berlin: Springer, 2000]).

For each big object in the catalog, the risk of collision can be studied deterministically, using the equations of motion. Given a patch in the sky and a time period, the equations may be solved to see if the object passes through. However, there are too many small objects to learn about and handle individually in this way. Another approach is needed, one that uses probability and statistics.

A PROBABILISTIC MODEL FOR SMALL OBJECTS

The small objects in low earth orbit make up a (changing) population, which can also be thought of as a population of orbits. It seems reasonable to assume that the orbits of this population are random because of the irregular character of things like launch times and places, mission goals, and accidents such as explosions. Figure 3 shows three orbits differing in initial conditions, with each path representing an object-orbit. One of these orbits passes through the patch of sky of interest. It makes sense to talk about the probability

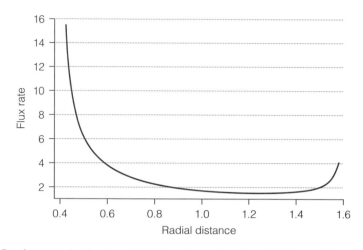

FIGURE 4 An example of a theoretical flux rate as a function of distance. The *x* and *y* units are arbitrary.

of some object-orbits passing through a specified patch of sky, and to compute such a probability we need to know the rate at which objects pass through the patch.

The number of objects passing through an oriented patch of unit area in a unit interval of time is called the flux rate. Using the equations of motion, the dependence of the flux on altitude may be shown to have the form illustrated in Figure 4. One sees a higher flux at the extremes of the orbit where the object is moving more slowly.

ESTIMATING THE FLUX RATE

In obtaining an approximation of the probability that an object will pass through a given patch, we must have an estimate for the flux rate. In fact, different values will be needed for different populations of objects and different time periods. To estimate the flux rate we need to learn more about the smaller objects. (As mentioned above the larger ones are tracked individually.) One way of learning more employs data collected from a radio astronomy facility near Boston, the Haystack Observatory. The facility is converted into a radar installation for certain time periods during which the radar beam is focused onto a specific cell. (The cell shape is the frustum of a cone.) Then, objects that are moving through the cell are detected.

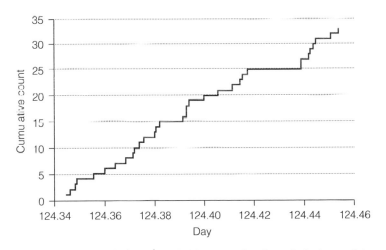

FIGURE 5 Cumulative count of number of objects passing through the beam of the radar observation volume for one time period

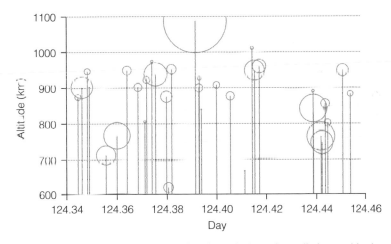

FIGURE 6 Representation of objects passing through the radar cell: times, altitudes, and sizes. The radius of circle is proportional to radar cross section (RCS).

Figures 5 and 6 provide examples of the data obtained by aiming the radar at a cell at height 900 km, width 200 m, the radar cone subtending a solid angle of 0.05 degrees. Figure 5 displays the cumulative count of the objects passing through the cell of sky as a function of time, measuring time in units of day from the beginning of the year. In this case 33 pieces of debris were detected in a time period of length 160.1 minutes. Figure 6 provides the times and also altitudes

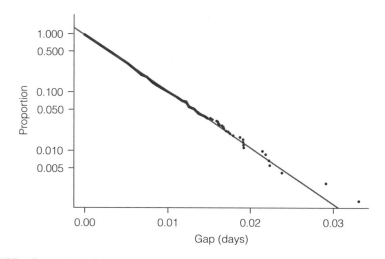

FIGURE 7 Proportion of time gaps greater than a given gap value plotted versus that value

and sizes. The estimated altitudes are indicated by the heights of the vertical lines and the estimated sizes by the radii of the circles. The units of the x-axis are days; the units of the y-axis, kilometers.

Such figures give one a feel for the data and in particular may be used to detect abnormal measurements and to suggest statistical models.

The flux is estimated by picking the altitude band and size interval of interest, dividing the cumulative count observed by the total staring time and the surface area of the cell at that altitude, and then scaling the data to one year. For the complete data set from which Figures 5 and 6 were derived, picking an altitude band of 800 to 1000 km, and objects of size greater than 5 cm, the cumulative count was eight, the staring time was 0.011864 yr., and the area of the surface of the frustum was 510.89 km². This led to a flux estimate of 1.320E-06 per m² per year. There is statistical uncertainty attached to this estimate that may be approximated using the Poisson distribution.

The probability that an object passes through the patch may be associated with the gaps in time between passages of pieces of debris having a so-called exponential distribution. Some support for this assumption is provided in Figure 7, which plots, for a larger data set, the proportion of gaps greater than a given value against the value. The plot would fluctuate about a straight line were the exponential assumption reasonable. It is seen to do so.

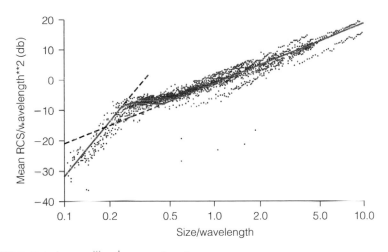

FIGURE 8 Data from calibration experiment

Source: Final Report of the Haystack Orbital Debris Data Review Panel, http://ston.jsc.nasa.gov/collections/ TRS/_techrep/TM-1998-4809.pdf.

Once flux has been estimated, there is still an important difficulty to overcome—namely, how to define the "size" of a space object. In terms of possible damage both the object's mass and its three-dimensional size (length, width, and height) are important. However, what is measured by the radar equipment is the so-called radar cross section (RCS). This is the strength of the reflected radar signal and an imperfect measure of physical size.

In order to proceed, we need a method to convert RCS to physical size, specifically a calibration relationship. To develop one, a calibration experiment was carried out for NASA. A satellite-like object was exploded. Afterward, the resulting pieces had their RCSs measured from many directions and for various radar wavelengths. The scatter plot, Figure 8, graphs the measured RCS values versus the average of the three greatest mutually perpendicular dimensions of the object divided by the radar's wavelength. One sees a bivariate distribution with considerable spread. The straight lines on the left and the right are derived from theory. The curved portion in the middle is a smooth joining of the lines.

A naive calibration would involve reading off an estimated size for an observed RCS value using the curve alone. However, because of the substantial scatter more needs to be done; statistical distributions need to be considered. It can be shown that the average value of the observed distribution of RCSs is the average of the desired size distribution, weighted by a calibration distribution. At this stage one has a so-called inverse problem. One has the problem of finding a useful solution of an empirical integral equation. A difficulty is that the solution is not unique. There are algorithms for finding a useful solution, and NASA incorporates the result in its models.

PUTTING IT ALL TOGETHER

Using the known orbits of large objects, the estimates of flux, and the information about the distribution of object sizes, NASA regularly estimates the risks of proposed flights and uses them to make decisions. For example, Figure 9 provides probabilities of collision over a one-year time period, for a 1 m^2 patch of sky for heights of 200 to 2000 km above Earth. One sees increased probability of collision around the altitudes of 900 and 1400 km. These are altitudes to avoid to the extent possible.

To make these estimates, NASA employs the program ORDEM2000 (ORDEM refers to "Orbital Debris Engineering Model"). This program describes the debris environment and

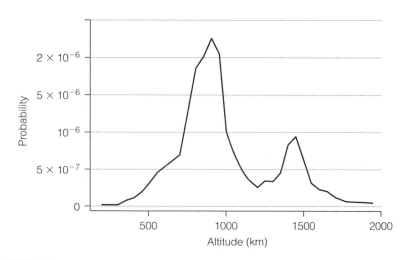

FIGURE 9 Passage probability of debris >5 cm, using NASA program ORDEM2000. Area = 1 m^2, time period = 1 year

predicts impact flux and velocity distributions of debris for populations of prescribed orbits and patches using extrapolations of future solar activity and launch rates. The program ORDEM2000 is available at http://sn-callisto.jsc.nasa.gov/model/modeling.html and was used to prepare Figure 9.

The larger objects are tracked individually, and if one seems likely to come close to the International Space Station in a given period the station is shifted to a safe location for that period.

DISCUSSION AND SUMMARY

We have been living in the space age for almost 50 years now, and our lives abound with technological spin-off. At the same time, there have been steady reminders of the risks: failed launches, reentry of radioactive materials, and, most recently, the shuttle *Columbia* disaster. In investigating space-debris risk, interesting statistical problems and methods arise, and statisticians and their methods add important information to what the engineers and scientists know and do on their own. To obtain effective estimates of space-debris risk, we use statistics to figure out how to sample the sky, and solve inverse problems to determine the size of objects in space. Then, we handle measurement error and provide uncertainty estimates. Finally, we extrapolate the distribution of the debris population to future dates.

It is noteworthy just how many difficult analytic problems NASA has found effective solutions for and how much of its work is available on the web.

ACKNOWLEDGMENTS

I thank Abdel El-Sharaawi of the Canadian Centre for Inland Waters and fellow member of the Haystack Review Committee. I thank the researchers at the Johnson Space Center in Houston. In particular, Mark Matney provided discussion of the problems and the data for Figures 5, 6, and 7.

Rick Kawin and Roger Purves helped by reading drafts of this essay and making important suggestions. The research was supported in part by National Science Foundation Grants DMS 99-71309 and DMS 02-03921.

REFERENCES

Brillinger, D. R. (2002). "Space Debris Risk." In *Encyclopedia of Environmetrics,* vol. 4, edited by A. H. El-Shaarawi and W. W. Piegorsch, 2036–40. New York: Wiley.

Johnson, N. L. (1998). "Monitoring and Controlling Debris in Space." *Scientific American* (August): 62–67.

Johnson, N. L., and D. S. McKnight. (1987). *Artificial Space Debris.* Malabar, Fla.: Orbit Book.

QUESTIONS

1. Visit the NASA space debris website, http://www.orbitaldebris .jsc.nasa.gov, and write a one-page report on one of the topics you find listed there.

2. How does the information from a radar telescope, which detects a "silhouette" view, get converted into information about the three-dimensional size of objects?

PART THREE

BIOLOGY AND MEDICINE

✺ : : ✺

Modeling an Outbreak of Anthrax
Ron Brookmeyer

The Last Frontier: Understanding the Human Mind
William F. Eddy & Margaret L. Smykla

Leveraging Chance in HIV Research
Charles M. Heilig, Elizabeth G. Hill & John M. Karon

Statistical Genetics: Associating Genotypic Differences with Measurable Outcomes
Rongling Wu & George Casella

DNA Fingerprinting
Bruce S. Weir

How Many Genes? Mapping Mouse Traits
Melanie Bahlo & Terry Speed

MODELING AN OUTBREAK
OF ANTHRAX

RON BROOKMEYER

*Department of Biostatistics, Bloomberg School of Public Health,
Johns Hopkins University*

———————— ❋ : : ❋ ————————

On October 2, 2001, a 63-year-old Florida man who worked as a
photo editor at a media publishing company was admitted to an
emergency department complaining of nausea, vomiting, and fever.
His symptoms began four days earlier on a recreational trip to North
Carolina. The man died shortly thereafter. An astute clinician
quickly made the surprising diagnosis of inhalational anthrax, which
is a serious and deadly disease. The diagnosis was surprising because
inhalational anthrax is extremely rare; only 18 cases were reported in
the United States between 1900 and 1978. Public health officials at
first believed that the Florida case was an isolated, rare event that
might have gone unnoticed except for the heightened state of public
health vigilance and alert following the catastrophic events of September 11, 2001. However, when a second case occurred in a
73-year-old man who worked at the same Florida media publishing
company and delivered mail to the first man, the coincidence seemed
remarkable. Employees of the company reported seeing a suspicious
letter on or about September 19, 2001, although that letter was never
found. Public health officials theorized that a letter contaminated
with deadly finely milled anthrax spores was the source of the disease.
Thus began the 2001 anthrax outbreak in the United States caused
by the intentional release of anthrax spores, an act of bioterrorism.

Anthrax is caused by the bacteria *Bacillus anthracis*. There are three types of anthrax. Gastrointestinal anthrax is caused by eating contaminated meats. Cutaneous anthrax is transmitted through the skin. Inhalational anthrax, the most serious form with the highest mortality rate, is transmitted when anthrax spores are inhaled deep into the lungs, where they may germinate and produce toxins. Inhalational anthrax is of most concern to public health and law enforcement officials because of its potential as a biological weapon. Although anthrax is not transmitted from person to person, anthrax spores have the potential to be turned into an aerosol and disseminated widely through the air, causing widespread disease. Antibiotics can prevent disease in people who inhaled spores, provided the antibiotics are circulating in the body when a spore begins to germinate.

Following identification of the two cases in Florida, public health officials closed the media publishing company and treated more than 1100 employees with antibiotics. On October 9, a letter addressed to Senator Thomas Daschle was processed at the Hamilton Post Office in New Jersey. The letter was then sent to the Brentwood Post Office in Washington, D.C., and ultimately to the Hart Senate Office Building. When a clerk opened the letter, a puff of white powder spread deadly anthrax spores through the air. The building was immediately closed, and workers were treated with antibiotics to prevent disease. At the time, it was not completely clear to public health officials that the tiny anthrax spores were smaller than the pores of an envelope and thus the spores could seep through a sealed envelope, putting at risk all people who came in contact with the letter even before it was opened. After postal workers at the Hamilton postal facility in New Jersey and the Brentwood facility in Washington became ill, all postal workers at those facilities were treated with antibiotics to prevent disease, and the buildings were closed.

While there were ultimately only 11 cases of inhalational anthrax nationwide in the 2001 U.S. outbreak, including five deaths, more than 10,000 people were treated with antibiotics to prevent disease. None of the people who received antibiotics came down with the disease. An important public health question concerns whether cases of disease were actually prevented by the use of antibiotics.

How big could the outbreak have been? Was the public health measure of giving antibiotics to exposed people effective? Unfortunately, there is relatively little direct data available to answer these questions. However, statistical models can help fill in the gaps of knowledge and tell the story of what happened and what could have happened.

THE ICEBERG PHENOMENON

A model can be developed to help answer the question of how many cases of anthrax were prevented by the antibiotics. The key idea is what we call the iceberg phenomenon. The cases that are symptomatic and come to public attention may be only the tip of the iceberg, as there may be many other potential cases that are also infected but their disease is still incubating. These cases, the ones below the tip, are the silent cases that have not yet become symptomatic. The iceberg phenomenon is illustrated in Figure 1. The question is, Can we estimate the entire size of the iceberg from just the tip?

A common-source outbreak is one in which the time and location of the exposure to the infectious agent is the same for all affected cases. When anthrax spores are disseminated into the air, it is believed that they remain suspended for at most one day, during which time they have the potential to be inhaled into a person's lungs. The clusters of anthrax cases in the New Jersey and Washington postal facilities and the Florida media publishing company can be considered common-source outbreaks. The time from when a person is

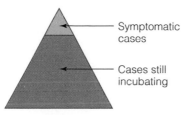

FIGURE 1 The iceberg phenomenon. If X cases become symptomatic t days after exposure, then $N - X/F(t)$ estimates the total number of cases in the common-source outbreak.

exposed to an infectious agent and the onset of symptomatic disease is called the incubation period.

Let's consider a simple numerical example. Suppose four cases of disease occur within 11 days of exposure in a common-source outbreak. Let's also suppose that the median incubation period is 11 days; that is, the probability is half that the incubation period is smaller or equal to 11 days and half that it is greater than 11 days. Thus the four symptomatic cases that occurred in the first 11 days following exposure to the infectious agent should represent only about half the total cases, as the other half are still incubating, Therefore, based on the size of the tip of the iceberg, which is four cases, and the incubation period, we estimate that the entire iceberg is about eight cases.

The cases of disease that initially occur when an outbreak begins to unfold are the cases with the shorter incubation periods. There may be many other people who are also infected but their disease has not yet produced symptoms. Thus, a central idea underlying the iceberg phenomenon is that there is variability in incubation periods and that incubation periods follow a probability distribution. The probability distribution of the incubation period $F(t)$ is the probability that the incubation period is less than t days. To illustrate the notation, $F(11) = 0.5$ means that the probability is 0.5 that an incubation period is less than 11 days. We shall define $F(t)$ only among people who received sufficient dose of the infectious agent to ultimately become ill with symptoms, and as such $F(t)$ is a proper distribution function; that is, $F(t)$ eventually approaches 1 after sufficient time has elapsed. In general, suppose X cases of disease occur within t days of the exposure in a common-source outbreak. We would like to estimate the total number of cases that would eventually occur, which is called N. The X cases are just the cases with incubation periods shorter than t days and are the tip of the iceberg. We can estimate the entire size of the iceberg N from the equation $F(t) \approx X/N$ if we had knowledge of the incubation period and specifically the value of $F(t)$. Thus, $N = X/F(t)$.

To illustrate these ideas, suppose the incubation period distribution is $F(t) = 1 - e^{-0.07t}$, which is an example of an exponential distribution. Suppose in a common-source outbreak we observe 10 cases of disease within seven days following the exposure. We would expect only about 39% of incubation periods to be less than seven days because $1 - e^{-0.07 \times 7} = 0.39$. Thus, the total number of cases we would eventually expect to see in this entire outbreak is about 26, because $N \approx 10/0.39 \approx 26$.

THE INCUBATION PERIOD OF ANTHRAX

If we want to gain understanding about the anthrax outbreak from the iceberg phenomenon we need to know something about the incubation period of the disease. There is some experimental data from monkeys but little data about the incubation period in humans. The largest human outbreak occurred in the city of Sverdlovsk, Russia, about 900 miles east of Moscow, in April 1979 (Meselson et al., 1994; Guillemin, 1999). Some public officials initially suggested that the Sverdlovsk outbreak was of the gastrointestinal type that resulted from eating contaminated meat. However, epidemiological investigation showed that the Sverdlovsk cases lived or worked in a narrow geographical band consistent with wind directions. Pathological reports also indicated that the outbreak was inhalational anthrax. Ultimately, Soviet officials confirmed that the outbreak occurred because of an accident at a military microbiology research facility in which a vent was accidentally left open on April 2, 1979, causing the dispersal of anthrax spores into the air. Soviet public health officials attempted to control the outbreak by distributing antibiotics and vaccine to exposed people in Sverdlosvk. However, it is not known how many people received antibiotics or vaccine, for how long, or the effectiveness of those antibiotics and vaccines.

It is believed that spores fall to the ground within one day and do not generally resuspend into the air, and thus it can be assumed that the Sverdlosvk outbreak was a common-source outbreak, with the exposure date of April 2, 1979. A histogram of the incubation periods of 70 cases from the Sverdlosvk outbreak shows a long tail, with incubation periods as long as 40 days. Can we obtain a credible estimate of the incubation period distribution from this data? An important caveat with a naive analysis of the data is that cases destined to have long potential incubation may be less likely to be included in the data set for the following reason: It took about two weeks following the accidental release of the anthrax spores for public health officials to begin disseminating antibiotics and vaccine. Thus, only people with potential incubation periods greater than two weeks would have had an opportunity to obtain antibiotics and perhaps have their disease prevented. Thus a naive analysis of the data might underestimate the true incubation period distribution because it might preferentially exclude some of the cases with long incubation periods. Statisticians call this phenomenon right truncation, a form of selection bias that occurs when larger observations may be less likely to be included in the data set.

A statistical model can help adjust the data for the bias introduced by right truncation. For example, it could be assumed that all cases that became symptomatic prior to the start of the public health control measures are included in the data set but only a random sample of cases with potential incubation periods beyond the start of the public health control program are included in the data set because disease in the remaining persons was prevented. That assumption, together with an assumed functional or parametric form for the

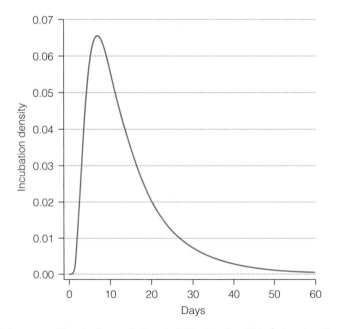

FIGURE 2 Lognormal incubation period probability density of inhalational anthrax

Source: Based on the analysis in Brookmeyer et al. (2001).

incubation period distribution, is sufficient to adjust the data for the bias introduced by right truncation. Sartwell (1950) pioneered the use of the lognormal distribution for the incubation distribution of infectious diseases. The distribution is lognormal if the logarithm of the incubation period follows a normal distribution, that is, if a logarithmic transformation produces the familiar bell-shaped curve. The lognormal distribution is right skewed, which means that it has a long right tail. An analysis that used the data from the Sverdlovsk outbreak, the lognormal distribution and a correction for the bias introduced by right truncation, showed that the median incubation period was about 11 days (Brookmeyer et al., 2001). The incubation period distribution based on this analysis is illustrated in Figure 2.

STATISTICAL MODEL FOR THE 2001 U.S. ANTHRAX OUTBREAK

The 2001 U.S. outbreak of inhalational anthrax occurred principally in three clusters: two cases in the Florida media publishing company, two cases among the postal workers exposed at the Hamilton postal facility in New Jersey, and four cases among the postal workers at the Brentwood postal facility in Washington D.C. (Jernigan et al., 2001). In total, eight cases of inhalational anthrax occurred from these three clusters. The data that is available for analysis, illustrated in Figure 3, shows the dates of onset of symptomatic disease among the cases, the dates antibiotics were distributed to all people in the

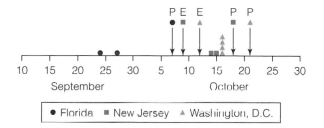

FIGURE 3 Dates of exposure to anthrax spores (*E*), and dates when people began to use antibiotic prophylaxis (*P*) in each of three clusters: workers at the Florida publishing company (circles), postal workers in New Jersey (squares), and postal workers in Washington, D.C. (triangles). Also shown are the dates that cases became symptomatic in each of the three clusters.

cluster, and the dates people were exposed to the letter containing the anthrax spores in New Jersey and Washington. The date of exposure is unknown in the Florida cluster because that letter was never found.

The total number of cases that could have occurred in the New Jersey and Washington clusters can be estimated by using the iceberg phenomenon, provided there is an independent estimate of the incubation period distribution such as that derived from the Sverdlovsk outbreak. The statistical model used to describe the number of cases that occurred in each cluster is the binomial distribution. The sample size parameter of the binomial distribution, N, represents the total size of the iceberg and is a parameter that is estimated. The "success" probability of the binomial distribution is the probability that the incubation period is less than the time interval between exposure to the infectious agent and when antibiotics were distributed. The modeling is a little more complicated for the Florida cluster because the date of exposure is uncertain, and so an additional parameter needs to be introduced into the model, the calendar date the Florida cases were exposed to the letter contaminated with anthrax spores (Brookmeyer and Blades, 2002; Brookmeyer and Blades, 2003).

The parameters of the statistical model were estimated using maximum likelihood methods. The maximum likelihood method is a statistical procedure to find estimates of the model parameters that are most consistent with the observed data. The results of the analysis are summarized in Figure 4, which shows the profile likelihood

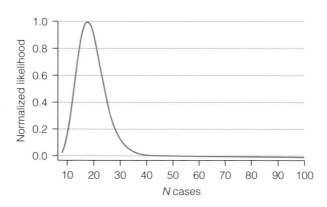

FIGURE 4 The likelihood function for the size of the 2001 U.S. outbreak (N) if antibiotics had not been initiated

(normalized to 1) for the total number of cases from all three clusters that would have occurred if not for the use of antibiotics. The profile likelihood is a useful graphical display of the evidence for how big the outbreak could have been. The likelihood function shows which values of N are best supported by the data. The mode of the profile likelihood in Figure 4 is $N = 17$ cases. Thus, while only eight cases of inhalational anthrax actually occurred in these three clusters, it is estimated that potentially there could have been 17 cases, suggesting that the public health control campaign to distribute antibiotics halved the number of cases. The date that the letter contaminated with anthrax spores arrived at the Florida publishing company estimated by the statistical analysis was September 18, 2001, only one day earlier than when employees reported seeing that suspicious letter.

QUANTIFYING UNCERTAINTY

Describing uncertainty in estimates from a model is an important component of a thorough analysis. The profile likelihood shown in Figure 3 not only displays the "best" estimate but also visually communicates uncertainty in that estimate. The estimate of the total number of cases that could have occurred was 17, but there is uncertainty in that estimate. The 95% confidence interval accounts for statistical error by giving a range of plausible values for N, and the confidence interval was between 10 and 28 cases.

However, that confidence interval does not actually account for all sources of uncertainty. It reflects only uncertainty due to sampling variation, and does not account for possible errors in the underlying model assumptions. For example, one underlying model assumption is the incubation period distribution. The incubation period was estimated from the Sverdlovsk outbreak, which is subject to both sampling and non-sampling errors. One way to address these additional sources of error is to perform a sensitivity analysis to different assumptions about the incubation period. For example, if the median incubation period is assumed to be 17 days instead of 11 days, then the potential size of the outbreak would have been 34 cases instead of 17 cases. If longer incubation periods are used in the statistical model, then the estimates of the numbers of cases that were prevented becomes larger. Intuitively that happens because if the incubation period is longer, more cases are incubating and thus the tip of the iceberg is a smaller part of the entire iceberg. Even with extreme assumptions about the incubation period distribution, the estimates of the potential size of the outbreak were always fewer than about 50 cases.

Another source of error concerns the statistical procedure itself. For example, the procedure that produced a confidence interval for N was an asymptotic procedure (actually called a likelihood ratio based confidence interval procedure). An asymptotic confidence interval procedure is supposed to "work" with the stated level of confidence when the sample size is large. However, the U.S outbreak was not what is typically considered large, with only eight observed cases and 17 potential cases. Simulation studies are useful to evaluate the performance of a statistical procedure in small sample sizes that theoretically works in large sample sizes. A simulation study revealed that the confidence interval procedure used in the analysis of the anthrax outbreak performed very well even though the sample sizes were small (Brookmeyer and Blades, 2003).

In summary, while models can help fill in critical gaps in knowledge, there are important sources of errors, including sampling error and model assumptions. A variety of approaches can be used to evaluate and assess uncertainty in the results, including confidence intervals, sensitivity analyses, and simulation studies. Bayesian methods can also be used to incorporate and tie together many of the sources of uncertainty.

MODEL IMPLICATIONS

More than 10,000 persons were treated with antibiotics to prevent disease in the fall of 2001 in the anthrax outbreak in the United States. Yet, public health officials did not have any idea if cases of disease were prevented by the public health control measures and if so, how many. The case study presented in this essay illustrates that a statistical model can piece together various sources of data into a coherent picture to answer these critical questions.

A by-product of the analysis was an estimate of the date that people in the Florida media publishing company were exposed to the letter contaminated with anthrax spores. If in a future outbreak the date of exposure is unknown, then statistical analyses can estimate the date of exposure and perhaps help identify the perpetrator of an act of bioterrorism or identify other people who may have also been exposed to the infectious agent and could benefit from medical care.

Several important messages follow from these analyses. Antibiotics halved the number of cases of disease, indicating that the public health control measures likely saved lives. Yet even without antibiotics, the 2001 U.S. outbreak was unlikely to have been greater than 50 cases. More than 10,000 persons were treated with antibiotics to prevent roughly a handful of cases of disease. Because widespread use of antibiotics could result in significant numbers of adverse reactions, an important question raised by our results is whether more targeted and limited distribution of antibiotics to those people at highest risk could have prevented roughly a similar number of cases of disease while minimizing unnecessary antibiotic use. Related questions are how long people really need to stay on the antibiotics to make sure they don't get sick, how quickly they must start antibiotics, and what is the role of vaccines in containing anthrax outbreaks. Models can help answer those questions, too (Brookmeyer, Johnson, and Bollinger, 2003; Wein, Craft, and Kaplan, 2003; Brookmeyer, Johnson, and Bollinger, 2004). The anthrax outbreak heightened awareness of the critical importance of rapid detection of outbreaks, together with effective and targeted control measures in protecting the health of the public.

Models similar to the kind used in this essay are also used to track other kinds of epidemics. For example, a method to predict future trends of the HIV/AIDS epidemic called back-calculation (Brookmeyer and Gail, 1994; Brookmeyer, 1991) is based on an idea similar to the iceberg phenomenon. One difference is that HIV infection is transmitted from person to person, and so infections can occur continuously in time rather than at a single time point, as in a common-source outbreak. HIV infection is characterized by a long and variable incubation period with a median of about 10 years. Similar ideas have also been used to model the epidemic of bovine spongiform encephalopathy among cattle in the United Kingdom, also known as mad cow disease, and the subsequent epidemic it spawned in humans called variant Creutzfeldt Jakob disease (Donnelly and Ferguson, 2000).

REFERENCES

Brookmeyer, R. (1991). "Reconstruction and Future Trends of the AIDS Epidemic in the United States." *Science* 253 (July 5): 37–42.

Brookmeyer, R., et al. (2001). "The Statistical Analysis of Truncated Data: Application to the Sverdlovsk Anthrax Outbreak." *Biostatistics* 2, no. 2:233–47.

Brookmeyer, R., and N. Blades. (2002). "Prevention of Inhalational Anthrax in the U.S. Outbreak." *Science* 295, no. 5561:1861.

———. (2003). "Statistical Models and Bioterrorism: Application to the U.S. Anthrax Outbreak." *Journal of the American Statistical Association* 98, no. 464:781–88.

Brookmeyer, R., and M. H. Gail. (1994). *AIDS Epidemiology: A Quantitative Approach.* Oxford: Oxford University Press.

Brookmeyer, R., E. Johnson, and R. Bollinger. (2003). "Modeling the Optimum Duration of Antibiotic Prophylaxis in an Anthrax Outbreak." *Proceedings of the National Academy of Sciences* 100, no. 17:10129–32.

———. (2004). "Public Health Vaccination Policies for Containing an Anthrax Outbreak." *Nature* 432 no. 7019:901–4.

Donnelly, C. A., and N. M. Ferguson. (2000). *Statistical Aspects of BSE and vCJD: Models for Epidemics.* Monographs on Statistics and Applied Probability 84. New York: Chapman and Hall/CRC.

Guillemin, J. (1999). *Anthrax: The Investigation of a Lethal Outbreak.* Berkeley: University of California Press.

Jernigan, J. A., et al. (2001). "Bioterrorism-Related Inhalational Anthrax: The First 10 Cases Reported in the United States." *Emerging Infectious Diseases* 7, no. 6:933.

Meselson, M., et al. (1994). "The Sverdlovsk Anthrax Outbreak of 1979." *Science* 266, no. 5188:1202–8.

Sartwell, P. E. (1950). "The Distribution of Incubation Periods of Infectious Diseases." *American Journal of Hygiene* 51:310–18.

Wein, L. M., D. L. Craft, and E. H. Kaplan (2003). "Emergency Response to an Anthrax Attack." *Proceedings of the National Academy of Sciences* 100 no. 7:4346–4351.

QUESTIONS

1. Why is it of public health interest to know how many cases of disease were prevented by the antibiotics program?

2. Suppose the incubation period distribution is $F(t) = 1 - e^{-0.03t}$ where t is the number of days since exposure. That is, $F(t)$ is the proportion of infected people who are symptomatic at or before t days after exposure to the infectious agent. Now, suppose you have seen a total of 16 cases by the seventh day following exposure. What would be your estimate of the number of people ultimately getting sick in this common-source outbreak?

3. What is the median incubation period for the distribution in Question 2?

4. Can the iceberg phenomenon be useful in other kinds of diseases? For example, can it be used with diseases with very long incubation periods? Very short incubation periods? Diseases where the incubation period is very variable? Diseases where there is little variation in the incubation period?

5. If we can detect an outbreak more quickly, will we be able to prevent more people from getting sick? If so, under what conditions?

THE LAST FRONTIER

Understanding the Human Mind

WILLIAM F. EDDY

Department of Statistics, Carnegie Mellon University

MARGARET L. SMYKLA

Department of Statistics, Carnegie Mellon University

It began like this. It was a cold and dreary afternoon in Pittsburgh in November 1993. I was sitting in my office reviewing census data when someone knocked. I looked up. It was a rather wild looking man, his long gray hair flying, shirt untucked, eyes ablaze with an intensity that was almost scary. "I'm a psychologist; I need some help," he said, oblivious to the irony: it was usually the other way around.

He set a black-and-white picture on my desk that he said was of a slice of a human brain. It was not a very sharp image. "I thought X-rays were a lot clearer than this fuzzy thing," I said. He explained that the picture was taken using magnetic resonance imaging (MRI), not X-rays, and that he was studying brain function, not brain structure.

He wanted my help as a statistician in an experiment he was conducting on short-term memory. He wanted to know: What areas of the brain are used for short-term memory? If the demand on your

---

This essay is a joint effort of both authors; the use of the first person singular refers to the first author.

memory is greater, will those areas show a higher level of activity? If so, what is the relationship between memory load and activity level in the brain? Does brain activity increase in a straight-line fashion as memory load increases?

For me this was a chance to work on what I had come to regard as one of the biggest remaining mysteries in science. During my lifetime, scientists have made extraordinary progress on three great problems—understanding the nature of matter, understanding the nature of space, and understanding the nature of life—but we still know comparatively little about a fourth great problem: the nature of mind.

DESIGNING THE EXPERIMENT

As a statistician, my biggest challenge in helping the psychologist would be to design a good experiment. That may not sound like a big challenge, but consider how much would be riding on it. We knew that the use of the equipment alone would end up costing around $10,000, not the kind of outlay that makes it easy to say, "Oops! That didn't give us the information we need. We'll have to rerun the experiment with a different protocol."

A typical experiment takes an hour and a half to run, per subject. Out of this time, data might be collected for no more than 30 or 45 minutes. The rest of the time is spent setting up and calibrating the equipment. Questions of greater complexity require more data, necessitating both more subjects and longer scanning times.

It is important to design experiments in such a way that the scientists can investigate particular questions of interest, while making efficient use of the time and equipment available. As statisticians, we can—and do—take an active role in this stage of the planning to help ensure that the resultant study will be sound.

In broad outline, the design challenge would involve three kinds of questions and three different subject areas. We had to decide which measurements to make, whether the response was a result of the stimulus, and how to manage variability. Very roughly, these three questions correspond—for this study—to radiology, cognitive psychology, and statistics.

Radiology: What Measurements to Make

The first challenge we faced is due to a simple fact: your skull may be hard, but your brain is soft. Until recent decades, the only way to get a picture of soft tissue was "the old fashioned way": you had to cut a person open and take a look. Volunteers for brain research were hard to find.

MRI changed all that. Compared with X-rays, MRI is both much more powerful and at the same time much more limited. An X-ray machine, like a camera with a fast lens, makes a quick picture. But the picture is basically just a shadow cast by bones or other dense objects. Your skull is hard enough to cast an X-ray shadow, but your brain is too soft. Trying to detect brain activity using X-rays would be completely hopeless. MRI solves the hard/soft problem, although the process is not nearly as fast. Whereas a high-resolution X-ray of your teeth is almost instantaneous, a high-resolution MRI of your brain takes many minutes. Nevertheless, the technology is truly impressive.

The underlying idea of magnetic resonance was discovered in 1946 by Felix Block and Edward Purcell, who were awarded a Nobel Prize for their work seven years later. Every atom spins on its axis, much the way the earth does. In a magnetic field the axes of rotation all line up with the field. If energy is added, the axes tip away from the direction of the field. Immediately afterward, the atoms "relax" (their axes return to alignment with the magnetic field) and emit microwave energy. If you put an object, such as your body, inside a magnetic field, you can use a scanner to record the microwave energy, and use the scan to create a picture. Although the applications of magnetic resonance to medical imaging have now been in use for many years, the Nobel Prize for pioneering contributions in this area was awarded only recently, in 2003, to Peter Mansfield and Paul Lauterbur. Meanwhile, MRI has become so common today that millions of MRI scans are performed annually to visualize the body's internal structures and diagnose a number of ailments, including tumors, stroke damage, heart and brain disease, and back and knee problems.

Back in 1993, although I knew MRI was used to pinpoint damage caused by a stroke or other calamity, my psychologist's claim that the technique could be used to make movies of the brain in action was news to me. After he left, I did a computer search and discovered that on July 1 of the prior year, a team of researchers from Bell Laboratories and the University of Minnesota reported the first demonstration that *activity* in the human brain could be detected by MRI.

Increased activity of the neurons in the brain induces an increase in blood flow to the region of activity. This increased flow leads to an increase of oxygenated blood in the small veins that drain the active region because the increased activity does not require much extra oxygen.

This technique, in which MRI detects the changes in blood oxygenation caused by brain function, became known as functional MRI, or fMRI. It is not invasive, and fMRI allows the study of high-level cognitive processes such as language, visual attention, and problem solving. With fMRI, a series of MR images is collected over time to gather information about brain activity during the course of the scan. While the scan is being performed, subjects may be asked to carry out various mental tasks; the images will then contain information about which regions of the brain were active and at what times.

These facts about fMRI determined the basics of what we would measure, but we still had many decisions to make. From the radiological point of view, for example, we were limited by an unavoidable trade-off between time and resolution. If we needed high resolution, it would take a long time to get an image. If we needed frequent images in order to track changes within the brain, we'd have to settle for fuzzier pictures. Following the advice of a radiologist, we settled on one image per second. Now we could turn to the next challenge, which was mainly a matter of cognitive psychology.

Cognitive Psychology: Isolating the Effects of Interest

At this point in our planning, my psychologist and I knew we would use fMRI to detect and record increases in brain activity. Our basic premise, as with all fMRI, was that small changes in the stimulus will cause small changes in the response of the subject. We also knew the two related questions we hoped to answer: Is there a particular location in the brain used for short-term memory? If so, is the level of activity related to memory load? We also knew, of course, that we would be using human subjects. The main thing left to decide at this point was what we should ask them to do while they were inside the scanner giving us pictures of their brains.

As a statistician I knew that our initial challenge was to *isolate the effects of interest.* Suppose we were to find an increase in activity, I asked myself. How could we be sure the activity was tied to short-term memory, and not to something else—like processing sounds, recalling vocabulary (long-term memory), making logical deductions, or just plain daydreaming?

Many experiments in cognitive psychology are based on the same general strategy. The idea is to invent a simple task that calls for a yes/no response, but is cleverly chosen to get the subject to engage in a particular kind of thinking. For example, if you wanted to know which parts of the brain are used for visualization, you might show subjects pictures of the letter R or its mirror image, rotated through various angles (0°, 30°, 60°, 90°, etc.), and ask whether the rotated image is an R (yes) or its mirror image (no). By requiring a response, you help ensure that subjects stay on task, and by keeping the tasks simple, you can know what answers to expect when subjects are following instructions. Any subject who starts daydreaming, for example, will almost surely start giving lots of wrong answers. Simple tasks with a built-in check of this sort help isolate the effects of interest, provided, of course, that the tasks themselves are chosen well.

The psychologist and I used a similar strategy to get our subjects to use short-term memory. We decided to show them a sequence of letters, one at a time, at a rate of one letter per second.

In the "Zero" level of the experiment, the subject was asked to push a button whenever the letter X appeared. No short-term memory for letters was required.

In the "One" level of the experiment, the subject was asked to push the button whenever the same letter occurred consecutively (twice in a row); this required the subject to briefly remember each letter in turn. Obviously, the memory load is greater than in the "Zero" level.

In the "Two" level of the experiment, the subject was asked to push the button whenever the same letter reoccurred two letters later (with one intervening letter); this required the subject to briefly remember two consecutive letters and to update those two letters, once per second. The memory load was clearly larger than in the "One" condition.

In the "Three" level the subject was asked to push the button whenever the same letter reoccurred with two intervening letters; again the memory load was increased.

In an ideal world, this set of tasks should make it easy to test the psychologist's theory, because it predicts that subjects would show the lowest levels of brain activity at level Zero, with the next lowest levels of activity for tasks at level One, and so on up to the highest level. In addition, the hypothesis predicts a straight-line relationship between the level of the task and the level of activity in the brain.

This might be what we would see in an ideal world, but as a statistician, I knew the world is far from ideal. In the real world, things vary.

Statistics: Managing Variability

No two people are alike. No two brains have the same size or the same shape. Even if size and shape are similar, what's inside will hardly be the same. Even the same person is not 100% the same now and five minutes from now. For example, heads don't stay perfectly still. Even when lying still inside the tunnel of an MRI with your head in a restraint, it will still move a little. Moreover, your brain moves inside your skull. It moves when you breathe. It moves every time your heart beats. Fresh blood brings new supplies of sugar and oxygen. Neurons fire, chemicals react, your brain changes. As the airlines put it, "Contents may shift while in transit."

To the psychologist and me, all this was a source of "noise," a form of statistical static that would make the scientific signal harder to detect and decode. We'd made good decisions about what measurements to make, and we'd made a good choice of tasks in order to isolate the effects of short-term memory, but without an effective plan for managing the statistical noise, we could still end up with a worthless experiment. In a real sense, noise management was the biggest challenge of the whole experiment. For sure it was the main reason why the psychologist needed the help of my brain.

In this phase of the planning, I relied on three principles that are often useful in statistical work: *use repetition, control what you can, and randomize the rest.*

1. *Use repetition, both to measure the size of variation, and to reduce it by averaging.*

    a. *Multiple subjects.* We knew that one person differs from the next, so we decided to use several subjects. This would let us quantify the differences: How far was the reading for Subject 1 from the reading for Subject 2? It would also let us combine the readings from all our subjects together, into a kind of average, to take advantage of the fact that averages tend to be less variable (less "noisy") than the individual values that go into those averages. Our choice of 10 subjects was a compromise. More would have been nice—from a statistical point of view, it's hard to have too many—but we were limited by practical considerations, such as time and cost.

    b. *Within-subject comparisons.* We knew that for most experiments involving human subjects, the differences from one subject to the next—so-called between-subject differences—would typically be a lot larger than differences from one occasion to the next for the same subject—within-subject differences. (The example of skull size gives an extreme illustration: a person's skull size hardly varies at all from one moment to the next, but the variation in skull size between you and other people is fairly substantial. For skull size, within-subject variability is tiny; between-subject variability is not.) Knowing this fact about the two kinds of variability meant that we wanted to design our experiment to permit within-subject comparisons: every subject should get sequences using all four memory loads.

    c. *Repetitions within subjects.* We knew that the same subject will be different at different times, so we used multiple sequences with each subject. In all, each subject got 12 sequences, three at each level. Just as our choice of 10 subjects was a compromise, so was our choice of three sequences per level. In principle, more would have been better, but in practice we didn't want to keep our subjects in the MRI longer than necessary. Multiple sequences per level offer the same sort of advantage we got from multiple subjects: we would be able to measure the typical size of the variability from one sequence to another for the same subject at the same level of memory load, and we would be able to reduce noise levels by working with averages.

d. *Long sequences.* Finally, we knew that our measurement process, the MR imaging, was noisy. By using 200 letters in each sequence, we would be able to get lots of scans per sequence, which we could use during the analysis to remove some of the noise by averaging. (In this context, "high resolution" and "low noise" mean essentially the same thing.)

A second principle that I relied on to manage variability is in fact closely related to "isolate the effects of interest":

2. *Control what you can.* Throughout the experiment, we tried to keep conditions as constant as possible, apart from varying the levels of memory load. We treated the subjects as much the same as we could. We used the same MRI protocol for all of them. We kept the conditions for the 12 sequences as uniform as possible. We also tried to choose subjects who could be expected to be similar in brain function. For example, because left and right sides of the brain have different functions, with some of those brain functions related to a person's dominant hand, we used only right-handed subjects. Also, because our task of recognizing letters was one that might well be affected by a person's native language, we used only native speakers of English as subjects.

3. *Randomize what you can't keep constant.* One set of effects we were unable to control was linked to time. Each subject was to complete 12 sequences of 200 letters, and we knew that over the course of the hour or so it would take, a number of things might change in ways we couldn't predict. Some subjects might become more relaxed as they got used to being inside the MRI machine. Others might start to feel more claustrophobic. Some subjects might start fresh but see their performance deteriorate as they tired. Others might actually get better as their brains got used to the memory task.

Rather than try to predict these effects, we decided to manage them by random assignment. Each subject would do three sequences at each level, but the order of the levels would

	SEQUENCE NUMBER											
SUBJECT	1	2	3	4	5	6	7	8	9	10	11	12
A	2	1	1	0	0	3	1	3	3	0	2	2
B	1	2	0	0	1	2	0	3	2	3	1	3
C	0	1	0	1	2	1	3	0	3	3	2	2
D	3	2	3	1	0	2	1	1	0	3	2	0
E	2	0	3	2	2	0	0	3	1	3	1	1
F	0	0	1	2	3	0	2	2	3	1	1	3
G	0	3	1	3	2	0	2	1	0	2	1	3
H	0	2	1	3	2	0	0	1	3	3	2	1
I	3	1	3	3	0	0	2	0	2	2	1	1
J	3	0	0	2	1	3	1	2	3	0	1	2

TABLE 1 Randomizing the assignment of memory loads to sequence numbers

be determined using a chance device, with a separate random order for each subject. Whatever the time-related effects, the use of a chance device would tend to balance those effects, just as a sequence of coin flips tends to balance half heads, half tails— even though the individual flips can't be predicted. All subjects started with a baseline scan that was used to determine where within the skull the other scans would be taken.

Table 1 is a schematic representation of the whole experiment. The numbers in the body of the table tell the level of memory load.

ANALYZING THE DATA

Two Kinds of Variation: Systematic and Random

The images we collected during the experiment had low spatial resolution and were fairly noisy, with two kinds of noise: systematic variation and random variation. Systematic variation is the sort that can be attributed to a specific source, such as mistimings in the electronics of the machine or motion of the brain associated with breathing. Random variation is the sort for which we have no explanation. We attempted to remove the systematic variation through a series of pre-processing steps. The specific sources are complex, and the pre-processing was correspondingly time-consuming. At the end of the pre-processing we had a collection of images that were as free of systematic noise as we could then achieve, apart from the systematic variation the experiment was designed to study. We were then ready to address the first question of experimental interest: Does increasing memory load increase brain activity?

Consider, for simplicity, two of the experimental conditions: the "Zero" level, which imposes no load on the short-term memory, and the "Two" level, which imposes a fairly heavy load.

To understand our next step, you must remember that MR images are black and white; that is, there is no color, only various shades of gray. The shade of gray corresponds to the strength of the signal at that location; black indicates no signal and lighter shades indicate stronger signals. Inside a computer, such an image is represented by numbers, zero for black and the lighter the gray, the larger the number. Thus an image can be thought of as an array of numbers, and we can perform arithmetic on images. We will think this way from now on.

If a "Zero" level image is subtracted from a "Two" level image, some pixels in the difference image would be positive and some would be negative and the rest would be zero. (A pixel—short for "picture element"—is the smallest unit of a picture that can be individually considered; typically, one measure of the quality of a digital image is the number of pixels.)

If there were no noise, the positive values would indicate regions of increased brain activity in the "Two" level; the negative values would indicate regions of reduced brain activity (that is, increases in the "Zero" level); and the zeroes would indicate regions of no change. Because there is noise we can't be certain that the positive values indicate increased brain activity; they might be positive simply by chance. So we need a formal method to decide if the values are greater than zero.

Random Variation: Is the Difference "Real" or Just Noise?

We now consider a much simpler but analogous situation. Suppose that I want to know whether cross-training will lower my resting heart rate; some studies have shown that it will. I first establish a baseline heart rate by making 100 repeated measurements of my pulse rate while sitting. The measurements will not all have the same value; they will vary about the typical or "true" value, with some values larger and some smaller. If a single value is needed, I would report the average as my best estimate of my resting pulse rate. Suppose now that after six weeks of aerobics and weight training, I record my resting pulse again. How can we decide if the new pulse rate is essentially the same as all of the ones I measured before I began exercising, or if, in fact, the exercise program has lowered my pulse rate?

One way would be to look at the individual values that I had recorded to get my baseline, and see if the additional, post-exercise measurement was "compatible" with them. To be specific, suppose that I had recorded 100 different pretreatment values. Suppose that the new, post-exercise measurement was smaller than all of those 100 recorded values. That would be quite surprising if the cross-training had had no effect on my pulse rate. We interpret a surprise to mean that cross-training did, in fact, lower my resting heart rate. A surprise is a real difference. Or, suppose instead that only one of my 100 baseline values was larger than the additional measurement; that would also be surprising but somewhat less surprising than in the first case.

To turn this into an operational procedure, all that is required is to decide, in advance, on an acceptable level of surprise. It is common, even traditional, to use five out of 100 as an acceptable level; if fewer than five are smaller (or larger) than the additional measurement it is deemed surprising and presumed to indicate a real difference.

Too Many False Alarms: Adjusting the Level of Surprise

We applied a slightly more complex form of this procedure to the differences ("Three" minus "Two"; "Three" minus "One"; "Three" minus "Zero"; "Two" minus "Zero"; "Two" minus "One"; and "One" minus "Zero") and made some pictures. It was immediately obvious that using five out of 100 as an acceptable level of surprise did not work. There were way too many surprises! (Remember we interpret a surprise to mean a real difference.) A little thought shows

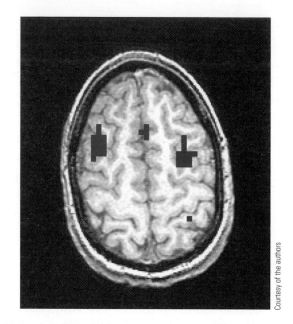

Courtesy of the authors

FIGURE 1 The high-resolution structural image of one horizontal slice of the brain of one subject. The image has had the surprising pixels (in the "Two" versus "Zero" comparison) from the functional images indicated by red regions. These regions are composed of large pixels, larger than the pixels in the brain image. This boxiness gives some sense of the difference in spatial resolution between the high-resolution (structural) scans and the low-resolution (functional) scans.

what the problem is: multiple comparisons. Suppose you were to perform the procedure described above 100 times. Then, just by chance you should expect that about five times out of the 100 you should be surprised by a result that really isn't a surprise; it is in fact a false alarm. Our images had 128 × 128 pixels, which meant that we were actually performing the procedure more than 16,000 times, and we should have expected about 820 false alarms. After we adjusted our level of surprise to account for the number of pixels, we made new pictures. Figure 1 shows the places where the "Two" minus "Zero" differences were surprising.

I'm not a neuroscientist but I know enough of brain anatomy to say that the two large contiguous regions of activity in Figure 1 are in the prefrontal cortex, exactly the area where short-term memory is believed to reside. The psychologist confirmed this, and the experiment was deemed a success. Increased memory load *does* cause an increase in brain activity.

How Does Brain Activity Relate to Memory Load?

You will remember that my psychologist friend had a second question, assuming that the answer to the first was yes. He wanted to know something about the nature of the relationship between activity and memory load. This is a more difficult question, and the answer is less certain and less clear. Part of the problem is that his question doesn't specify a definition of brain activity. I could, for example, count the number of pixels that are identified as active in each experimental level, but that would depend on, among other things, the chosen level of surprise and the size of the brain of the subject. Alternatively, I could compute the average pixel value of the "active" pixels at each experimental level. This would still depend on the chosen level of surprise, and it doesn't tell us what do with the small number of apparently unrelated pixels that seem active (that is, the ones that are not contiguous to lots of others). Without going into detail of how we got there, I'll just jump to the decision we ended up with after a complicated analysis: The active regions were determined by counting a pixel as active if it was active in all three comparisons to level "Zero." This eliminated all the isolated pixels and left us with contiguous regions of activity.

Another complication in the analysis was that the psychologist didn't say whether he meant the relationship would be a straight line for each person or just that when averaged over all 10 individuals in some way the relationship would be a straight line.

It would take too much space to describe our procedure for deciding the straight-line question, but Figure 2 presents some of the data for the 10 subjects; we computed the average value of the pixels in the active regions for each combination of subject and memory load, and then plotted those averages versus the load separately for each subject. As shown in Figure 2, for some of the subjects the plots bend up, and for others the plots bend down, but there are about half in each group, and the average across all 10 subjects is almost a straight line. The answer to the second question is yes; the relationship between

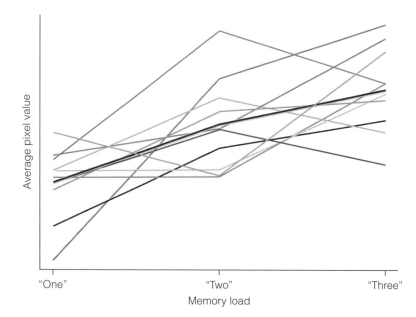

FIGURE 2 A plot of the average pixel value for the contiguous regions that were "surprising" for each comparison of a level to "Zero." The values for each subject are connected by a line so that the values can be compared across levels. The red line joins the average pixel value averaged across subjects. There is a clear upward trend, and one could argue that it is nearly a straight line.

increased memory load and brain activity can be described by an upward sloping straight line.

WHAT HAVE WE LEARNED?

It turned out that my psychologist was right. Increased memory load causes an increase in brain activity. Moreover, the relationship between the two appears to be reasonably well described by a straight line. Our success in establishing these two conclusions illustrates how fMRI can be used to learn about brain activity and, more specifically, to study the relationship between the level of a stimulus (memory load) and level of the response (brain activity).

The experimental design was a success as well. Since that first study, we have designed other fMRI experiments using the same set of key features: multiple scans, fuzzy and frequent, rather than a smaller number of higher resolution scans; multiple subjects, to measure subject-to-subject variability and to reduce it by averaging; several trials per subject, to permit within-subject comparisons; and randomized assignment of conditions to trials within each subject, to protect against bias due to time order.

Using this approach, we have learned a great deal about the detailed organization and function of short-term memory: for colors, numbers, locations, and so on. Some studies have involved subjects who are psychologically normal, but others have looked at autistic adults and still others at schizophrenics. For these studies, our goal has been to find ways in which the brain activity of subjects with autism or schizophrenia differs from the activity of subjects who are psychologically normal. Another set of studies is looking at concussion injuries and how brain function is different for those with sports-related concussions and a control group of other athletes. In all these studies, the ultimate goal is to understand brain function in order to develop more effective treatments that help brains to function normally.

Some of the studies involve as many as 300 subjects. Each subject generates a few hundred megabytes of brain images; combining all the resulting data into one single analysis is currently not feasible because of the sheer size. And so statistical challenges remain. These experiments generate huge quantities of data, with myriad sources of variability, some of it systematic, and some chancelike. We are still far from knowing the most efficient way to combine a set of brain scans in order to get the clearest, most informative possible pictures, those that have eliminated all extraneous differences, leaving only the one difference we want to know about.

QUESTIONS

1. The design of the MRI experiment had to allow for a possible effect of the response to stimulus changing as the subject tired of the measurement process. What strategy was used for this, and why would this strategy reduce the bias that the changes due to tiring might have caused?

2. How would you check whether the design chosen (see Table 1) achieved its objective of balancing out the order effect, if there is an order effect?

3. To control variability in the experiment, only right-handed, native English speakers were chosen as subjects in the experiment. How might this affect the conclusion of the experiment?

4. In Figure 2, six of the profiles are concave down and four are concave up. Why does the author suggest that the relationship is approximately a straight line, when some of the profiles are decidedly bent?

LEVERAGING CHANCE
IN HIV RESEARCH

CHARLES M. HEILIG

*Division of Reproductive Health, National Center for Chronic
Disease Prevention and Health Promotion, Centers for Disease
Control and Prevention*

ELIZABETH G. HILL

*Department of Biostatistics, Bioinformatics, and Epidemiology,
Medical University of South Carolina*

JOHN M. KARON

Emergint Technologies, Inc.

———— ❋ :: ❋ ————

The first cases of what came to be known as acquired immune deficiency syndrome (AIDS) were reported in June 1981 in the Centers for Disease Control's *Morbidity and Mortality Weekly Report* (30, no. 21:250). Human immunodeficiency virus (HIV) infection, the cause of AIDS, is now the fifth leading cause of death for Americans between the ages of 25 and 44. Men who have sex with men have the greatest risk of infection (*Morbidity and Mortality Weekly Report* 50, no. 21:440); many of them are first infected in their teens and early twenties.

Another group at high risk of HIV infection is children born to women with HIV. The mother-to-child transmission of HIV is particularly acute in resource-limited countries. According to UNAIDS—a joint United Nations program on HIV/AIDS—more than half a million children in sub-Saharan Africa were infected with HIV in 1999. In the United States, one of the most encouraging achievements in AIDS research has been the development of therapies that dramatically reduce the number of mother-to-child transmissions. Some U.S. doctors say pediatric AIDS is no longer a crisis in the United States ("Moms Passing HIV to Kids Is Urgent Topic," *San Francisco Chronicle,* July 14, 2000). But these treatments are too expensive and impractical to be widely adopted by many countries. This compelling public health need has motivated researchers to seek alternative ways to reduce the mother-to-child transmission of HIV. Carefully designed clinical trials have made possible a comparison of various affordable drugs and treatments for pregnant women with HIV.

To reduce the number of new HIV infections, local and national HIV prevention efforts need to be developed and targeted to groups at the highest risk of infection. In this essay, we focus on two such groups: young men who have sex with men, and children born to mothers with HIV.

But it is quite a challenge. For each risk group, data are needed to plan and evaluate interventions, but getting good data is neither easy nor straightforward. For example, to learn more about the incidence of HIV and the factors that govern its spread among the young men of interest, we need to find and study a sample that is representative of the population to ensure that the study findings could be generalized to the whole population. Researchers at the CDC employed an innovative statistical sampling technique to find a representative sample of this hard-to-reach population. In our second example, the Uganda study of mother-to-child transmission, investigators confronted the problem of designing an experiment such that the findings could support scientifically valid conclusions about the effectiveness of new treatments. In both cases, randomness plays an important, but different, role in the solution to the problem of controlling the spread of HIV.

THE YOUNG MEN'S SURVEY (YMS)

One approach to controlling the spread of HIV is through behavior change. Unprotected sex is a common risk behavior among men who have sex with other men; efforts might be directed, for example, toward increasing condom use. Any systematic effort to change risk behaviors is difficult, however, because such a program must first identify the people who are likely to engage in risky behaviors and then persuade them that it is worthwhile to avoid risky behavior. If you are going to administer this kind of program, you have to know which groups to target, how and where to reach them, what behaviors to try to change, and what to say to be the most persuasive. The more you know, the more efficient and effective you can be. Good data are essential to these efforts.

Many standard survey sampling methods involve constructing a list of all members in the population or a list of all households in which members of the population are known to reside. The advantage of these lists is that a chance mechanism can be used to select members of the population where every member will have a known probability of being chosen for the sample. When analyzed appropriately, results obtained from this probability-based design generalize to the population of interest. But here we have no such lists.

The population of young men who have sex with men is not readily identified, so traditional sampling designs are unsuitable. Random digit dialing would be ineffective in this situation because few calls would result in someone who belongs to the population. And then there is the problem of getting honest answers on such a sensitive subject from a young man whom you have reached by calling his home. But selecting individuals from those who have, say, recently visited a clinic for treatment of a sexually transmitted disease might yield a biased view of the population. For example, the prevalence of a risky behavior, such as unprotected sex, within the past six months among those seeking treatment at a clinic is unlikely to be the same as the prevalence in the general population.

Fortunately, a previous study pointed to a way of reaching the target population and obtaining a probability sample: in a door-to-door survey in San Francisco, it was found that 91% of homosexual and bisexual men aged 18 to 23 reported attending gay bars in the past six months. It appears that by visiting public places known to be frequented by the young men of interest we can reach nearly all of the population. These public places, called *venues*, may be bars, cafés, dance clubs, parks, business establishments, street locations, and social organizations.

In an effort to measure prevalence of HIV infection and associated risk behaviors among young men who have sex with men, the CDC designed and funded the Young Men's Survey (YMS)—a study based on a probability sample of the population of men aged 15 to 22 who have sex with men and who frequent public venues. The sampling method is a special type of *venue-based* sampling called *space–time* sampling.

Space–Time Sampling

The task of reaching people at appropriate venues presents several challenges. We first must find these places and figure out the best time to visit them, and once there, apply a consistent way of choosing people and approaching them to participate in the survey. We must include all types of venues on the list to avoid *selection bias*. For example, the individuals who frequent social organizations may be different in terms of age, education level, or other characteristics than those who visit dance clubs.

The YMS was conducted in seven large cities in the United States. In each city, the survey team constructed an exhaustive list of public venues frequented by the population of interest. In addition, they figured out which days of the week and times of the day the venue was attended by members of the study population. This process was labor-intensive but crucial to both the operational efficiency of the study and the overall validity of survey findings. Field-workers interviewed business proprietors, examined local publications, consulted representatives of community-based organizations, and conducted focus groups, all for the purpose of constructing a comprehensive list of appropriate venues that were well-attended by the population of interest. This list is known as the *sampling frame,* and each venue in the frame is a *primary sampling unit.*

For each primary sampling unit, the survey team identified those four-hour blocks of time for each day of the week that had a minimum

Stage 1: Randomly select public venues frequented by young men who have sex with men	Stage 2: For each venue, choose at random a four-hour time period to visit	Stage 3: Sample young men from randomly selected time periods
Café	Café:	Deploy four workers to
Dance club	Thursday 7 P.M. to 11 P.M.	the café on the third
Street location	Dance club:	Thursday,
Bar	Friday 9 P.M. to 1 A.M.	7 P.M. to 11 P.M.
Clothing store	Main Street:	Count apparent 15- to
...	Sunday 5 P.M. to 9 P.M.	22-year-olds
	...	Approach and briefly interview
		Invite eligible men
		Interview, draw blood

FIGURE 1 Depiction of the three stages in space–time sampling for the Young Men's Survey

number of members of the study population attending the venue. Only those venues with at least one qualifying four-hour block of time were included in the frame. The process of constructing the sampling frame and identifying the busy time periods for each venue was repeated every month because each venue's attendance could vary over time. Moreover, business establishments close, new ones open, and social organizations may not meet during certain months. This illustrates the need for ongoing formative research to adapt the sampling scheme to reflect the social organization of the population of interest.

There were three stages to selecting people for the survey, each of which was conducted on a monthly cycle (Figure 1). In the first stage, 12 to 14 venues were randomly selected without replacement from the list of primary sampling units. In the second stage, time periods to visit the sampled venues were chosen. To do this, the selected venues were sorted according to their number of available time blocks. Starting with the venues with the fewest time blocks, one time period was randomly selected with equal probability from those available for the venue. A visit by the survey team was scheduled for a matching day and four-hour time slot that month. By sampling time periods in this manner, scheduling conflicts were minimized.

In the third and final stage of the selection process, field-workers arrived at the venue on the selected date and time to enroll participants. Some field-workers counted the number of men who appeared to be eligible (between the ages of 15 and 22), while others intercepted and briefly interviewed all of the enumerated young men. Those determined to be eligible were asked to participate in the survey and complete a face-to-face, confidential interview with a staff member using a standardized questionnaire. Participants answered questions pertaining to demographics (such as race, income, education), how often they attend the venue, and their HIV-related risk behaviors. In addition, field-workers provided counseling and testing services, as well as referrals for medical and social support.

This is a complex survey design involving random selection of venues and time periods and selection of all individuals in the venue/time-period combination. The random selection of venues gives each place an equal probability of selection regardless of its estimated attendance. This process avoids unintentional bias that may arise if the survey staff chooses venues and time slots at their discretion. In addition, allowing staff to select participants or participants to self-select could result in *selection bias* because those who are more approachable or more willing to participate may differ from the overall study population in nontrivial ways. On the other hand, randomly selecting individuals to interview could cause confusion and lead to errors. Selecting the entire group at the sampled time and place avoids these problems. Although such a complex sampling scheme helps avoid biased samples, the design must be kept in mind when the data are analyzed to avoid misleading results.

Analysis and Related Issues

Once the data are collected, the final stages of a survey include analysis and interpretation of the data, and identifying action items (if necessary) related to study findings. A fundamental analysis issue associated with this survey is that of *duplication*. For example, a participant who frequented five venues in the frame was more likely to be selected for the sample than one who attended a single listed venue. Furthermore, individuals have different attendance records, so someone who goes to a café once a month is less likely to be chosen for the

Race/ethnicity	Number	HIV prevalence	Odds ratio (95%CI)
White	1246	3.3%	1.0
Black	587	14.1%	4.8 (3.3, 7.1)
Asian/Pacific Islander	203	3.0%	0.9 (0.4, 2.1)
Hispanic	1027	6.9%	2.2 (1.5, 3.2)
Mixed or other	380	12.6%	4.2 (2.8, 6.6)

TABLE 1 HIV prevalence estimates from the young men's survey by race/ethnicity, with unadjusted odds ratios

Source: Valleroy et al., 2000.

sample than another who visits the café every week. Ignoring such differences could introduce bias in the survey findings because not everyone has the same chance of being chosen for the survey, but the chance mechanism in the YMS design allowed analysts to account for these differences: the data are weighted by the inverse of their selection probabilities. In the hypothetical café example, the person who visits the café weekly is four times as likely to be sampled than the monthly visitor, and as a result, in the analysis, the weekly visitor's response counts only one-fourth that of the monthly visitor.

Between 1994 and 1998, the YMS enrolled 3492 eligible men from different metropolitan areas. These men were a subset of the 38,622 young men enumerated in the sampled areas during 1592 sampling events at 194 venues. Among those entering the sampled areas, 6866 were eligible; 4272 enrolled; and 3492 were unique (nonduplicate) participants who reported having ever had sex with a man.

In addition to good estimates for HIV prevalence and risks in the population, researchers also need to understand the variability in these estimators in order to direct their decision making. To illustrate, consider the data in Table 1 from the survey. These findings show that the prevalence of HIV is about 14.1% among young black men, while the estimate for young white men is 3.3%. The odds ratio comparing HIV prevalence in young black and white men is 4.8. If there were no difference in prevalence, the odds ratio would be 1.0. Without an estimate of sampling variation, it is difficult to assess whether there is a true difference between population subgroups. If there is truly a greater prevalence in one group compared with another, it would make sense for public health officials to target behavioral interventions toward the group with a higher prevalence. Thus, we need a measure of the uncertainty associated with these estimates to determine what action to take. With the additional information that the standard error of the logarithm of the odds ratio is about 0.19, we get a 95% confidence interval for the population odds ratio of (3.3, 7.1). We conclude that prevalence of HIV among young black men in the population is statistically significantly different from that among young white men in the United States.

Specialized software, such as SUDAAN (RTI International, http://www.rti.org), provides reliable estimates of variability in survey samples because it accounts for the sampling design. For example, in the Young Men's Survey, the individuals are sampled in clusters because all of those present at a venue in a particular day/time period are eligible for the study. Generally speaking, subjects sampled within the same venue and time period are likely to exhibit a greater degree of similarity than subjects sampled from different venues and time periods. If the data are analyzed ignoring the potential for homogeneity among subjects within primary sampling units, researchers risk underestimating variance, which can lead to drawing false conclusions about the study population.

THE HIVNET CLINICAL TRIAL IN UGANDA

In the early 1990s, a study in the United States and France found that the transmission of HIV from a mother to her baby was dramatically reduced when drugs were given to these women during pregnancy and delivery and to their newborns. The cost of the drug regimen was about $800. It was started early in the pregnancy and required a serious commitment on the part of the mother—she had to take pills five times a day for five months of her pregnancy and give her baby medicine four times a day for six weeks after birth. She also had to bottle-feed her baby to avoid transmission through breast milk. Most industrialized countries have adopted an approach like this to controlling mother-to-child transmission of HIV, and it has paid off. Prior to the advent of effective prevention measures, 15% to 40% of U.S. mothers infected with HIV transmitted the virus to their newborns. Now, the transmission rate for women who follow this regimen is less than

5%. Unfortunately, there are many reasons why this drug regimen will not work for mothers in countries such as Uganda.

To be considered effective, a drug regimen must be able to prevent or reduce the likelihood of the transmission of the virus from mother to infant. It must also meet two other challenges: the regimen must be affordable and it must be relatively easy to take correctly. If the drug is too expensive in the country where it will be used, it cannot be effective there. Moreover, it won't be effective if patient *compliance* is low. In Uganda, for example, pregnant women often do not seek medical care until the beginning of labor, if at all. UNAIDS estimates that only about 40% of births in Uganda are attended by skilled health personnel. Further, many mothers have no choice but to breast-feed their children. The cost of infant formula may be prohibitive, the lack of clean water may lead to fatal infectious diseases, and even if safely used, bottle-feeding may lead to stigma and rejection by the community.

Controlling the spread of HIV from mother to child in Uganda needs a different treatment strategy. Alternative regimens had been proposed that involve short courses of treatment during labor and immediately following delivery. Before these new drug regimens could be recommended for widespread adoption, however, they needed to be tested on smaller groups of people to determine their clinical effectiveness. The HIV Network for Prevention Trials (HIVNET, now HPTN, http://www.hptn.org) at the National Institutes of Health sponsored a clinical trial in Kampala, Uganda, during 1997–99. This trial compared short regimens of zidovudine (ZDV), the first anti-HIV drug approved by the U.S. Food and Drug Administration, and a newer drug called nevirapine (NVP).

Researchers design *clinical trials* for testing new treatments. A clinical trial is a controlled intervention where volunteers are screened, enrolled in the program if eligible, and given the drug regimen. The subject's response to the treatment is measured, and based on these data, either the new treatment is recommended for adoption or not. (In this example, the response is whether or not the infant contracted HIV.) Good designs are required in order to give good data. From good data we can draw valid inferences about the effectiveness of a treatment and generalize the findings to the population. Two basic questions arise: How can we ensure that the study results are generalizable? How can we tell if the treatment works?

Controlled Intervention

A study's *generalizability* is ensured in part through selecting appropriate subjects because the scientific findings may only directly apply to people who resemble those in the study. In the Uganda study, potential subjects were required to have documented HIV infection, to be at a late stage of pregnancy (at least 32 weeks), and to satisfy basic healthfulness criteria across a range of laboratory tests. Eligibility is also linked to ethical considerations. Women were excluded from the study for uncontrolled hypertension and chronic alcohol use, because these conditions could have compromised the safety of the treatment and put the women and their infants at greater risk. Furthermore, for their safety, women could be taken out of the study and given appropriate medical treatment if they showed certain life-threatening conditions during labor, whether or not the conditions could be directly attributed to the study.

Clinical trial methods resemble methods for designed experiments in other fields such as agriculture and manufacturing, with the key difference that the units of observation are human beings rather than crops or widgets. This affects both the logistics and the ethics of research. For this reason, only women who were of the age of legal consent and who gave written informed consent participated in the trial. The women also received counseling about their infection status. Further, eligible women had the freedom to participate in the study or not, to adhere to the rules of the study or not, and to show up for evaluations or to drop out before the end of a study. To ensure high compliance, clinical trials must be designed with the risks and potential benefits for the participants carefully weighed and controlled to the extent possible. For these and other reasons, clinical trial methods stand apart from other experimental design methods.

To be considered effective, the treatment must be shown to reduce HIV infection. But to determine whether or not the transmission of HIV is less likely, we need a benchmark for comparison; that is, we need a comparison group, called a *control group,* of HIV-infected pregnant women who would not receive the new treatment (NVP). The women in the control group in the Uganda study received a short regimen of the older drug, ZDV. Several trials for preventing mother-to-child HIV transmission used control groups in different ways.

Date	Location	Drug	Control	Blind	Before delivery	During labor	Infant dose	Estimated cost (USD)
1991–93	U.S. and France	ZDV	Placebo	Yes	Oral 4 months	IV	Oral 6 weeks	$800
1996–97	Bangkok, Thailand	ZDV	Placebo	Yes	Oral 4 weeks	Oral	None	$50
1997–99	Kampala, Uganda	NVP	ZDV	Open label	None	Oral 1 dose NVP/ every 3 hr. ZDV	Oral 1 dose NVP/ 7 days ZDV	$4

TABLE 2 Three trials to reduce mother-to-child HIV transmission

Note: In the Uganda study, the ZDV regimen consisted of several doses given orally to the mother from onset of labor through delivery, followed by syrup administered orally to infants twice daily for the first seven days. The NVP regimen consisted of a single dose given orally to the mother at the onset of labor, followed by a single dose given orally to the infant within three days after birth.

The Uganda trial was originally intended to have a third group, one that did not receive any drug, but that portion of the design was dropped after the results of a clinical trial in Thailand became known. The Thailand trial took place in Bangkok during 1996–97. It compared a short ZDV regimen that cost $50 to no treatment. The comparison in this study was particularly controversial, given that a longer ZDV regimen had been proven safe and effective. In a resource-limited setting, however, it was thought that no treatment was the more feasible alternative to the short, inexpensive ZDV treatment. Table 2 presents the details of the regimen for the Uganda and Thailand trials as well as for the U.S./France trials that established the effectiveness of the more expensive, longer ZDV treatment. The U.S./France trial had a control group that received no treatment, because at the time there was no existing treatment for mother-to-child transmission of AIDS.

When we have a control group, we are bound to observe some difference between the responses of the mothers in the two groups. We also need to figure out whether or not the observed difference between the two groups can be attributed to the treatment rather than some other factor. For example, suppose the women in the control group were sicker than those receiving the treatment. That would make for an unfair comparison. Many unforeseen differences between these two groups could bias the results and yield misleading conclusions. Several precautionary measures have been put in place in clinical trials to help avoid these potential pitfalls. One important precautionary measure involves *randomization;* that is, each mother enrolled in the study is assigned to the treatment or the control group according to a chance mechanism, as determined by something like the flip of a coin. The random assignment of participants to the groups means the two groups are highly likely to be very similar to each other in many ways, especially in unforeseen ways that may affect the transmission of the virus. This randomization allows the researcher to analyze the variability of results in a meaningful way, and to attribute large differences between the study groups (larger than might be expected by chance alone) to the difference in treatments rather than to some other factor.

An additional precaution, the use of a *placebo,* ensures the comparison groups are as similar as possible. When the alternative to the new treatment is no treatment, as was the case in the U.S./France and Thailand trials, a placebo is often given to the control group; that is, both groups received a drug, but the control group's drug was fake. In the U.S./France study, the treatment drug was ZDV, and the placebo was an inactive compound made to resemble ZDV and given on the same dosing schedule. In this way, the patient did not know whether she was receiving the active or inactive treatment. When the treating clinician also does not know whether the patient is receiving the active or inactive treatment, the study is called *double-blind.*

Unlike the U.S./France and Thailand trials, the Uganda study was an *open-label* randomized trial, meaning that it was not blinded to either the subject or to her clinicians. The two-dose NVP regimen in the Uganda study costs about $4, lower still than the short ZDV regimens in this study and the Thailand trial. Although infants were examined up to eight times by 18 months of age, the study end point focused on the chance of infection by six months of age, because 98% of the babies in the study were breast-fed.

Analysis and Related Issues

The target size of each study was set in order to achieve specified error rates; that is, based on assumptions about the expected reductions in HIV between groups, the sample size would guarantee a limit on the probability of an incorrect conclusion about the effectiveness of the new treatment. This is a minimum requirement for carrying out a credible, ethical efficacy trial.

Many clinical trials monitor accumulating data for early, persuasive evidence of harm or efficacy to determine whether the study should end before its planned completion. Such a finding would permit the ethical choice to avoid harming further subjects by giving them an inferior treatment. To this end, the Uganda study planned for an *interim analysis* by a *data monitoring committee*. This interim analysis set stricter limits on the probability of an incorrect conclusion, because repeated analysis of the interim data inflates the chance of observing persuasive evidence when in fact the treatment is not effective. In other words, it is more likely that one of many analyses, in comparison to one analysis alone, will incorrectly provide convincing evidence of efficacy, and the error for each intermediate analysis must be reduced to preserve the overall planned error limits. In addition, earlier analyses use relatively less data (because they have yet to be gathered), and we must compensate in the interim analyses for the additional variability in the results due to fewer data. The U.S./France trial showed signs of effectiveness at the first analysis, contributing to a decision to halt the study and conclude that ZDV was superior to the placebo. The other two studies carried through to their planned completions.

The efficacy results in Table 3 are expressed in terms of the estimated probability of HIV infection among infants in each study group by a specified age. These analyses needed to account for the possibility of incomplete study follow-up while avoiding bias or unnecessarily discarding information. Generally speaking, early withdrawals of HIV-uninfected infants were analyzed as *censored* observations because the time of HIV infection in those infants was not observed. Censoring, a concept from *survival analysis*, represents a special kind of missing data for which specific analytic techniques have been developed. We use the Kaplan-Meier estimator for time to HIV infection,

	U.S./France	Thailand	Uganda
Planned total sample	636	392	556
Number screened	Not stated	1140	2125
Number enrolled	477	423	626
Number analyzed	363	392	616
HIV prevalence			
New treatment	8.3 (3.9, 12.8)	9.4 (5.2, 13.5)	13.1 (9.1, 17.1)
Control group	25.5 (18.4, 32.5)	18.9 (13.2, 24.2)	25.1 (19.5, 30.8)
Relative difference	67.5 (40.7, 82.1)	50.3 (15.4, 70.6)	47.8 (20.0, 64.0)

TABLE 3 Data from the U.S./France, Bangkok, and Uganda studies of HIV transmission from mother to child

Note: The U.S./France results were reported at 18 months of age, Thailand at two months, and Uganda at 14 to 16 weeks. Ranges in parentheses represent 95% confidence intervals.

a standard survival method, to retain some information even from missing observations.

The Uganda study demonstrated a 48% reduction by 14 to 16 weeks of a two-dose NVP regimen compared with a simple ZDV regimen in a breast-feeding population (Table 3). That is, 13.1% of the infants receiving NVP and 25.1% of the ZDV-treated infants contracted HIV, the relative difference being $(25.1 - 13.1)/25.1 \times 100\% = 47.8\%$.

A 95% confidence interval for this relative difference is (20%, 64%). (In comparison, the U.S./France study estimated a 67% relative reduction of an intensive ZDV regimen compared with placebo, and the Bangkok study showed a relative reduction of 50% for a simpler, less expensive ZDV regimen.) As a result of the Uganda trial, the NVP regimen was widely adopted in Uganda for HIV-infected pregnant women and their newborn children.

CONCLUSION

Since the first cases of AIDS were reported, statistical methods have proved invaluable in characterizing the epidemic and in measuring evidence about the ability of interventions to prevent or treat HIV and AIDS. The unique features of HIV have prompted statistical innovations, as seen in our first example about sampling hard-to-reach populations. The YMS found high HIV prevalence in young homo-sexual men, especially in young men of color, and that most of the infected men were unaware that they were infected. As a result, the CDC increased both the emphasis on young men of color in preven-tion programs and the funds available for those programs, and also initiated new studies to understand the disproportionately high HIV infection rates in these populations.

Other statistical methods in HIV epidemiology have often reflected the applications of statistics in broader public health research. Our second example concerns three international clinical trials of drugs intended to reduce the risk of mothers transmitting HIV to their infants during labor and delivery. Each of these studies has proven influential in its own time. The first trial prompted new treatment guidelines in the United States, contributing in turn to a dramatic decline in mother-to-child transmission of HIV. As a result of the Bangkok trial, the Thai government adopted a national policy implementing the short-course ZDV regimen, also leading to a sig-nificant decline in transmission. At the time this essay was drafted, the NVP regimen was beginning to become more widely available in several resource-limited countries, including Uganda.

REFERENCES

Connor, E. M., et al. (1994). "Reduction of Maternal-Infant Transmission of Human Immunodeficiency Virus Type 1 with Zidovudine Treatment." *New England Journal of Medicine* 331, no. 18:1173–80.

Gehan, E. A., and N. Lemak. (1994). *Statistics in Medical Research.* New York: Plenum.

Guay, L. A., et al. (1999). "Intrapartum and Neonatal Single-Dose Nevirapine Compared with Zidovudine for Prevention of Mother-to-Child Transmission of HIV-1 in Kampala, Uganda: HIVNET 012 Randomised Trial." *Lancet* 354:795–802.

MacKellar, D., et al. (1996). "The Young Men's Survey: Methods for Estimating HIV Seroprevalence and Risk Factors among Young Men Who Have Sex with Men." *Public Health Reports* 111 (Supplement 1): 138–44.

McFarland, W., et al. (2001). "HIV Incidence among Young Men Who
Have Sex with Men—Seven U.S. Cities, 1994–2000." *Morbidity and
Mortality Weekly Report* 50, no. 21:440–44.

Shaffer, N., et al. (1999). "Short-Course Zidovudine for Perinatal HIV-1
Transmission in Bangkok, Thailand: A Randomized Controlled Trial."
Lancet 353:773–80.

Valleroy, L. A., et al. (2000). "HIV Prevalence and Associated Risks in
Young Men Who Have Sex with Men." *Journal of the American
Medical Association* 284, no. 2:198–204.

QUESTIONS

1. What is the main problem in the design of the Young Men's
   Survey, and how is it solved?

2. Explain why "selected venues were arranged in order according
   to their number of available time blocks" in order that
   "scheduling conflicts were minimized"?

3. Why was it decided to use all eligible men at a venue (in a time
   block) rather than a random selection of eligible men?

4. Why is it necessary to weight the data of a subject who frequently
   uses a certain venue with a lower weight than a subject who rarely
   uses the venue (but is nevertheless caught by the survey)?

5. The odds ratio of blacks to whites among the population of
   young men who have sex with men was estimated to be 4.8.
   Why is it useful to know something about the size of error in
   this estimate?

6. Why is the North American solution to pregnancy transmittal of
   AIDS not appropriate for Uganda?

7. What is the "double-blind" strategy and why was it necessary in
   the Thailand study?

8. Were prospective subjects in the U.S./France study required to
   agree to the placebo treatment?

9. Why was NVP recommended for Uganda as a result of the
   Uganda study?

STATISTICAL GENETICS

Associating Genotypic Differences with Measurable Outcomes

RONGLING WU

University of Florida

GEORGE CASELLA

University of Florida

———— ❊ : : ❊ ————

During World War I, the French army dispatched ammunition-carrying Briard dogs to the front lines. The Red Cross turned to the shaggy-haired sheepdogs to haul first-aid supplies. And now Briards once again have been called to service in medical research. For the first time ever, animals that were born blind gained the ability to see after undergoing gene therapy, opening the door to the development of treatment for people with a rare, inherited eye disorder.[1]

Gene therapy is a treatment designed to correct defective genes responsible for disease or illness. There are several approaches currently under study, but the most common (and the one used to reverse blindness in the Briard dogs) involves introducing a "normal" gene to replace the abnormal gene that causes the disease. This is a

---

[1] Based on research from the University of Florida, Cornell University, and the University of Pennsylvania, published in the May 2001 issue (vol. 28, no. 1) of *Nature Genetics*.

complicated process, and its success depends on knowing which gene (or combination of genes) is responsible for the disease. Identifying the offending gene can be difficult—one statistician, Mark Yang of the University of Florida, described finding a particular gene as similar to finding a particular one-yard long segment along a 170-mile-long highway. Fortunately, with the aid of statistical analyses, it is now possible to attempt to locate genes (see the essay "How Many Genes? Mapping Mouse Traits" by Melanie Bahlo and Terry Speed, included in this book, for a description of this process).

The success of the gene therapy experiment involving the Briard dogs represented a huge advance for gene therapy research. Briards are susceptible to a genetic disease that causes blindness. Because researchers were able to identify the gene responsible for the disease (a defect in the RPE65 gene), they were able to devise a treatment and test it on Lancelot, a four-month-old Briard puppy who had been blind from birth. Normal RPE65 genes were injected into the sub-retinal areas of Lancelot's right eye. About three months later, vision was detected in his right eye, and after 10 months, Lancelot could see well enough with his good eye to negotiate a crowded room without bumping into anything. The success of this experimental therapy has medical researchers optimistic about the prospect of developing a treatment for a similar disorder, Leber's congenital anaurosis, which causes blindness in humans, and for which currently there is no cure.

Conditions or diseases that are a consequence of defects in a single gene, as is the case for blindness in Briard dogs, are the best candidates for gene therapy. Unfortunately, some common conditions, such as heart disease, arthritis, obesity, and infection rate, are complex genetic traits. Such traits are controlled by many genes, and each one may have only a minor effect on the trait. Moreover, these traits are often sensitive to both genetic and environmental variation, so it is difficult to isolate the genetic influence. Hence, there is a need for statistical analysis to help make valid conclusions from the data.

This success of gene therapy illustrates the type of applications we see today in the science of genetics. Like many other branches of the biological sciences, genetics has been energized by recent advances in *genomics* (the study of the molecular structure and function of genes). The results of genetic research have started to be felt in every aspect of our daily lives. For example, genetically modified crops provide us with higher yielding and more nutritional grains, and personalized drugs, designed specifically to correspond to a patient's genetic makeup, are more effective in curbing diseases. Genetics and genomics are even joining in the fight against bioterrorism.

To draw valid inference from genetic data, we strive to understand the genetic architecture of a complex trait, such as the number of genes, gene effects, and interactions between different genes or between genes and the environment. The *phenotypic variation* of any trait (the variation in the observed outcome, such as growth, amount of infection, etc.) can be partitioned into its genetic and environmental components through statistical analysis.

Today, we are most interested in finding the connections between the variability that we see in the phenotype (the observed outcome) and the individual genetic components, referred to as *quantitative trait loci* (QTL). (A *locus*, or point, leads us to a place on the *genome*, the string of genes, and identifies a particular gene that is influencing the trait.) This work has been greatly aided by two different developments. First, advancements in molecular technologies in the mid-1980s led to the generation of a virtually unlimited number of *markers* (loci on the genome that the scientist can identify exactly) that specify the genome structure and organization of any organism. Second, and almost simultaneously, improved statistical and computational techniques were developed, which gave us the computational power needed to tackle statistically complex genetic problems.

MENDELIAN GENETICS

Arbitrarily pick out two individuals from a population. You will find considerable differences between these two individuals in various traits such as body size and form, eye color, and behavior. Some of the things causing these differences are *genes* (the physical entities transmitted from parent to offspring that determine the inheritance of a trait). The set of genes present in an individual constitute its *genotype*, and the physical expression of the genotype is called the *phenotype*. Genes can exist in different forms or states, called *alleles*.

Genes are lined up in a *chromosome*, and the location of the gene along a chromosome is the *locus* of the gene. (You may recall the famous work of James Watson and Francis Crick, who showed that the

FIGURE 1 Possible offspring from heterozygous parents

chromosomes in an organism are arranged in a double helix.) If we have two alleles, symbolized by *A* and *a*, at a gene locus, then different combinations of these alleles in a *diploid* (pair) form three possible genotypes, including two homozygotes (*AA*, *aa*) and one heterozygote (*Aa*), each corresponding to a particular phenotypic value for a trait.

Gregor Mendel was the first man to postulate that genotypes could influence phenotypes. He observed an almost perfect 3:1 segregation ratio for all seven characteristics that he measured in hybrid plants derived from a series of crosses for the garden pea. For example, he found that 5474 seeds were round and 1850 seeds were wrinkled, and 5474/1850 = 2.96. These data allowed Mendel to hypothesize that an offspring obtains one allele randomly from each parent.

Mendel's hypothesis became *Mendel's first law*—a gene will segregate during the formation of the reproductive cells (*meiosis*), thus passing into different *gametes* (reproductive cells). Figure 1 illustrates Mendel's first law, where crossing two heterozygous parents, *Aa*, generates $\frac{1}{4}$ homozygote *AA*, $\frac{1}{2}$ heterozygote *Aa*, and $\frac{1}{4}$ homozygote *aa*. If allele *A* dominates allele *a*, we will have $\frac{3}{4}$ dominant phenotype (a mix of *AA* and *Aa*) and $\frac{1}{4}$ recessive phenotype, the 3:1 ratio (see Question 1).

Based on his pea hybrid study, Mendel developed a second law, which says that if there are two or more pairs of genes on different chromosomes, they segregate independently (combine independently in the offspring). However, in his study of fruit flies in the 1930s, Thomas Morgan found that genes that were close together on the same chromosomal region (two linked genes) will not segregate independently. Morgan's finding laid a solid foundation for modern linkage analysis and the construction of genetic maps from molecular markers.

LINKAGE ANALYSIS

The ultimate goal of complex trait dissection is to identify the actual genes involved in the trait and to understand the cellular roles and functions of these genes. As a first step, a linkage analysis seeks to associate marker genotypes with measurable traits.

Molecular markers are segments of the genome that exhibit heritable variation, that is, the alleles that can be determined. We try to detect the location of a QTL (recall that a QTL is a locus on the genome that identifies a particular gene that is influencing a trait) by finding markers that are close to the QTL and are associated with the trait of interest. To do this we measure the phenotypic trait of members of a randomly segregating population and also determine their molecular genotype. We then try to determine if an association exists between any of the markers and the quantitative trait. Markers that are close in location to the QLT should have traits that are associated with the trait being studied.

A common method of determining the association is by analyzing phenotypic and genotypic data with standard statistical methods such as a one-way analysis of variance or regression analysis. If the genotype class is found to be significant, then the molecular marker used to define the genotype class is considered to be associated with a QTL. This idea is later illustrated in Figure 2, which displays loci that are associated with differential growth in poplar trees.

As a simple example, suppose that we have data on the heights of 12 tomato plants (in cm)—79, 82, 85, 87, 100, 101, 102, 103, 124, 125, 126, 127—and we want to see if we can associate height with a particular gene. At each marker we can group the above data according to the genotype of the marker for each particular plant. A grouping at one marker might be giving some evidence that genotype *AA* is associated with increased height.

	GENOTYPE		
	aa	Aa	AA
Data	79, 82, 85, 87	100, 101, 102, 103	124, 125, 126, 127
Mean	83.25	101.5	125.5

On the other hand, at another marker the data might group as follows:

	GENOTYPE		
	bb	Bb	BB
Data	79, 87, 102, 125	82, 100, 103, 126	85, 101, 124, 127
Mean	98.25	102.75	109.25

And here we have less evidence that tomato plant height is associated with this gene.

Remember that the genotypes that we use are those of the marker, for we do not know the location of the QTL. The accuracy and precision of locating QTL depends, in part, on how close the markers are to one another, and of course it is better to have the markers close together. In the next section we will look more closely at how we can link the markers to the QTL.

QTL: KEYS TO AGRICULTURAL PRODUCTIVITY

Most of the traits important to agriculture are quantitatively inherited. In other words, they are shaped by a number of different genes that interact with each other and the environment. Yield, flavor, and stress adaptation, including drought and salinity tolerance, are examples of economically important traits that can be called quantitative.

In agriculture, a breeder will cross two parents and practice selection until advanced-generation lines with the best phenotype for the quantitative trait under selection are identified. This type of program, though, requires a large input of labor, land, and money. Therefore plant breeders are interested in identifying the most promising lines as early as possible in the selection process. Another way to state this point is that the breeder would like to identify as early as possible those lines that contain those QTL alleles that contribute to a high value of the trait under selection.

The molecular dissection of complex traits into individual QTLs needs two steps: (1) the construction of a genetic linkage map (a map that tells us the ordering of the genes) using *polymorphic markers* (which have more than one allele in a population), and (2) the identification of QTLs for the traits on the genetic map.

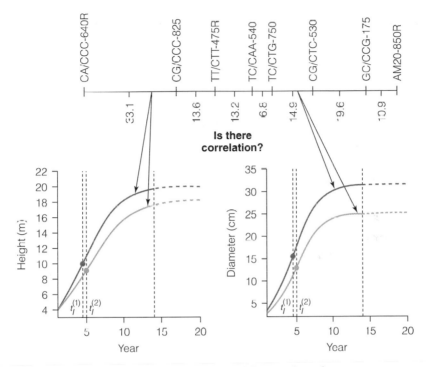

FIGURE 2 Possible locations of QTL for height and diameter in poplar trees

In the first step, all individuals in a controlled population are geno-
typed for different polymorphic markers and measured for a complex
trait of interest. The distances between each pair of the markers are
calculated, and a genetic linkage map composed of these markers is
constructed using statistical algorithms. Of course, the higher the density of the markers on the map, the more precise the location of the
potential QTL.

The top panel in Figure 2 depicts a linkage map. There we see the
markers arranged along a line (the linkage group). The distance on
the line between two markers is representative of the distance between the two markers on the chromosome.

In the second step, the markers are systematically associated with
the phenotypic values of the trait, the QTL underlying the trait are
identified on the map, and the number of the QTL and their genetic
effects are estimated.

The bottom panel of Figure 2 shows two quantitative traits
(growth trajectories for height and diameter) in poplar trees. The two

QTLs exert increased effects on stemwood growth when trees develop, but the QTL effects will be constant after age 10 years or so. The goal of the analysis is to find loci on the linkage map (top panel) for which different genotypes results in differential growth trajectories, therefore allowing the breeder to ultimately control the growth of the poplar tree by specifying the QTL genotype. (Ma, Casella, and Wu [2002] describe such an analysis in detail.)

GENETICS AND OBESITY

Obesity is a major health problem in the United States and most industrialized civilizations. It is estimated that 50% to 60% of Americans over age 30 are overweight and 25% to 30% are clinically obese. More worrisome is obesity in children, where rates in developed nations worldwide have risen two- to fourfold in the last 10 to 25 years. Obesity is a significant medical concern because numerous studies link obesity with increased risk of cardiovascular disease, metabolic disorders (such as type 2 diabetes mellitus and lipid abnormalities), and some forms of cancer. Obesity is thought to be a complex disease, influenced by numerous genes, along with other demographic and environmental factors like diet, age, physical activity, sex, and ethnicity.

It is now estimated that 40% to 70% of the variability in body weight is related to genetic factors, and obesity represents a classic example of the negative interaction between genetics and modern lifestyles. The ability to store energy when food was available represented a selective advantage for times when food was scarce. Only in recent times, when there is an overabundance of food and when many people are not engaged in physical labor, has obesity begun to rise and represent a significant health problem. It is also now clear that obesity is *polygenic* (influenced by multiple genes) in nature, and it is likely that populations in different geographic areas were under different environmental pressures, resulting in different sets of genes that contribute to fat/energy storing efficiency. Given the considerable rise in obesity rates worldwide, there is substantial interest in attempting to unravel its genetic basis.

Current approaches for genetic studies of obesity in humans broadly include (1) twin studies used to estimate the heritabilities of fat mass and other obesity-related traits, (2) animal models characterizing spontaneous single-gene mutations or candidate genes to understand obesity in humans, (3) gene association studies with obese cases and non-obese controls, and (4) a genomewide scan for obesity traits using linkage analysis, aimed at characterizing candidate QTL carrying human obesity genes.

Some polymorphic sites (QTL) predisposing to human obesity were identified using data from 643 women drawn from the Women's Ischemia Syndrome Evaluation (WISE) study (see Wu et al., forthcoming b). This study found that different QTLs are responsible for two different aspects of obesity—the amount and the distribution of body fat. It appears that the obesity QTLs are population dependent. For example, more QTLs for obesity were detected in the black population than white population, and there is no common QTL between these two populations. Such a population difference suggests using different strategies to prevent or reduce obesity in different populations.

GENETIC DIFFERENCES IN HIV DYNAMICS

The way in which the human immunodeficiency virus type 1 (HIV-1) particles that cause AIDS change over time varies considerably from patient to patient, and it is possible that such variation in HIV-1 behavior is related to an individual's genetic makeup. Traditional genetic analysis of HIV-1 infection has been based on various biochemical approaches, but it has not yet had great success. This is because HIV-1 dynamics, as a complex trait, is under polygenic control (influenced by more than one gene) and, moreover, seems sensitive to environmental changes. Moreover, at this time, there is a lack of data measuring both the amount of virus, and the genetic markers, on a population large enough to detect any such correlation.

However, there are statistical *models* that can help us. Using such a model, we have found statistical evidence for the possible existence of a host major gene responsible for HIV dynamics. The three genotypes at the detected major gene displayed marked differentiation in their viral load trajectories (Figure 3). The heterozygote (*Aa*) and one

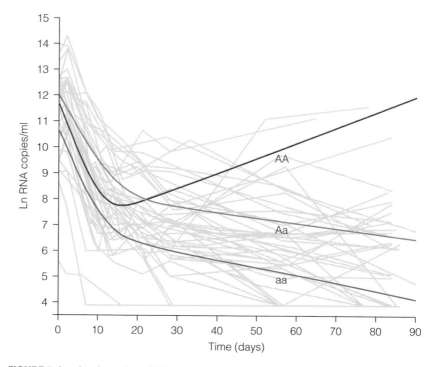

FIGURE 3 Load trajectories of HIV-1 virions (measured in viral RNA copies) for 53 patients (with shadow curves). The x-axis is time (in days) and the y-axis is a measure of the viral load of the patient. A high viral load implies that the drug therapy is not effective. The three thick curves each represent a different genotype (denoted by *AA, Aa,* or *aa*) at a possible major gene detected by the statistical model.

homozygote (*aa*) that together account for an overwhelming majority of patients (94%) were found to decline consistently with time in viral load after initiation of antiviral drugs. For the second homozygote (*AA*) in the frequency of 6%, viral load turns out to increase from day 20 following a short period of decline after the treatment of antiviral drugs.

For the staggering number of people infected with HIV (40 million), it is imperative to develop the most potent drugs to quickly eliminate all HIV from the blood and from the body. The possibility of a major gene for HIV dynamics suggests that HIV infection may be more efficiently prevented and curbed using personalized strategies of gene therapy targeted for the gene a particular patient carries.

CONCLUSIONS

The past 20 years have witnessed great progress and promise in genetics and genomics, which shapes our daily lives. The ability to gather data at the genome level, combined with statistical algorithms and computing power, has opened new doors in the analysis of biological systems. The use of this information, both as a diagnostic tool and a prescriptive tool, has the promise of greatly increasing the power of modern medicine. It can lead us to not only more accurate diagnoses of disease but also to the use of personalized drug therapies that will provide ever more effective treatments, helping us to cure diseases that, as of now, remain without cure.

REFERENCES

Fisher, R. A. (1918). "The Correlation between Relatives on the Supposition of Mendelian Inheritance." *Transaction of the Royal Society of Edinburgh* 52:399–433.

Ma, C. X., G. Casella, and R. L. Wu. (2002). "Functional Mapping of Quantitative Trait Loci Underlying the Character Process: A Theoretical Framework." *Genetics* 161, no. 4:1751–62.

Wu, R. L., et al. (Forthcoming a). "A Major Gene Detected for HIV-1 Dynamics." *Journal of Theoretical Biology*.

———. (Forthcoming b). "Genetic Variants for Human Obesity as Revealed by Linkage Disequilibrium Mapping." *Pharmacogenetics*.

ADDITIONAL READING

The Story of DNA and the Double Helix

Watson, J. D., and L. Bragg. (1991). *The Double Helix: A Personal Account of the Discovery of the Structure of DNA,* reissue ed. New York: New American Library.

Statistical Genetics

Doerge, R. W., Z. B. Zeng, and B. S. Weir. (1997). "Statistical Issues in the Search for Genes Affecting Quantitative Traits in Experimental Populations." *Statistical Science* 12, no. 3:195–219.

Lynch, M., and B. Walsh. (1997). *Genetics and Analysis of Quantitative Traits*. Sunderland, Mass.: Sinauer Associates.

Mendelian Genetics

Weir, B. S. (1996). *Genetic Data Analysis*. Sunderland, Mass.: Sinauer Associates.

Gregor Mendel

Henig, R. M. (2001). *The Monk in the Garden: The Lost and Found Genius of Gregor Mendel, the Father of Genetics*. New York: Mariner Books.

The educational website http://www.mendelweb.org covers the material in Mendel's 1865 paper "Experiments in Plant Hybridization."

QUESTIONS

1. Explain why the reproduction scheme portrayed in Figure 1 leads to a 3:1 ratio of dominant phenotype to recessive phenotype. Hint: To see all possible combinations of offspring, fill in the table.

PATERNAL GENOTYPE	MATERNAL GENOTYPE	
	A	a
A		
a		

2. How does the association of a marker trait with the trait of interest help to detect the location of the gene that controls the trait of interest?
3. What are the implications of Figure 3 for the effectiveness of the drug? That is, if a patient's genotype is known, how does Figure 3 advise the doctor?

DNA FINGERPRINTING

BRUCE S. WEIR

North Carolina State University

———— ❧ :: ❧ ————

In 1892 Francis Galton wrote a book intended to provide a sound statistical basis for the growing use of fingerprints in forensic science. He wrote, "They have the unique merit of retaining all their peculiarities unchanged throughout life, and afford in consequence an incomparably surer criterion of identity than any other bodily feature." But he could not have known that what he was saying about fingerprints would be relevant, in 1995, to the famous trial of O. J. Simpson, where "a forensic scientist testified that DNA tests showed a genetic match between Simpson's blood and stains found near the bodies of his former wife and her friend" (http://www.courttv.com).

Galton, a scientist, statistician, and cousin of Charles Darwin, went on to write, "They may be made to throw welcome light on some of the most interesting biological questions of the day, such as heredity, symmetry, correlation, and the nature of genera and species." Galton was writing about fingerprints at a time when they just beginning to be used for individual identification, but the same words were relevant a hundred years later when DNA profiles were just beginning to be used. In striking anticipation of the debates of the mid-1990s, Galton went on to describe "an attempt to appraise the evidential value of finger prints by the common laws of probability, paying great heed not to treat variations that are really correlated, as if they were independent."

In these few words, Galton captured the essence of the subsequent issues surrounding the evidential value of DNA profiles: they are heritable, they do not change over a person's lifetime, they are subject to detection uncertainty, and they serve to distinguish one person from another. There is now as much acceptance of DNA "fingerprints" as there is of dermal fingerprints, and this acceptance owes something to statistical calculations about match probabilities. Moreover, there are situations, such as the current efforts to identify remains from the World Trade Center disaster of September 11, 2001, where statistical calculations remain an integral part of the procedure.

This essay describes some of the statistical issues associated with the forensic use of DNA profiles by referring to the most hyped trial in recent years: *People v. Simpson.*

THE SIMPSON MATTER

O. J. Simpson was a legendary college football player, winning the Heismann trophy at USC in 1968, and then going on to a successful professional football career followed by an acting career. He was both rich and famous.

In June 1994, the bodies of Nicole Brown Simpson and Ronald Goldman were found outside Simpson's apartment in Los Angeles. Subsequently, Nicole Simpson's estranged husband, O. J. Simpson, was charged with the murders and brought to trial, a trial that was televised on Court TV and captivated the United States. On October 3, 1995, an estimated 142 million people worldwide either watched on TV or heard on the radio as he was acquitted of these charges. A year later he was found liable for wrongful death in a civil suit brought by the families of the deceased.

The cases against Simpson were circumstantial and relied heavily on the DNA profiles determined for bloodstains found at the scene of the crime, in Simpson's automobile, and at Simpson's house. Some of these profiles can be used to illustrate the statistical issues surrounding the interpretation of DNA profiles, although a full analysis of all the profiles will not be given.

DNA PROFILES

The genetic instructions needed to make us human are encoded in our chromosomes, and we receive one set of chromosomes, or genome, from each of our parents. The chromosomes consist of long molecules of DNA that would measure about two meters if they were straightened out and which can be represented as sequences of letters (bases). Although there are only four base types (denoted by A, C, G, and T), the total human genome sequence has more than 3 billion bases. Any two of us are the same at 99.9% of the positions in the sequence, but the remaining 0.1% still has more than 3 million bases—more than enough to distinguish between any two people except identical twins. The sequence positions where people in a population do not all have the same base are said to be polymorphic and are often referred to as genetic markers. There is currently an international effort under way to find the base types at 600,000 of these markers in samples of people from Africa, Asia, and Europe (see http://www.hapmap.org). Forensic scientists use only a small number of polymorphic markers, and the commercial genetic-typing kit known as Polymarker used only half a dozen.

Population geneticists refer to the type that a person has on a chromosome at a marker position as an allele, and it is conventional to give these alleles alphabetical labels A, B, and so on. As we each have two copies of a chromosome, one from each of our parents, we have two alleles for each marker, and the pair is referred to as a genotype. For a marker with alleles A, B there are three genotypes: AA, AB, BB. People with the same genotype at one marker are said to match.

The Polymarker system consists of five separate markers. When we look at many people we find there are two alleles in the population, A and B, for the markers called LDLR, GYPA, and D7S8, whereas the markers HBGG and Gc have three alleles: A, B, and C. The first three markers therefore have three genotypes and the other two have six genotypes: AA, AB, BB, CC, AC, and BC. Altogether there are $3 \times 3 \times 3 \times 6 \times 6 = 972$ different possible Polymarker profiles. Modern profiling systems use many more than five markers, and each has at least 10 alleles, so that the number of possible profiles is far greater than the number of people on earth. (The system used by the FBI allows at least 10^{24} possibilities; see Budowle and Moretti [1999].)

The Polymarker profiles for just two of the 45 bloodstains in the Simpson case will be considered: one ("Bundy") found on the ground next to the bodies, and one ("Bronco") found in Simpson's car. Along with these two evidentiary stains, profiles were determined for the two victims and the accused. All these profiles are

	BLOODSTAIN		INDIVIDUAL[a]		
Marker	Bundy	Bronco	NB	RG	OJ
LDLR	AB	AB	AB	AB	AB
GYPA	BB	AB	AB	AA	BB
D7S8	AB	AB	AB	BB	AB
HBGG	BC	ABC	AB	AA	BC
Gc	BC	ABC	AC	AA	BC

TABLE 1 Polymarker profiles of interest

[a]NB = Nicole Brown Simpson; RG = Ron Goldman; OJ = O. J. Simpson

shown as the columns in Table 1. The genotype for each marker has been written with the two alleles in alphabetical order. There is no implication that the first allele written for each marker came from the same parent; for the markers LDLR and D7S8, for example, we do not know if Simpson received the A alleles from one parent and the B alleles from the other, or if he inherited alleles A at LDLR and B at D7S8 from one parent and the remaining pair from his other parent.

All three people must have received the A form of the LDLR marker from one of their parents and the B form from the other. We cannot distinguish between the three profiles at this marker. For the GYPA marker, however, all three people have different genotypes, as is also the case for HBGG and Gc. The NB and OJ profiles are the same for marker D7S8. The Bundy profile has a C allele for the HBGG marker, which means that it cannot represent blood from either victim, as they did not have that allele. O. J. Simpson does carry the C allele, and so he cannot be excluded as a source of the Bundy profile on the basis of the information in Table 1. Can you see that markers GYPA, D7S8, and Gc also exclude Ronald Goldman from being a source of the Bundy stain, and that GYPA, HBGG, and Gc exclude Nicole Brown Simpson? O. J. Simpson has exactly the same genotype as the Bundy profile at all five markers.

The Bronco stain has three alleles at two of the markers, which means that it must contain DNA from at least two people. One person has only two alleles per marker. All three of the NB, RG, and OJ profiles are contained within the Bronco profile, meaning that the table contains no information to exclude any of those people. Note that the AB type for GYPA in the Bronco stain could arise by any combination of AA, AB, and BB genotypes. It is only the allelic types that are recorded in the table, not the amount of each type. If all three of the NB, RG, and OJ profiles are added together, then the Bronco profile is obtained, but so is it if just the RG and OJ profiles or just the NB and OJ profiles are added together. Can you see that the Bronco profile cannot be obtained by adding together the NB and RG profiles? This combination is ruled out by the HBGG and Gc markers.

Once statements about exclusion have been made, quantification of the strength of the evidence requires that alternative propositions be articulated and probabilities of the evidence calculated under each alternative.

HYPOTHESES

In the Simpson case, the prosecution believed that O. J. Simpson cut himself while committing the murders and then transferred both his and his victims' blood to his car. These beliefs can be translated into prosecution propositions H_p for each item of evidence.

Bundy H_p: OJ is the source of the stain.
Bronco H_p: OJ and one or both of RG and NB are the sources of the stain.

If the first of these prosecution propositions is correct, then the Bundy and OJ profiles should be the same. (They are.) If the second proposition is correct, then the alleles at each marker of the Bronco profile should include all those of the OJ profile and all those of at least one of the NB and RG profiles, but no other alleles. (They do.)

A typical defense approach is to deny that the defendant is a contributor to the evidentiary stain, and that approach will be taken here. Purely for the purposes of this discussion, the defense propositions H_d will be taken as follows:

Bundy H_d: OJ is not the source of the stain.
Bronco H_d: OJ is not a source of the stain.

In the Simpson case, moreover, the defense challenged the collection and handling of the evidence. From that view, not accepted by the prosecution, whether or not Simpson was the source of the stains was moot. It will be necessary to explore the consequences of the defense making statements about NB and RG for the Bronco stain.

QUANTIFYING EVIDENCE

The probabilities of evidence E given each of two competing hypotheses are compared by means of a likelihood ratio LR:

$$LR = \frac{P(\text{Evidence given } H_p \text{ is true})}{P(\text{Evidence given } H_d \text{ is true})}$$

A large ratio means that the evidence is much more likely if the prosecution is correct than if the defense is correct. Large ratios support the prosecution view of the case. How large it needs to be to lead to conviction is a matter for the jury members to decide. Ratios less than 1 support the defense.

For a single-contributor stain, such as the Bundy example, the evidence is the pair of profiles of the stain and the suspect. We will also assume that

1. The probability that the suspect has a particular DNA profile does not depend on the hypotheses H_p, H_d.
2. In the absence of any errors in the DNA typing, the stain and suspect profiles should have the same profile if H_p is true.

These assumptions lead to us being able to write the likelihood ratio as 1

$$LR = \frac{1}{P\left(\begin{array}{l}\text{Stain profile, given that } H_d \text{ is true and} \\ \text{the suspect profile has already been seen}\end{array}\right)}$$

where the denominator is the match probability or the probability that some unknown person has a DNA profile matching that of the suspect, given that the profile in question has already been seen. It is important to distinguish this from the profile probability, which is the probability that an unknown person has that particular profile without regard to the profile of anyone else.

Match probabilities allow for the effects of a relationship between the source of the crime stain and the defendant (under the defense hypothesis that they are not the same person), whether that relationship means membership of the same family or of the same population.

Evidently, the important question is not the profile probability (What is the probability that someone has this DNA profile?) but rather the match probability (Given that the defendant has this profile, what is the probability that someone else has it?). The numerical difference between these two probabilities can be enormous if "someone else" means the defendant's brother, and it can still be substantial if "someone else" means a person in the same population as the defendant.

DNA DATA

Match probabilities are estimated using data collected from convenience samples, such as blood banks or paternity testing agencies. More extensive data sets are being amassed as banks of profiles from convicted offenders, although that source is not generally used for probability estimation.

All the statistical calculations in this discussion will rest on FBI data (Budowle and Moretti, 1999). The FBI determined DNA profiles of about 200 people in each of the self-declared groups of African American, Caucasian, or (Southwest) Hispanic. The Polymarker genotype counts for these people are shown in Table 2, but it is allele counts or proportions that are needed in forensic calculations. There is a simple counting procedure that converts genotype counts to allele counts: for a marker with alleles A and B

No. A alleles = 2 × No. AA genotypes + No. AB genotypes

No. B alleles = 2 × No. BB genotypes + No. AB genotypes

and the Polymarker allele counts are shown in Table 3.

If, as is usually the case, the profile in question does not appear in a frequency database, the profile probability has to be estimated by means other than simply counting occurrences of the profile. Instead, we use allele frequencies and assume that all alleles in the

Marker	Genotype	SAMPLE[a]		
		AfAm	CA	SW
LDLR	AA	6	34	53
	AB	68	105	111
	BB	132	60	44
GYPA	AA	37	55	83
	AB	122	102	98
	BB	47	42	27
HBGG	AA	39	54	24
	AB	33	85	87
	BB	16	60	89
	AC	66	0	4
	BC	33	0	4
	CC	19	0	0
D7S8	AA	92	63	73
	AB	89	101	98
	BB	25	35	37
Gc	AA	1	19	9
	AB	28	24	20
	BB	117	4	15
	AC	6	60	48
	BC	47	29	62
	CC	7	63	54
Sample size		206	199	208

TABLE 2 FBI polymarker genotypic counts

[a]AfAm = African American; CA = Caucasian; SW = Southwest Hispanic

profile are independent. We then multiply together the corresponding allele proportions.

At markers for which the two profile alleles are different, an additional factor of two is needed. This is because it is not known which of the two alleles is maternal and which is paternal. Both combinations are assumed to be equally probable. Table 4 reflects this adjustment.

Marker	Allele	SAMPLE[a]		
		AfAm	CA	SW
LDLR	A	80	173	217
	B	332	225	199
GYPA	A	196	212	264
	B	216	186	152
HBGG	A	177	193	139
	B	98	205	269
	C	137	0	8
D7S8	A	273	227	244
	B	139	171	172
Gc	A	36	122	86
	B	309	61	112
	C	67	215	218
Sample size		412	398	416

TABLE 3 FBI polymarker allelic counts

[a]AfAm = African American; CA = Caucasian; SW = Southwest Hispanic

PROFILE	SAMPLE[a]		
	AtAm	CA	SW
NB	0.00348	0.00990	0.01370
RG	0.00001	0.00057	0.00016
OJ	0.00148	—	0.00023

TABLE 4 Estimated profile probabilities using FBI samples

[a]AfAm = African American; CA = Caucasian; SW = Southwest Hispanic

PROFILE PROBABILITIES

Profile probabilities are easy to calculate if the two alleles for each marker in the profile are assumed to be independent of each other and independent of all the other alleles in the profile. At a marker for which the profile has two copies of allele A, for example, the probability is the square of the allele proportion p_A under this independence assumption. If the profile is AB, the match probability is twice the product of allele frequencies, $2p_A p_B$. These calculations are then multiplied over markers. The probability of finding the OJ profile in the African American population is therefore estimated as twice the product of allele proportions for the A and B alleles at LDLR, multiplied by the square of the allele proportion for the B allele at GYPA, and so on:

$$P(OJ - AfAm) = \left(2 \times \frac{80}{412} \frac{332}{412}\right) \times \left(\frac{216}{412}\right)^2 \times \left(2 \times \frac{98}{412} \frac{137}{412}\right) \times \left(2 \times \frac{273}{412} \frac{139}{412}\right)$$
$$\times \left(2 \times \frac{309}{412} \frac{67}{412}\right)$$
$$= 0.00148$$

The NB, RG, and OJ profile probabilities under the assumption of complete allelic independence have the estimates shown in Table 4. Unless some adjustment is made for the missing HBGG-C allele in the Caucasian sample, the OJ profile probability cannot be estimated from that sample.

MATCH PROBABILITIES

Providing numerical estimates of match probabilities requires some statement of the relationship between the people named in the competing hypotheses H_p and H_d about the sources of the crime scene stains. For single-contributor stains, where H_p states that suspect is the source and H_d states that an unknown person is the source, either evolutionary or family relationships, or both, between the suspect and the unknown person need to be specified.

For an unrelated suspect and an unknown person, the match and profile probabilities are the same, because knowledge of the suspect's profile does not alter the probability of an unknown person having the profile. The likelihood ratios for the Bundy stain are therefore just the reciprocals of the OJ profile probability: 675 for the African American population. We can say that the Bundy profile is 675 times more likely to have the type seen in Table 1 if it came from O. J. Simpson than if it came from some member of the African American population unrelated to Simpson. We have to be careful to avoid the "prosecutor's fallacy" and not transpose this statement by saying that it is 675 times more likely that Simpson, rather than some unrelated African American, was the source of the Bundy profile. Unfortunately, media reports virtually always make this mistake. It is obvious that "the probability of an animal having four legs if it is a dog is 675 times greater than the probability of it having three legs if it is a dog" and "the probability of an animal being a dog if it has four legs is 675 times greater than the probability it is a dog if it has three legs" are different statements. It is unfortunate that newspaper reporters and editors do not see the difference between statements such as "the probability of the Bundy profile if it came from O. J. Simpson is 675 times greater than the probability of the profile if it came from someone else" and "the probability that the Bundy profile came from O. J. Simpson is 675 greater than the probability it came from someone else."

In the case where suspect and unknown person are related because of the past evolutionary history of their population, good approximations for the match probability (Balding and Nichols, 1994) have been expressed in terms of a parameter θ that can be thought of as the correlation of allele frequencies within a population. In human populations θ is of the order of 0.01, and forensic scientists often take a conservative value of 0.03. The match probability for profile AA is the quantity

$$[3\theta + (1 - \theta)p_A][2\theta + (1 - \theta)p_A] / [(1 + \theta)(1 + 2\theta)]$$

which is larger than p_A^2. For a profile with different alleles, AB, the match probability is

$$2[\theta + (1 - \theta)p_A][\theta + (1 - \theta)p_B] / [(1 + \theta)(1 + 2\theta)].$$

Using the African American allele frequencies and $\theta = 0.03$, the match probability for the O. J. profile is 0.00195. The likelihood ratio has been reduced to 512, thereby reducing the quantitative strength of the evidence against Simpson.

Results are also available for the case when the unknown person is thought to be in the same family as the suspect (Evett and Weir, 1998). In the case of father and son, for example, an AA profile has match probability p_A whereas an AB profile has match probability $(p_A + p_B)/2$. The match probability for the OJ profile for the case of the Bundy stain being from a child of O. J. Simpson, using African American allele frequencies, is 0.01706. In this case where family relatedness is considered, the likelihood ratio has been reduced to 56:

	SAMPLE[a]		
	AfAm	CA	SW
Match probability	0.00195	0.00031	0.00065
Confidence interval	(0.00124, 0.00281)	(0.00021, 0.00043)	(0.00042, 0.00100)

TABLE 5 Estimated match probabilities (with 95% bootstrap confidence intervals) for Bundy profile, using FBI samples and assuming $\theta = 0.03$

[a]AfAm = African American, CA = Caucasian, SW = Southwest Hispanic

a match is less unusual if the alternative source of a profile is related to the suspected source. (A further reduction follows if both evolutionary and family relatedness are considered.) This example should not be taken to suggest that O. J. Simpson's son was in any way connected with the crime. If a close relative is ever suggested in an actual situation, the best procedure is to obtain a DNA profile from that other person. The person will either be excluded or found to have the same profile as the accused, in which case the evidence points equally to accused and relative.

SAMPLING VARIATION

No matter how profile or match probabilities are calculated, the fact that they depend on sample allele frequencies means they are subject to sampling variation. There are many methods for quantifying this variation, resulting in confidence intervals for the match probabilities (or likelihood ratios, if preferred). Rather than delve into those details (see Weir, 2003), we just present the 95% bootstrap confidence intervals for the Bundy profile match probabilities in Table 5.[1] The upper and lower limits of these intervals differ from the estimate by about a factor of 1.5. For current DNA profiles, the factor is more likely to be 100 or more.

Another source of uncertainty in a match probability is the choice of population from which to sample. To which population does the unknown contributor in the defense proposition belong? Some indication

[1]The bootstrap is a computationally intensive numerical method that avoids distributional assumptions and that can be used to construct confidence intervals.

of the likely effects of population choice is given by presenting the results from several populations, as has been done in Table 5. It would be conservative to report the largest of the estimates.

ALLELIC INDEPENDENCE

The main statistical issue in calculating profile or match probabilities has been that of independence of alleles within or between markers. The Hardy-Weinberg law, published in 1908, showed that the genotypic probabilities at a marker were the products of allele probabilities under the idealized conditions of infinite populations mating completely at random and not under the influence of any disturbing forces such as mutation, natural selection, or migration. Clearly, human populations do not meet these idealized conditions, but just as clearly it has been difficult to detect departures from the Hardy-Weinberg relation in forensic databases. (For more on the best statistical testing strategy, see Weir [1996].) Appropriate tests were conducted for the Simpson case and failed to find substantial indications of allelic dependence within markers or between pairs of markers.

MIXED PROFILES

How should evidence such as the Bronco stain be quantified? It is first necessary to make the alternative propositions more explicit, and possibly consider many pairs of propositions. For illustration, consider the following pair:

H_p: OJ and RG are the contributors of the Bronco profile.
H_d: Two unknown people are the contributors of the Bronco profile.

The likelihood ratio still requires the probabilities of the evidence under each of these alternatives, and it is easiest to calculate these under the assumption of allelic independence (i.e., assuming no family or evolutionary relatedness). For the Bronco stain at marker LDLR, under H_p there are two profiles, OJ and RG, from known sources. The probability of the evidence is 1 under this hypothesis.

Under H_d, however, the two unknown people must have alleles A and B between them, and so they could be AA, BB or AA, AB or BB,

AB or AB, AB. The first three of these could also have been written in reverse order. The probability of the evidence under this hypothesis is therefore the sum of the probabilities of these four pairs of genotypes, and the four values need to be multiplied by 2, 4, 4, or 4 to account for the orders of the different genotypes and for the orders of alleles within heterozygous genotypes. The resulting likelihood ratio may not be much larger than 1, and it is certainly less than the values obtained for single-contributor stains. Mixed DNA profiles tend to have less evidential value than single-contributor profiles.

The defense may well choose to raise other propositions that could include one or more of the three principals and/or different numbers of unknown contributors. In the absence of other information about the numbers or identity of the contributors it would be conservative to report the smallest likelihood ratio found from different propositions. The prosecution may be required to perform calculations for any propositions offered by the defense.

DISCUSSION

DNA profiles are now widely used in human identification. They are used to associate suspects with crimes, children with their alleged fathers, and remains with the families of missing people after wars or mass disasters. When large numbers of genetic markers are used, DNA profiles can be highly discriminating—to the point that coincidental matches between even close relatives are unlikely. However, it can still be useful for courts of law to hear about the numerical strength of DNA evidence, and the statistician has a crucial role to play in determining numerical values. Attention must be made to the principles of forensic science and population genetics as well as to the usual statistical issues such as independence and sampling variation.

DNA evidence has now resulted in the conviction of many criminals, as well as the exoneration of many people who were falsely convicted but for whom DNA evidence was subsequently obtained and found to exclude them as the source of an evidentiary stain. The Simpson criminal jury was apparently convinced that the DNA evidence had been mishandled; however, the civil jury believed that the evidence pointed to Simpson's role in the murders.

REFERENCES

Balding, D. J., and R. A. Nichols. (1994). "DNA Match Probability Calculation: How to Allow for Population Stratification, Relatedness, Database Selection and Single Bands." *Forensic Science International* 64, no. 2–3:125–40.

Budowle, B., and T. R. Moretti. (1999). "Genotype Profiles for Six Population Groups at the 13 CODIS Short Tandem Repeat Core Loci and Other PCR-based Loci." *Forensic Science Communications.* http://www.fbi.gov/programs/hq/lab/fsc/backissu/july1999/budowle.htm.

Evett, I. W., and B. S. Weir. (1988). *Interpreting DNA Evidence.* Sunderland, Mass.: Sinauer.

Weir, B. S. (1996). *Genetic Data Analysis II.* Sunderland, Mass.: Sinauer.

———. (2003). "Forensics." In D. J. Balding, M. Bishop, and C. Cannings, *Handbook of Statistical Genetics,* 2nd ed., 830–52. New York: Wiley.

QUESTIONS

1. The introductory paragraphs of the essay discuss fingerprints and DNA. What is the connection between fingerprints and DNA, in the context of this essay?

2. Describe how you would systematically list all 972 Polymarker profiles in such a way that the list would be complete and without duplicates the first time you wrote it down?

3. Find a newspaper story that reports probabilities for DNA evidence in a trial. Does the story contain the prosecutor's fallacy?

4. Why is the probability that O. J. Simpson has the Bundy profile not the most useful probability for establishing his guilt or innocence?

5. Based on Table 4, the probability of a random African American having the OJ profile is 0.00148. But the OJ profile was found to match the Bundy profile. Why does this imply that the Bundy profile is much more likely to occur if it is O. J.'s than if it is some other African American's?

6. Does the conclusion of Question 5 show that the Bundy stain is much more likely to be O. J.'s than some other African American's?

HOW MANY GENES?

Mapping Mouse Traits

MELANIE BAHLO

Walter & Eliza Hall Institute of Medical Research, Australia

TERRY SPEED

University of California, Berkeley
Walter & Eliza Hall Institute of Medical Research, Australia

———— ⁑ : : ⁑ ————

In biomedical research, the laboratory mouse is an extremely important "model organism." That is, regarding mice as "little people" and studying them tells us a lot we can't learn any other way about the causes, nature, prevention, and treatment of human disease. Laboratory mice come from inbred strains, which means that they always breed true: the offspring of matings within the strain are always identical to their parents, apart from minor variations. There are many different inbred strains of laboratory mice, and between strains, there can be great differences in characteristics such as coat (hair) color, size, behavior, disease susceptibility, and in many other qualitative and quantitative traits. Such measurable characteristics are called phenotypes. Are these differences genetic or environmental, or a mix of both? These are important questions for biomedical research.

Many modern-day lab mouse strains descend from mice bred as pets by mouse fancier Abbie Lathrop early in the 20th century. This strongly suggests that the visible differences we see between strains are indeed genetic, and raises the question of identifying the genes responsible for such differences. Genes are specific regions along chromosomes, and for much of the 20th century, it was a real challenge to identify and locate genes causing phenotypic differences. This process is called gene mapping. With the genetic engineering revolution of the 1970s and 1980s, new genetic tools became available that greatly facilitated this task. Although there are visual aspects to genetics that involve light and electron microscopes and elaborate staining techniques, and many biochemical aspects (such as those associated with Pauling, Watson, Crick, and others), statistical notions remain essential to the task of identifying the DNA variation that causes phenotypic variation.

In this essay, we discuss a simple example associated with the coat color of mice. The second generation of planned matings between an inbred albino strain of mouse and an inbred black-coated strain found 29 mice with black coats, 29 albinos, and 72 agouti mice (a grizzled color). We see that of these 130 mice, roughly $\frac{1}{2}$ are agouti, $\frac{1}{4}$ black and $\frac{1}{4}$ albino. Data such as these led Gregor Mendel to discover the principle that a parent passes on one of its two genes to its offspring, where each gene is passed on to an offspring with probability $\frac{1}{2}$ independent of the other offspring. The fact that not exactly a quarter of the mice are albino reflects the variability in this chance process.

We will soon see that the search for the albino trait is more complicated. This example gives us a glimpse of how statistics is used in contemporary mouse genetics. With the data collected on these mice we illustrate the first step in most gene mapping exercises, which leads to an approximate location of the gene associated with the albino trait. The creation and widespread use of statistical methods to map genes causing phenotypic differences is one of the great modern success stories of statistics.

GENETIC BACKGROUND

Loci and Markers

Hereditary information is encoded in the long DNA molecules known as *chromosomes*. The full set of an organism's genes is called its *genome*. Each chromosome contains many genes, and each particular gene has a specific location on one of the chromosomes. The mouse genome consists of 19 distinct non-sex chromosomes (compared with 22 in humans) and two sex chromosomes, an *X chromosome* and a *Y chromosome*. Each mouse has two versions of each of these 19 chromosomes and two sex chromosomes; female mice have two X chromosomes, and male mice have an X and a Y chromosome.

The notion of a *gene* has evolved considerably since Mendel discovered the now-accepted principle that every individual inherits one copy of each gene from each parent. Genes can cause visible traits called *phenotypes* to manifest themselves. Variation in some visible traits is determined by variation at single genes (this was the case with all the traits Mendel studied), while variation in others is determined by two or more genes. The idea that genes are located at specific positions, called *loci* (singular *locus*), along the chromosomes became accepted early in the 20th century, and it was later accepted that different variants of genes, called *alleles,* correspond to variant forms of the *DNA sequence* at the gene's locus. It is always the case (on the non-sex chromosomes) that each individual has two alleles at a given locus, and they are usually designated by symbols such as a, b, etc., while genotypes are unordered pairs of alleles, which we will write in the form a/a, a/b to emphasize the different parental origins of the alleles. An individual is *homozygous* at a non-sex chromosome locus if the two alleles at that locus are the same, a/a, and *heterozygous* otherwise, a/b.

Inbred Strains and Their Crosses

The main players in our example will be two inbred strains of mice that are created by brother–sister mating over many generations. As a result they are, for all practical purposes, homozygous at every locus. If a male and a female from the same inbred strain mate, then their offspring will be genetically identical to them, apart from random mutations that will occur from time to time. One strain is a robust, black-coated mouse that is used worldwide as an experimental organism, not just for genetic purposes. By contrast, the second strain is a fairly delicate, diabetes-prone albino mouse, widely used for research on type 1 diabetes. For details on these and other strains, see http://www.informatics.jax.org.

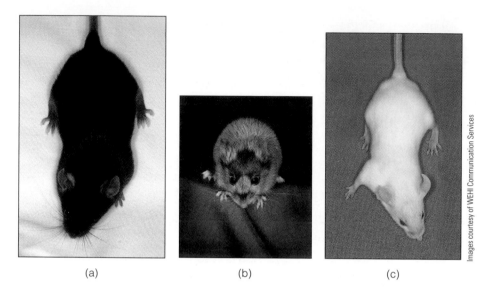

(a)   (b)   (c)

FIGURE 1 A black mouse (a), an agouti mouse (b), and an albino mouse (c)

The *coat color* of a normal (non-black, non-albino) mouse is called agouti, meaning that the fur has a grizzled color resulting from each hair having alternating dark and light bands. There is a locus called *agouti* on mouse chromosome 2, with a variety of alleles, including *dominant* agouti alleles and a *recessive* non-agouti allele (see http://www.mendelweb.org for details about dominance and recessivity). The mice in the black strain are homozygous non-agouti at the *agouti* locus. By contrast, the albino-strain mouse is homozygous for an agouti allele, and so should have an agouti-colored coat, but it is also homozygous for the recessive albino allele at the *albino* locus on chromosome 7, and so no coat color manifests itself. We will call the dominant allele at the albino locus the full color allele. See Figure 1 for photographs of the mouse strains.

A cross in the present context is a series of one or two planned matings, beginning with inbred strains. Here we discuss one type of cross: the intercross. We start with two different inbred parental strains and mate a male from one strain with a female from the other. The resulting offspring will be denoted by H for heterozygote. H has traditionally been called the first filial generation, while Mendel called them simply hybrids. Since the male and female parents are homozygous at each locus, the H offspring will be heterozygous at the loci where these two strains have different alleles. The offspring of all such matings will be genetically identical, with the exception of the sex chromosome. The *intercross* is the mating of an H male with an H female.

Data

The data we consider are for 132 female mice resulting from an intercross beginning with black and albino parental strains. That is, black strain mice and albino-strain mice are mated to produce H offspring, and then H offspring are mated. Coat color and genotypes at 156 other marker locations spanning all 19 autosomes were recorded for each of the 132 female mice resulting from this intercross. Only female mice were used in the actual study, as its main focus was on immunological phenotypes, and these are more variable for male mice because of their tendency to fight one another.

Mendel's First Law for One Locus

We have already mentioned Mendel's conclusion that offspring receive a maternal and a paternal allele at every locus. Mendel expressed this in his law of *segregation*, which says that every individual has two alleles at each locus, and passes on one to each of its offspring. More precisely, Mendel's first law is the following statistical statement: an individual passes on to its offspring either its paternally or its maternally inherited allele, each with probability $\frac{1}{2}$, with the inherited allele for different offspring being independent.

We can check the validity of this law with the data from our 132 intercross offspring. At any one of the 156 markers that have been typed on these mice, we expect to see evidence in the 132 offspring of independent, equally frequent segregation of the alleles in the heterozygous parents. Consider a particular marker on chromosome 7. If we denote the genotype homozygous for the albino-strain marker allele by A (representing a/a), and that homozygous for the black-strain allele by B (representing b/b), and the heterozygotes by H (representing a/b), then our intercross data includes 24, 29, and 67 offspring with genotypes A, B, and H, respectively, at this marker. These add to 120, indicating that there were 12 incomplete or missing genotypes at this locus. These counts are compatible with the

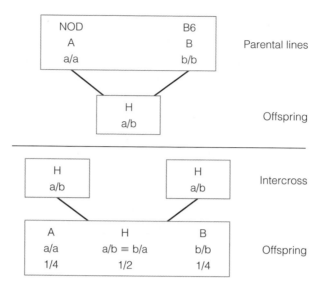

FIGURE 2 The intercross

Mendelian frequencies of $\frac{1}{4}$, $\frac{1}{4}$, and $\frac{1}{2}$ for genotypes A, B, and H (see Figure 2). According to Mendel's law for one locus, we would expect 120/4 = 30 offspring with genotype A, 30 with genotype B, and 60 with H. A χ^2 goodness-of-fit test, which compares the expected frequencies with those actually observed, confirms that the observed deviations from what we would expect are well within the bounds of what one might expect due to chance.

The alleles agouti and non-agouti at the *agouti* locus should also be segregating in a Mendelian fashion in this cross, as should the alleles albino and full color at the *albino* locus, but the situation is complicated by the fact that we cannot tell the status of the albino mice at the *agouti* locus. The 132 intercross offspring contained about 72 agouti, 29 black, and 29 albino. These figures are consistent with Mendelian segregation at the *albino* locus.

Mendel's Second Law for Two Loci: Recombination and Linkage

Mendel formulated a (second) law asserting the *independent* segregation of alleles at two or more loci. If alleles a_1 and b_1 are segregating at locus 1, and a_2 and b_2 are segregating at locus 2 in an individual,

Mendel proposed that not only should a_1 and b_1 segregate with equal probabilities of $\frac{1}{2}$, as we have just described, and similarly for a_2 and b_2, but also the combinations a_1b_1, a_1b_2, a_2b_1 and a_2b_2 should segregate jointly with equal probabilities $\frac{1}{4}$. This conclusion was entirely consistent with the pairs of loci Mendel discussed in his paper, and helps explain why the term *recombination* is used in this context. Our doubly heterozygous hybrid father has genotype a_1/b_1 at locus 1, and a_2/b_2 at locus 2. According to Mendel's second law, all four combinations, a_1a_2, a_1b_2, b_1a_2, and b_1b_2, are equally likely to be passed on by him to his offspring.

Mendel's first law of segregation has been found to be generally true at single locus, but his second law holds only for certain pairs of loci, and is not generally true. The reason goes back to the behavior of chromosomes during *meiosis*. Meiosis is a complex process, but there are only a few features of it we need to know. First, although a father (and mother) passes on to every one of his offspring a copy of each non-sex chromosome and a sex chromosome, these copies are only rarely (perhaps never) precise copies of one of those he received from his (her) parents. Each chromosome passed on by a father in a sperm cell to one of his offspring is a mosaic of his two parental chromosomes, the result of a random process that takes place during meiosis. This random process is repeated independently for each chromosome. The same is the case for mothers and egg cells, although there are important differences in detail that need not concern us here.

The mosaic chromosome 1 (for example) that a father passes on to his offspring results from a process whereby his two chromosome 1s are aligned alongside each other and replicated, so that there are now 4 versions of chromosome 1—two copies of his paternal chromosome 1 bundled together and two copies of his maternal chromosome 1 bundled together, as illustrated in Figure 3a. The two chromosomes in a bundle, which are just copies of one another, are called "sisters."

These chromosome pairs line up with each other gene by gene. The adjacent non-sister chromosomes twist around each other and then break and exchange segments (called chiasmata formation), resulting in new allele combinations. That is, rather than the alleles from each parent staying together, new combinations are formed, as shown in Figures 3b and 3c.

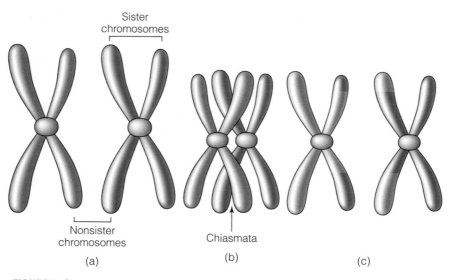

Sister chromosomes

Nonsister chromosomes

Chiasmata

(a)   (b)   (c)

FIGURE 3 Chromosomes crossing over during meiosis

At the end of this process, there are now four versions of chromosome 1—two that were inherited from a parent (called the parental) and two that are a combination of paternal and maternal alleles (called recombinations). These four versions of chromosome 1 now separate, and each is passed to a sperm cell. Thus, a sperm cell might contain a parental or a recombination version of chromosome 1. The positions of the break points are "random," and this entire process appears to take place independently for the different chromosomes.

With this background, we can now begin to understand the meaning of the term *linkage*. For two loci nearby on the same chromosome, there will be a tendency for the alleles from one parent to be transmitted together to a sperm cell. In fact, the closer the loci are to each other on the chromosome, the more likely it will be that these alleles are transmitted together. All of this is a clear violation of Mendel's second "law." On the other hand, when two loci are on different chromosomes, or far apart on the same, long chromosome, the joint probabilities of co-segregation of the four combinations of alleles turn out to be close to $\frac{1}{4}$ all around. When two loci are on different chromosomes the choice of transmitting one allele or another at one locus is independent of the choice of allele at the other locus because meiosis treats different chromosomes independently. When the loci are well separated on the same chromosome, a similar conclusion holds approximately, as multiple breakage and rejoining events effectively randomize the chromosomal segment between the loci. It is this relationship between the distance between two genes and the chance that they are transmitted together that allows us to use information about how frequently recombination occurs to estimate the distance between two genes. If recombination occurs as often as $\frac{1}{2}$ the time, it is an indication that the two genes under study are either located on different chromosomes or that they are very far apart on the same chromosome. If recombination rarely occurs, it is an indication that the two genes are close together on the same chromosome.

GENETIC MAPPING

Genetic mapping is the positioning of genes along chromosomes. Mapping can involve ordering loci in relation to one another and to landmarks (a marker whose location is known) along chromosomes, and the estimation of genetic (also called map) distances between loci. Although other mapping methods exist, we restrict ourselves here to mapping that is based on recombination, that is, to what is known as linkage mapping.

Why do we create genetic maps? The answer is that maps are a natural way of organizing information about chromosomes, much the same way that ordinary maps organize geographic information. And like ordinary maps, genetic maps have many uses. For example, the human genetic map has been helpful for genetic counseling, and, more recently, for assisting researchers in locating and cloning disease genes. And, if two loci are linked, knowing the genotype at a locus that is close to a locus of interest is sometimes as good as knowing the genotype at the locus itself.

Recombination Fractions and Genetic Distance

The recombination fraction between two loci is a simple measure of the extent to which they are linked. Let us go back to our doubly heterozygous father, who has genotype a_1/b_1 at locus 1 and a_2/b_2 at locus 2. If he obtained the pair a_1a_2 from one parent, and b_1b_2 from the other, we call these *parental* pairs, and the other two, a_1b_2 and

	LOCUS 2			
LOCUS 1	A	H	B	Total
A	26	10	0	36
H	10	46	9	65
B	0	5	23	28
Total	36	61	32	129

TABLE 1 Two-locus genotypic frequencies

b_1a_2, are called *recombinant* pairs. The recombination fraction, denoted by r, between locus 1 and locus 2 is simply the fraction of recombinant pairs passed on to offspring. For two loci that independently segregate, $r = \frac{1}{2}$. This would occur for loci on different chromosomes or for loci that are far apart on the same chromosome. The distance between loci on the same chromosome can be estimated by the corresponding recombination fraction.

Now let's use the intercross mouse data and consider two of our marker loci known to be relatively close on the same mouse chromosome; we denote them by 1 and 2. Table 1 gives the two-way table of their genotypes, where we use the A, B, and H symbols for the genotypes at both loci. Thus A means homozygous for the albino-strain allele, B homozygous for the black-strain allele, and H heterozygous for these two alleles. This is a widely used and convenient notation, and should cause no confusion as long as we can rely on context to tell us which marker is under discussion. In Table 1, the rows are labeled by locus 1 genotypes, and the columns by locus 2 genotypes, and so 26 of the 129 offspring in this intercross are homozygous for the albino-strain allele at both marker loci, et cetera.

Can we estimate the recombination fraction between these two marker loci? The answer here is certainly yes, but the task is not as simple as we might wish. In particular, it is not as simple as just counting recombinant pairs in the offspring. Why not? Because recombination may occur in the father or the mother or both, and all we see are the genotypes of the resulting offspring. In most cases, the parental origin of any possible recombinations can be identified, but not in every case. This is what makes the problem a little tricky.

Recall that this intercross resulted from crossing the progeny from the cross of two inbred strains. If we denote the albino-strain alleles

FEMALE	MALE			
	a_1a_2	a_1b_2	b_1a_2	b_1b_2
a_1a_2	$\dfrac{(1-r)^2}{4}$	$\dfrac{r(1-r)}{4}$	$\dfrac{r(1-r)}{4}$	$\dfrac{(1-r)^2}{4}$
a_1b_2	$\dfrac{r(1-r)}{4}$	$\dfrac{r^2}{4}$	$\dfrac{r^2}{4}$	$\dfrac{r(1-r)}{4}$
b_1a_2	$\dfrac{r(1-r)}{4}$	$\dfrac{r^2}{4}$	$\dfrac{r^2}{4}$	$\dfrac{r(1-r)}{4}$
b_1b_2	$\dfrac{(1-r)^2}{4}$	$\dfrac{r(1-r)}{4}$	$\dfrac{r(1-r)}{4}$	$\dfrac{(1-r)^2}{4}$

TABLE 2 Probabilities of parentally transmitted allele combinations

by a_1 and a_2 and the black-strain alleles by b_1 and b_2, then all the offspring of the first mating received pairs a_1a_2 on one chromosome and b_1b_2 on the other from their inbred parents. Denoting the unknown recombination fraction between the two loci by r, an H parent will produce the pair a_1a_2 with probability $\frac{(1-r)}{2}$, and similarly for b_1b_2, while the pairs a_1b_2 and b_1a_2 are transmitted with probability $\frac{r}{2}$. Where do the factors of $\frac{1}{2}$ come from? To see this, recall that each recombination event involves *two* non-sister chromosome pairs breaking and rejoining differently, but the sperm or egg receives just one of these, chosen at random. Thus if r is the probability of recombination, $\frac{r}{2}$ is the probability of an offspring receiving a *particular* one of the two recombinant chromosomes. Similarly we have $\frac{(1-r)}{2}$ for the parental chromosomes.

The transmission events in the different parental meioses are independent, and so we can easily build up Table 2 from the products of the $\frac{r}{2}$ and $\frac{(1-r)}{2}$ probabilities. For example, the father will pass on the recombinant pair a_1b_2 with probability $\frac{r}{2}$, and, independent of this, the mother will pass on the non-recombinant pair a_1a_2 with probability $\frac{(1-r)}{2}$, and so these two events will occur together with probability $\frac{r(1-r)}{4}$, which leads to the entry in row 2, column 1 of the table.

Terms in this table can now be summed to allow us to build up Table 3 of probabilities for the nine different two-locus genotypes. For example, we observe A($= a_1/a_1$) at locus 1 and H ($= a_2/b_2$) at locus 2 if and only if the transmitted male and female combinations are a_1a_2 and a_1b_2 or a_1b_2 and a_1a_2, leading to a combined probability of

LOCUS 2	LOCUS 1		
	A	H	B
A	$\dfrac{(1-r)^2}{4}$	$\dfrac{2r(1-r)}{4}$	$\dfrac{r^2}{4}$
H	$\dfrac{2r(1-r)}{4}$	$\dfrac{2[r^2+(1-r)^2]}{4}$	$\dfrac{2r(1-r)}{4}$
B	$\dfrac{r^2}{4}$	$\dfrac{2r(1-r)}{4}$	$\dfrac{(1-r)^2}{4}$

TABLE 3 Probabilities of two-locus genotypes

$\dfrac{2r(1-r)}{4}$. The remaining terms are derived in the same way, with the most complex case being the probability for the two-locus genotype HH, where four different terms need to be summed.

Notice that the probabilities in Table 3 are simple quadratics in r. For $r = \frac{1}{2}$, the probabilities along the rows would be $\frac{1}{16}, \frac{1}{8}, \frac{1}{16}, \frac{1}{8}, \frac{1}{4}$, $\frac{1}{8}, \frac{1}{16}, \frac{1}{8}$, and $\frac{1}{16}$. According to the multinomial probabilities, which is a generalization of the binomial probabilities, the chance of observing the counts in Table 1 when $r = \frac{1}{2}$ is

$$\left(\frac{1}{16}\right)^{26} \times \left(\frac{1}{8}\right)^{10} \times \left(\frac{1}{16}\right)^{0} \times \cdots \times \left(\frac{1}{16}\right)^{23}$$

(We ignore the factorials in the probability because they are the same for all values of r.) Each value of r yields its own multinomial probability,

$$\left(\frac{(1-r)^2}{4}\right)^{26} \times \left(\frac{2r(1-r)}{4}\right)^{10} \times \left(\frac{r^2}{4}\right)^{0} \times \cdots \times \left(\frac{(1-r)^2}{4}\right)^{23}$$

This expression is called the likelihood function, $L(r)$, and we are interested in the value of r for which the observed data is most likely to have occurred. To find this value, we maximize the likelihood function with respect to r.

Inferring Linkage and Mapping Markers

We now turn to the principal methodological problem: deciding when two loci are linked, and in that case, estimating the map distance between them.

Suppose that we have a two-way table of frequencies like Table 1, but that we don't know whether or not the loci are linked (i.e., whether or not the recombination fraction r between them is equal to $\frac{1}{2}$). A natural way to address this question is to carry out a formal test of the null hypothesis that $r = \frac{1}{2}$, against the alternative that $r < \frac{1}{2}$. And, given that we used likelihood methods to estimate r, it seems natural to carry out a likelihood ratio test to decide between these two hypotheses. This practice entered genetics in the 1950s and has been firmly entrenched there ever since. The test statistic almost always used in this context is denoted by LOD, and it is the base 10 logarithm of the ratio of the likelihood at the maximum likelihood estimator $\hat{r}$ of r to that at the null, $r = \frac{1}{2}$,

$$LOD = \log_{10}\left\{\frac{L(\hat{r})}{L(1/2)}\right\}$$

Notice that the LOD will be near 0 when $\hat{r}$ is near $\frac{1}{2}$, and it will be large when it is unlikely that $r = \frac{1}{2}$. Normal statistical practice would have us determine a cutoff for the test that would limit the chance of a type 1 error under the null hypothesis, that is, the chance of rejecting the hypothesis that $r = \frac{1}{2}$ when it is in fact correct. This practice is rarely adopted in genetics, where tradition there dictates the use of more stringent thresholds that take into account the multiple testing common with linkage mapping. In genetics, we rarely carry out a single test of the kind just described; rather, many similar tests are conducted with the same data, some of which are correlated, and so new methods for controlling the overall error rate have evolved.

For example, to create a genetic map of markers from a cross like ours, we would begin by computing all pairwise recombination fractions and the associated LOD scores. In our case, this would be $156 \times 155/2 = 12{,}090$ calculations. We would then set a LOD cutoff, for example, 3, and a recombination fraction cut-off, for example 0.3, and declare as linked any pair of markers whose recombination fraction was less than 0.3 and whose LOD exceeds 3, or is connected by a sequences of marker pairs each satisfying this threshold. We could then use these results to group the markers, putting markers that were thought to be linked into the same group. These groups are called *linkage groups.* Let's suppose that we did this, and obtained 20 linkage groups. (If we didn't get something like the expected number, here the number of non-sex chromosomes plus 1 for the X chromosome, we might try again with modified thresholds.) The next step would be to order the loci within each linkage group. It would take us too far afield to describe this in detail, but we do need to say at this point that multilocus mapping, which means considering more than two markers at once, is essential to getting the best ordering from the data. Our 156 markers are spread over 20 chromosomes, and so there is on average eight per chromosome. Ordering eight markers might involve doing $8! = 40,320$ eight-locus likelihood calculations, and choosing the order with the largest likelihood. There are less computationally intense methods that are needed with more than 20 markers.

The mapped locations of the 156 markers in the mouse data are shown in Figure 4.

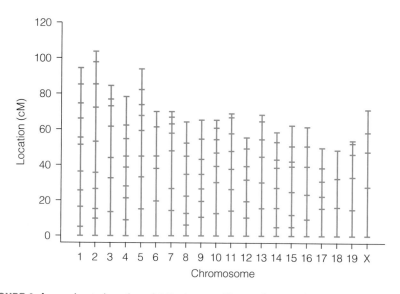

FIGURE 4 Approximate location of 156 microsatellite markers on the 19 autosomes and the X chromosome. The unit of map distance shown on the y-axis is a centiMorgan, cM, this being one one-hundredth of a Morgan. The name honors Thomas Hunt Morgan, a pioneer of Drosophila genetics.

Mapping the Albino Trait

Now that the markers are mapped, we next show how to map the *albino* locus. In order to do so, we need to know (or guess) a genetic model connecting phenotype with genotype. The albino trait is easily seen to segregate as a recessive gene, so we postulate the existence of a locus having a recessive and dominant allele, with the homozygous recessive case leading to the albino phenotype. Naturally in our cross, the albino-strain mouse will be homozygous recessive, and the black-strain mouse homozygous for the dominant (i.e., full color) allele. Our analysis seeks markers closely linked to this *albino* locus, and so we calculate the LOD score for each marker in relation to the trait.

We begin by classifying each intercross mouse as agouti, black, or albino (white), and building up the tables cross-classifying by coat color and genotype at any locus. We give three such, using three markers, one on chromosome 12, another on chromosome 2, and a third on chromosome 7, the first unlinked with either trait, the second closely linked to the *agouti* locus, and the third closely linked to the *albino* locus (Tables 4, 5, and 6).

	CHROMOSOME 12			
COAT	A	H	B	Total
Agouti	19	35	18	72
Black	8	18	3	29
Albino	9	12	7	28
Total	36	65	28	129

TABLE 4 Coat color against genotype for a marker on chromosome 12, unlinked to either trait locus

	CHROMOSOME 2			
COAT	A	H	B	Total
Agouti	24	46	2	72
Black	0	1	28	29
Albino	5	14	6	25
Total	29	61	36	126

TABLE 5 Coat color against genotype for a marker linked to *agouti* on chromosome 2

| COAT | CHROMOSOME 7 | | | |
	A	H	B	Total
Agouti	3	47	19	69
Black	0	19	10	29
Albino	21	1	0	22
Total	24	67	29	110

TABLE 6 Coat color against genotype for a marker linked to *albino* on chromosome 7

| COLOR | LOCUS | | |
	A	H	B
Albino	$\dfrac{(1 - r)^2}{4}$	$\dfrac{2r(1 - r)}{4}$	$\dfrac{r^2}{4}$
Full color	$\dfrac{1}{4} - \dfrac{(1 - r)^2}{4}$	$\dfrac{1}{2} - \dfrac{2r(1 - r)}{4}$	$\dfrac{(1 - r^2)}{4}$

TABLE 7 Probabilities of coat color genotype combinations

How can we use frequency tables like these to map *albino*? As shown in the previous section, we need to build up a table of probabilities. For mapping *agouti* we would combine the black and albino phenotypes into non-agouti. We need expressions for six probabilities of phenotype-genotype combinations. Table 7 is a collapsed version of Table 3. It is obtained by identifying the A row there with the albino genotype, and identifying and aggregating the H and B rows there with the dominant full color genotypes.

To pass to a likelihood analysis, we collapse tables such as 4, 5, and 6 to two rows: albino and full color, and use the probabilities in Table 7 to find the likelihood function $L(r)$, which we maximize and then use to find the LOD score.

In our present context, with 152 markers, we carry out what is known as a genomewide scan, or simply genome scan; that is, we compute the LOD score at every marker locus. We have done this and plotted the result in Figure 5. The result is striking and clear-cut: a single high peak above 20 on chromosome 7 at the locus corresponding to the counts in Table 6. It turns out that this marker is just 6 cM from the *albino* locus. Since most mouse chromosomes have length of the order of 1 Morgan, or less, a distance of 6 cM is quite small.

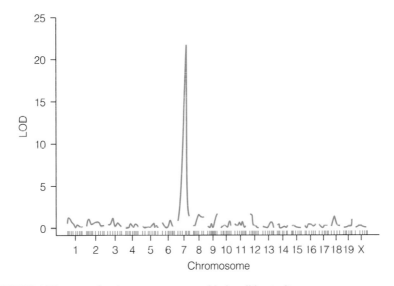

FIGURE 5 LOD scores for the genome scan with the albino trait

D7M25	–	B	H	A	A	A	A	A	–	A	A	A	A	A	–	B	H	H	H	H	H	A
D7M84	–	A	H	H	H	A	A	–	A	A	A	A	A	A	A	B	H	H	H	H	H	A
D7M126	–	A	A	A	–	A	A	A	–	A	A	A	A	A	–	A	A	A	A	A	A	H
D7M101	–	A	A	A	A	A	A	A	A	H	H	H	H	H	H	A	A	A	A	A	H	B
D7M71	–	A	A	A	A	H	H	H	H	B	B	H	H	H	A		H	H	B	H	B	B
D7M242	A	A	A	A	–	H	–	A	A	H	H	H	H	B	–	A	H	A	H	A	H	B
D7M292	–	A	A	A	–	H	A	A	A	H	H	H	H	B	H	A	H	A	H	A	H	B
D7M189	–	A	A	A	A	H	H	A	A	H	H	H	H	B	H	A	H	A	H	A	H	B

FIGURE 6 Allele pairs of the albino animals in the neighborhood of the marker with the highest LOD score in Figure 5. The A represents homozygous albino; B, the homozygous black strain; and H, the heterozygote. Genotypes are read down.

In order to check this conclusion, it is helpful to examine the genotypes of the 22 albino intercross progeny in the neighborhood of this marker. They should all, or almost all, be A, as albino mice are homozygous albino. This is found to be the case (see Figure 6): there is a single heterozygote, presumably resulting from a recombination in the 6 cM between the marker and the *albino* locus, and there are four mice whose genotype is missing at the marker.

Similar results are obtained for the *agouti* locus, although the peak is less striking, due to the previously mentioned fact that agouti status of albino mice cannot be determined.

CONCLUSION

We have tried to show how simple models can be put to work to map a gene contributing to phenotypic differences (here, coat color) between inbred strains of lab mice. The background required is substantial. We need to appreciate the ideas underlying an intercross between two inbred strains of mice. To obtain the precise form of the multinomial models, we need to understand some details of genetic linkage and recombination. To have a sense of the type of data we will get, and its quality, we need to become familiar with the molecular genetic markers used to identify sequence differences between our strains. When we have understood these topics sufficiently, it is almost a relief to turn to the likelihood methods used to carry out the statistical analysis. For the most part they are familiar, though sometimes presented in a different language or framework. One feature of the use of these methods in gene mapping is the larger scale than usual. In a genomewide analysis, we do not carry out one but 150 significance tests—or more. This demands novel methods to control type 1 error rates. For these and many other applications of statistics in mouse genetics, we refer to Silver (1995) and to more recent issues of the journals he cites.

Here in mouse genetics, we get a glimpse of how statistics can serve the cause of biomedical research. There are many human disease and disease susceptibility traits that can be studied in the mouse and the knowledge transferred back to humans. And there is no other general way to localize genes contributing to such traits than to use the statistical methods we have outlined or their analogues.

REFERENCES

Broman K. W., et al. (2003). "R/qtl: QTL Mapping in Experimental Crosses." *Bioinformatics* 19, no. 7:889–90.

Freedman, D., R. Pisani, and R. Purves. (1997). *Statistics,* 3rd ed. San Francisco: Norton.

Lange, K. (2001). *Mathematical and Statistical Methods for Genetic Analysis,* 2nd ed. New York: Springer-Verlag.

Makino S., et al. (1980). "Breeding of a Non-Obese Diabetic Strain of Mice." *Jikken Dobutsu* 29, no. 1:1–13.

MendelWeb. http://www.mendelweb.org/Mendel.html.

Silver, L. M. (1995). *Mouse Genetics.* New York: Oxford University Press.

QUESTIONS

1. Explain why the offspring of a black-strain $\times$ albino-strain cross will all have agouti coat colors.

2. Explain why the intercross offspring of the result of a black-strain $\times$ albino-strain cross can have albino, agouti, or black coat color.

3. The alleles agouti and non-agouti at the *agouti* locus should be segregating in a Mendelian fashion in this cross, as should the alleles albino and full color at the *albino* locus. The 132 intercross offspring contained about 72 agouti, 29 black, and 29 albino. Are these figures consistent with Mendelian segregation at each of the *agouti* and the *albino* loci? Explain.

4. Do the data in Table 2 support Mendel's second law; that is, do they support the hypothesis of a recombination fraction of $\frac{1}{2}$?

PART FOUR

BUSINESS AND INDUSTRY

✳ : : ✳

To Catch a Thief: Detecting Cell Phone Fraud

Diane Lambert & José C. Pinheiro

Reducing Junk Mail Using Data Mining Techniques

R. D. De Veaux & H. Edelstein

Improving the Accuracy of a Newspaper: A Six Sigma Case Study of Business Process Improvement

Ronald D. Snee

Assuring Product Reliability and Safety

Necip Doganaksoy, Gerald J. Hahn & William Q. Meeker

Randomness in the Stock Market

Richard J. Cleary & Norean Radke Sharpe

Advertising as an Engineering Science

William Kahn & Leonard Roseman

TO CATCH A THIEF

Detecting Cell Phone Fraud

DIANE LAMBERT

Bell Laboratories, Lucent Technologies

JOSÉ C. PINHEIRO

Novartis Pharma

————— ❋ : : ❋ —————

Millions of calls are placed on a telephone network every day. Most of the calls are legitimate, but some are placed by people or businesses that do not intend to pay, which constitutes fraud. According to an August 6, 2002, report by the U.S. Federal Communications Commission, wireless phone fraud in the United States accounts for losses of more than $150 million each year.

Figure 1 shows the calls for a wireless account attacked by fraud. Each vertical line in the plot corresponds to one call at the time shown. The length of the line represents call duration, and its color denotes call direction (incoming or outgoing). The calling behavior for this account changed dramatically on March 27. Before then, most calls were short and fell during business hours. But calls after March 27 were considerably longer and fell on any day of the week or time of day. Although the fraud is easy to spot visually, this is not a practical way to detect fraud when millions of accounts have to be monitored simultaneously. In fact the fraud pictured in Figure 1 was

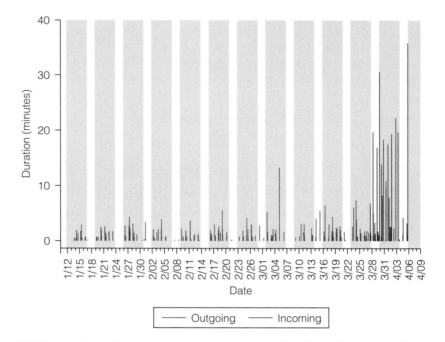

FIGURE 1 Calling activity on a wireless phone account hit by fraud. Each vertical line is a different phone call. Unshaded bars represent weekends.

not detected and stopped for almost two weeks. By then, many long-duration, costly fraudulent calls had been placed. An efficient fraud detection system would have detected the fraud much earlier. Statistical thinking plays a critical role in the design of an automatic, algorithm-based fraud detection system.

The key to detecting fraud is to find significant changes in calling behavior. A wide array of call characteristics, or *variables,* can be computed from wireless call records. These include the variables illustrated in Figure 1 (date and time, duration and direction) and many others as well, such as the use of call waiting or whether the call was made while roaming. The call characteristics help to identify fraud. For example, the account in Figure 1 never used call waiting until after fraud started.

Given the large number of accounts, it is clear that a method for reducing the available call data to a numerical summary that can be used to detect fraud automatically, without visualizing the data, is needed. The basic questions regarding automated fraud detection are statistical: what summary of call data best describes calling behavior, how should the summary be updated as new calls are made, and which calls should be judged atypical?

DEFINING AN ACCOUNT SIGNATURE

Summarizing call data well is particularly important because changes in behavior are impossible to detect reliably without a good baseline description of calling for the account. A single phone call can be represented as a collection of call characteristics such as duration, time of day, direction, and so forth. The baseline description of an account then summarizes which values of these characteristics or variables are typical for the account and which are atypical. (If the atypical values start to become frequent, then that might suggest that the account has been hit by fraud.) A probability distribution is an especially useful way to convey such information. For example, the probability distribution of the variable "call direction" specifies the probability p that a call is incoming and the probability $(1 - p)$ that a call is outgoing. Before fraud began on the account shown in Figure 1 the probability p of an incoming call was 0.15, but that probability rose substantially after fraud started. Similarly, before fraud started, the probability distribution of call duration assigned high probability to short calls, since about 90% of calls were less than two minutes and only 10% were longer than two minutes, but that distribution changed dramatically after the fraud started. The complete calling behavior for an account is described by a single probability distribution that represents *all* call variables simultaneously. That single distribution can be complicated because there are many different call characteristics and some may be interrelated. For example, if outgoing calls tend to be longer than incoming calls, then the probability distribution for call duration cannot be specified without considering call direction.

A natural way to build the probability distribution that describes the calling behavior for an account is to work in stages. Initially we can form the probability distribution of a single variable; for example, call direction would be described by the proportions of incoming calls and outgoing calls. Then, the distribution of a second characteristic, perhaps call duration rounded to intervals (such as 0–1 minutes, 1–2 minutes, 2–4 minutes, 4–8 minutes, 8–12 minutes, 12–20 minutes, greater than 20 minutes), is introduced, with separate distributions of call duration for incoming and outgoing calls. Incoming calls, for example, might tend to be shorter than outgoing calls. The distribution of duration for incoming calls and the distribution of duration for outgoing calls are together known as the conditional probability distribution of call duration given call direction (or sometimes as the probability distribution for call duration conditional on call direction). The use of conditional probability distributions allows us to capture the relationship between call duration and call direction accurately. Once the probability distribution of call direction and the distribution of call duration conditional on direction are specified, the process of building the complete probability distribution of calling behavior continues by specifying a probability distribution for another call characteristic conditional on both call direction and call duration. This process of specifying the probability for a call variable conditional on all variables that have already been considered continues until all call characteristics of interest have been considered.

A complete probability distribution for all call variables can have a huge number of probabilities that need to be specified, but it is often possible to reduce the task. For example, the probability of using call waiting may not depend on call direction and call duration. In that case the proportion of calls that use call waiting need only be specified once (instead of separately for each possible combination of call direction and call duration). We call the final set of probability distributions that are used to describe an account's behavior the *account signature,* and we call each probability distribution that is included in the signature a *signature component.* The same signature components are used for each account. Then the only differences from account to account are the numerical values of the probabilities for the different characteristics. It is these probabilities that reflect the behavior for the account.

For example, Figure 2 shows a simple signature for three different accounts. There are four signature components: (1) a probability distribution for call direction (incoming or outgoing); (2) a probability

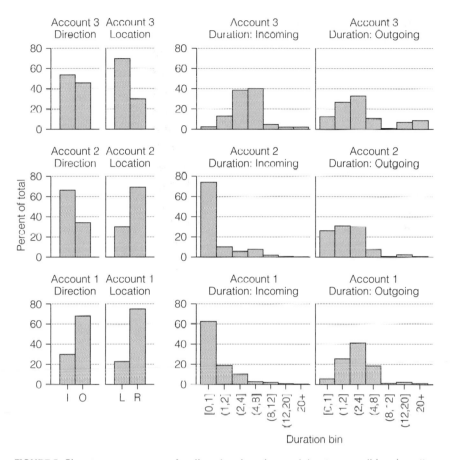

FIGURE 2 Signature components for direction, location, and duration conditional on direction for three accounts. Account 1 has the lowest proportions of incoming and local calls, for example. In all three accounts, outgoing calls tend to be longer than incoming calls.

distribution for call location (roaming or local); (3) a probability distribution for the duration of incoming calls; and (4) a probability distribution for the duration of outgoing calls. All three accounts have these same four signature components, but their probability values differ. For example, Account 1 is more likely to make outgoing calls than to receive incoming calls, Account 2 is more likely to receive incoming calls, and Account 3 is nearly equally likely to make or receive calls. The second signature component (location) indicates that Accounts 1 and 2 make more roaming calls, whereas Account 3 makes more local calls. All three accounts tend to talk longer when they initiate the call. Account 2 has the largest difference in call duration between incoming and outgoing calls, with about 60% of incoming calls lasting less than one minute and about 60% of outgoing calls lasting more than two minutes.

DESIGN OF AN EFFICIENT SIGNATURE

Smaller account signatures, that is, signatures based on fewer variables or with fewer conditional distributions, are preferable because they require less space to store (an important consideration when each account needs its own signature). Just as important, there is a strong statistical argument for smaller signatures. The more signature components that need to be filled in, the more data are needed to estimate all the probabilities well. For example, suppose a signature has the three components shown in Figure 2. Then, the outgoing calls cannot be used to estimate the probability distribution for the duration of incoming calls. The more a variable is broken down by conditioning variables, the more signature components it contributes, and the more data are needed to estimate the probabilities in the signature. Because there may not be many calls before fraud starts, signatures with lots of components are unappealing.

Signature design is based on analyzing all the calls for thousands or even millions of accounts with no fraud during a fixed period (60 days, for example). The call data are used to identify important characteristics to include in the signature and to determine when it is important to use a conditional distribution to capture the interrelationships among variables. The key point is that a well-designed signature allows an enormous amount of data to be summarized in a relatively small amount of computer disk space. In one application the signatures computed from 33 gigabytes of raw wireless call records for about 1.5 million accounts required only about 300 megabytes to store!

Once the signature structure is set, it is easy to compute a signature for existing accounts. The required probabilities are just estimated by the fraction of calls in the account's call history that have a particular characteristic. For example, the probability of incoming calls in the signatures shown in Figure 2 can be estimated from the proportion of incoming calls among the historical calls for each of the accounts (approximately 0.30 for Account 1, 0.65 for Account 2, and 0.55 for Account 3).

STARTING A SIGNATURE

Assigning a signature to a new account quickly is an important step in fraud detection because there are new accounts every day. Waiting until an account has a month or more of calls is not a viable option because fraud may start at any time. In fact, a customer may not intend to pay from the start, so all calls on a new account are fraudulent (this is known as subscription fraud). Delaying the start of a signature for these accounts for a month could lead to significant losses in revenue. An alternative is to use signatures on existing accounts with no fraud to start the signature for a new account. For lack of better information, we might assume that a new account is "average" and initialize the new account signature with the average of all signatures in the historical data. This starts the account with a baseline signature derived from legitimate calls, making it easier to detect fraud early on.

Of course not all accounts are the same, and some differences may be evident early enough to improve the initial signatures. Better initial signatures are possible if accounts in the historical data can be differentiated using their first few calls. For example, accounts could be divided into segments based on whether their first few calls are made during business hours (perhaps indicating a business phone) and the average duration of their first few calls. Then, a new account would be assigned to a segment after it has three calls, matching it to the segment appropriate for the average duration of its three calls and the number of these calls during business hours. The average signature of all accounts in the appropriate segment would then be assigned as the initial signature for the new account.

KEEPING A SIGNATURE UP-TO-DATE

It is important to keep the signature for an account up-to-date because the starting signature may be a crude estimate of the behavior of the account. Fortunately, statistical methods can be used to estimate an account's signature again every time there is another call on the account. This allows an account with many calls to move far from its initial signature, which was based on the experience of other accounts. Eventually, after the account makes many calls and the starting signature is updated many times, the account will have its own "personalized" signature. Updating an account signature can also incorporate honest changes in calling behavior, ensuring that there is always an up-to-date standard against which fraud can be assessed. Additionally, call-by-call updating is computationally efficient because it avoids accessing the account's past calls. Computational efficiency is important because processing has to be much faster than data collection to keep up with the flow of calls.

Updating a signature amounts to reestimating the probability distributions for the account. One approach relies on a technique called exponentially weighted moving averaging. In this approach an updated probability distribution is a weighted average (a kind of combination) of the current estimated probability distribution and the information in the new call. The relative contributions of the new call and the existing probability distribution are determined by a weight w (a number between 0 and 1, typically taking on a small value). Specifically, weight w is assigned to the new call, and weight $(1 - w)$ is assigned to the current probability distribution. For example, consider the probability that a call's duration is less than two minutes. This is just a small part of one signature component, but discussing this piece in detail helps demonstrate the approach. If the current estimate of the probability of a short call is p, then the updated estimate is either $(1 - w)p + w$ if the new call is shorter than two minutes or $(1 - w)p$ if the new call is at least two minutes. Because w is a number between 0 and 1, the revised probability estimate is higher than the earlier estimate if the new call is short, and the revised probability estimate is lower than the earlier estimate if the new call is not short. How much higher or lower depends on the weight w. One way to determine a reasonable value of w is to consider how the impact of a call wanes over time. The call that happened 10 calls ago has weight $(1 - w)^{10}$ relative to the new call; this exponential weighting is what gives the method its name. For $w = 0.1$ the older call has weight $(0.9)^{10} = 0.35$ relative to the new call, or, put in other words, 35% of the weight of the new call. For $w = 0.02$ the older call still has 82% of the weight of the new call. With small w, older calls continue to make large contributions to the signature component. Consequently, the smaller w is the more slowly the account signature changes over time.

Not all signature variables can be updated exactly as shown here, because some call characteristics, such as day of week, are not observed in random order. All the Tuesday calls are made, then all the Wednesday calls are made, and so forth. If we update as shown here, and ignore the lack of randomness, then the probability distribution for day of week will change constantly even if there is no change in calling pattern. Fortunately it is possible to update the estimated distributions for such quantities using an approach that is similar in spirit and in amount of computation to exponentially weighted moving averaging.

SCORING CALLS FOR FRAUD

Not all calls that are inconsistent with an account signature indicate fraud. A short call to a never-before-dialed local number need not be a sign of fraud—it may be a new friend! False alarms are a big concern in fraud detection. Just as summarizing the calling pattern for an account in a probability distribution is key to detecting fraud, summarizing the calling pattern for fraud in a probability distribution, or *fraud signature*, is key to limiting false alarms. Typically, a fraud signature has the same signature components as an account signature, because it simplifies the comparison with account signatures and because there is often too little fraud data for a separate, detailed analysis of interactions between variables. Generally we collapse all the fraud records into one set of target data that is then used to estimate the probabilities needed in the fraud signature. If several different kinds of fraud can be distinguished, then a fraud signature can be developed for each kind.

Scoring a call for fraud reduces to comparing the call's probability under the account signature to its probability under the fraud signature (or to its probability under several fraud signatures, if there is more than one). A *call score* is based on the ratio of the probability of a call calculated from the fraud signature to the probability of the call calculated from the account's (non-fraud) signature. It is common to take the logarithm of this ratio, so a ratio greater than 1 (indicating possible fraud) becomes a positive score and a ratio less than 1 becomes a negative score. The higher the call score, the more suspicious the call. For a call to obtain a high score, it has to be unexpected for the account. Calls that are not only unexpected for the account but also expected under fraud are considered more suspicious than calls that are unexpected for both the account and fraud. Thus, some departures from the signature are more interesting than others from the perspective of fraud management. So, just calling a new friend is not likely to trigger a fraud alert.

Call scores serve two purposes. One is to give information that can be used to identify *accounts* that may have fraudulent activity. The other is to identify *calls* that may be suspicious and so should not be used to update the signature. For example, a negative score suggests that the call is not fraudulent, so it can safely be used to update the account signature. Calls with high positive scores raise concerns about fraud, though, and to be safe should not be used to update the signature. Small positive scores are ambiguous. They might suggest a slight change in calling pattern or they might suggest that there is a subtle case of fraud. This ambiguity suggests the following procedure: update the signature if the call score is negative, do not update the signature if the call score is high, and choose whether to update randomly if the call score is positive but small. In the last case, a signature is updated with a probability that depends on the call score, varying from probability 1 for a call score that is less than or equal to 0 to probability 0 for a call score that is sufficiently high.

FLAGGING ACCOUNTS FOR FRAUD

Egregious cases of fraud may generate calls with scores so high that a service provider is willing to declare that fraud has occurred after only one call marked as fraud. Usually, however, the evidence from one call alone is not sufficient to identify fraud. One way to build up evidence of fraud is to monitor the cumulative score of all calls that exceed a threshold. Monitoring the rate of increase of the cumulative call score can also be important because fraudsters may try to make

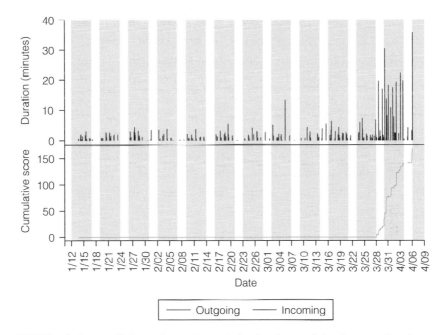

FIGURE 3 Call record information and cumulative fraud score (of calls exceeding the threshold) on a wireless phone account hit by fraud

as many calls as possible before being detected. A low threshold will identify most cases of fraud but increase the risk of mistakenly claiming fraud (a false alarm). A high threshold on the other hand risks missing cases of fraud by declaring that nothing unusual has occurred (a false negative). The need to choose an optimal threshold by trading off the two types of errors is common in decision problems. Here the threshold can be chosen to minimize the possibility of a false alarm subject to a specified limit on the proportion of false negatives.

We say that an account that exceeds the cumulative score thresholds is *flagged*. Figure 3 shows the cumulative call scores for the account considered in Figure 1. Around March 27, the cumulative score increases sharply. A signature-based fraud detection system would have flagged the account for investigation by an analyst close to March 27, and the fraudulent activity could have been detected and stopped soon after it began.

The need for statistics in fraud detection does not end once accounts are flagged. Flagged accounts have to be prioritized, preferably in terms of the losses that will be averted if the account has fraudulent activity and it is shut down. This amounts to estimating the severity of a suspected case of fraud and the rate at which fraudulent phone use on the account is occurring. The performance of a fraud detection system also has to be monitored. Performance is ultimately determined by the losses a service provider is able to prevent, but measuring averted losses (which by definition don't occur) is clearly difficult if not impossible. Instead, service providers use measures like the number of investigated cases that have fraud, average time to detection after fraud starts, and average number of fraudulent calls prior to detection. Finally, there is a continual need to learn from missed cases of fraud, to understand when a new type of fraud has started and a new fraud signature is needed.

SUMMARY

Detecting telephone fraud requires that fraudulent calling behavior be distinguished from ordinary account behavior for millions of accounts making dozens or even hundreds of calls each month. Probability and statistics tools are featured prominently in fraud detection algorithms. The use of probability distributions to characterize call behavior allows for a quantitative measure of how unusual a particular call appears relative to an account's typical behavior; the probability distribution combines information from a variety of call characteristics, including call duration, call direction, date/time, and phone features used. Statistical thinking requires that the call also be compared with the typical behavior of past fraudulent calls. Statistical methods are used to determine which call characteristics should be included in the account signature and to estimate the needed probability distributions. Thresholds for identifying troubled accounts are set by considering the relative cost of the two possible errors (identifying typical use as being fraudulent or identifying fraudulent use as being typical). The use of statistical methods to combine information from past call data and current call activity allows phone service providers to quickly and accurately identify fraud.

ADDITIONAL READING

Bickel, P. J., and K. A. Doksum. (1976). *Mathematical Statistics.* San Francisco: Holden-Day.

Chen, F., D. Lambert, and J. C. Pinheiro. (2000). "Incremental Quantile Estimation for Massive Tracking." *Proceedings of Knowledge Discovery in Data (KDD) 2000,* 516–22. Boston: Association for Computing Machinery.

Chen, F., et al. (2000). "Reducing Transaction Databases, without Lagging Behind the Data or Losing Information." Technical memorandum. Bell Labs, Lucent Technologies.

Devore, J. L. (2000). *Probability and Statistics for Engineering and the Sciences,* 5th ed. Belmont, Calif.: Wadsworth.

Federal Communications Commission. (2002). "FCC Continues National Fraud Awareness Week Activities: Day Two, Cell Phone Fraud." *FCC News,* August 6.

Gibbons, P. B., Y. E. Ioannidis, and V. Poosala. (1997). "Fast Incremental Maintenance of Approximate Histograms." *Proceedings of the 23rd International Conference on Very Large Data Bases,* 466–75. San Francisco: Morgan Kaufman.

Lambert, D., J. C. Pinheiro, and D. X. Sun. (2001). "Updating Timing Profiles for Millions of Customers in Real-Time." *Journal of the American Statistical Association* 96:316–30.

Yu, Y., and D. Lambert. (1999). "Fitting Trees to Functional Data, with an Application to Time-of-Day Patterns." *Journal of Computational and Graphical Statistics* 8:749–62.

QUESTIONS

1. It is obvious from Figure 1 that a change in usage pattern has occurred around March 28. Why is it suggested that a numerical summary is needed to detect fraud?

2. What are typical components of an account signature?

3. Why is it important to estimate a signature based on only a few days of account experience?

4. For Account 1 in Figure 2 the probability of an incoming call is 0.30. Suppose we use the exponentially weighted moving

average with $w = 0.10$ to update this probability. What would the updated probability of an incoming call be after a new incoming call is observed? What would the updated probability of an incoming call be after five new incoming calls are observed? How would the answers change if $w = 0.02$ was used?

5. Explain why a negative call score suggests the call is not fraudulent.

6. The essay describes the trade-off involved in selecting a threshold for including calls in the cumulative score. If the threshold is too high, then we will miss fraudulent activity. If the threshold is too low, we may identify legitimate activity as being fraudulent. Consider another problem requiring a decision (e.g., whether to cancel a picnic in the face of a bad weather forecast or how much deductible to accept on your auto insurance) and describe the trade-off involved in making the decision.

REDUCING JUNK MAIL
USING DATA MINING
TECHNIQUES

RICHARD D. DE VEAUX

Williams College

HERB EDELSTEIN

Two Crows Corporation

———— ❧ : : ❧ ————

Junk mail generates 4 million tons of paper waste every year at an annual cost of $550 million just to transport it. According to the Houston-Galveston Area Council, more than half of the unsolicited mail is thrown out without being read or opened. In fact, the overall response rate to junk mail is less than 2%. Companies don't make money by sending you mail that you throw out. How can statistical methods help them decide who should get what mail?

Today, companies collect an incredible amount of information about their customers. Much of this information is *transactional data,* data generated automatically by purchasing, ordering, billing, website transactions, calls to the 800-number call center, and other interactions with the organization. For a credit card account at a bank, the customer record may result in dozens or even hundreds of entries a year. For a philanthropic organization, the transactional data might consist of whether you responded to its recent solicitations, and if you responded, how much you gave. Separate from this trans-actional database is a database of personal information about the cus-tomer. The customer usually supplies this information when opening an account or when registering on a website. In addition, the com-pany may buy *demographic* and *behavioral* information about the in-dividual, including age, value of house, make of car owned, or club memberships. Neighborhood demographic data on a zip code or finer level can be purchased from agencies such as the U.S. Census Bureau or a variety of commercial organizations, and can provide valuable information about the area in which the customer lives.

These disparate databases are gathered in a central repository, called a data warehouse, where they may be used in a variety of query, reporting, and analytical applications. Analysts may want to use the data warehouse to build *predictive models* that can identify who will respond to a direct mail campaign or credit card solicitation. The types of analyses that they perform fall under the general term *data mining*. Data mining techniques are being used on large databases in myriad applications, from customer relationship management to genomics to fraud detection. Common themes of data mining are exploring relationships and building predictive models from very large sets of data.

The obstacles facing the data miner are formidable. If you think about the three different data sources for a minute, you can see that one of the huge challenges of the data warehouse is to *merge* these disparate data sources. One problem is that the data in each database are collected on a different temporal or spatial scale. Transactional data are collected every time a transaction is made. For a bank, a new item is stored whenever you use your ATM card, write a check, call the 800 number, or make a direct deposit. Your customer informa-tion, however, is likely to stay the same until you move, open a new account, or change your name. Demographic information changes slowly but only over a larger area, and does not pertain to only one customer. For a small-town bank, the same demographic informa-tion may apply to all its customers. Another challenge is that the same term may be used differently in the different databases. A variable called *age* may refer to the customer's age in one, the age of the account in another, and the average age of the residents within a zip code in a third. Reconciling these different meanings and main-taining the data warehouse is a major effort.

A DIRECT MAIL EXAMPLE

The Paralyzed Veterans of America (PVA) is a congressionally chartered organization serving veterans of the armed forces who have experienced spinal cord injury or dysfunction. A large part of their funding comes from proceeds from direct mail solicitations. With a mailing list of more than 13 million donors, the PVA is one of the largest direct mail fundraisers in the country.

The PVA estimates that 34% of its revenues are spent fundraising. For people on its mailing list, a solicitation accompanied by personalized address labels and/or greeting cards usually comes every four to six weeks. Certainly, not everyone who receives address labels or greeting cards in the mail returns money back to the PVA every time. It costs the PVA about $0.68 to produce and send a solicitation to each person. If it could send mail only to the people who send back more than $0.68, it stands to reduce its costs substantially, thereby having more money to assist paralyzed veterans. Furthermore the amount of unwanted solicitations—junk mail—sent is reduced.

This is a real example of a potential application of data mining. After discussing what data mining is, and a little about how it works, we will return to this example and show how data mining techniques were used to attack it.

WHAT IS DATA MINING?

Once a company has invested in creating and maintaining a data warehouse, it's only natural that it would want to get more value out of the information contained in it. The sheer size of the data warehouse makes that a daunting task. Databases today can range in size into the terabytes (TB)—more than 1,000,000,000,000 (1 trillion) bytes of data. For example, the tracking database of the United Parcel Service is estimated to be of the order of 16TB, roughly the digital size of all the books in the Library of Congress. Information of strategic importance may lie hidden within this massive collection of data. How can we find the needle of information in this huge haystack of data?

The first step of turning data into information is to give context to the data values. For example, a customer record of ("male," "123456," "123456") is essentially meaningless until we convert it into the information "customer number 123456 is male and made purchases this year of $1,234.56." In this case it is the data structure that supplies the context. Given this context, we can use a sequence of queries to deduce facts about customer behavior, asking specific questions of the data. Guided by their domain knowledge, analysts traditionally used this query-driven approach to deduce patterns or to test hypotheses using the data. This approach is often called online analytical processing, or OLAP. While it can answer a specific question like "how many customers under the age of 25 in the Midwest made more than two purchases last year?" analysis of this type is limited and does not build a predictive model.

By contrast, data mining moves beyond simple queries and fact collection, using a variety of data analysis tools to discover patterns and relationships in data that often are used to build useful models and make predictions. Data mining uses both the knowledge of the domain expert and the data itself to identify the best predictors and to uncover useful patterns. It is a general principle that statistical methods work best when combined with sound subject-matter knowledge. This subject-matter knowledge formulates the problem and guides data collection; statistical methods then provide answers from the data, and typically reveal unforeseen patterns or behaviors, which refine and advance the subject matter knowledge. The new knowledge, or domain theory, leads to further data collection to answer additional questions, and so the overall process of scientific inquiry continues in an iterative fashion. In this case, a data mining tool might discover that education and age are good predictors for sales of a product and that there is a segment of women, say in a particular geographic region and age range, with whom the product is particularly popular. While this might have been found by a set of queries, the initial result of the data mining effort is a model that predicts the success of targeted sales initiatives. Further data might then reveal the degree of success of this targeted sales initiative, produce additional learning, and generate additional questions to be answered.

Data mining also addresses another key problem in trying to make predictions that the descriptive approach fails to address. Even after a pattern is unearthed through a series of queries, the analyst doesn't know whether the pattern holds true for anything other than the collection of data used to find it. Yet often the analyst is more interested in generalizing his or her conclusions to other data. For example, the analyst might build a profile of customers who bought certain products and then want to use it to target potential future customers. But will this profile apply to people who are not yet customers? A pattern that says recent purchasers of our products are good candidates for future promotions may not be useful in a campaign to acquire new customers.

It should be emphasized that it is almost always easy to make accurate prediction on the data *at hand*. Given a data set and enough predictor variables, we can construct a model that will predict the response we want perfectly. It's easy to predict last year's stock market moves once they've happened. The challenge is to do well on data that we haven't seen. Next year's stock prices are more difficult to predict. As Niels Bohr, the Nobel Prize–winning physicist, once said, "Prediction is difficult, especially about the future." To account for this, data miners typically split the data into two (or three) sets, building the model on a *training set* and testing it on a *test set*. Sometimes a third set, a *validation set,* is withheld for a final evaluation.

Common Myths about Data Mining

Data mining can provide effective models, but it is a set of tools, not a magic wand. Many people buy data mining tools in the hopes that the tools will passively interact with databases and spew knowledge with little work on the user's part. But today's data mining tools won't sit in your database watching what happens and send you e-mail to get your attention when it sees an interesting pattern. They won't eliminate the need to know the business, to understand the data, or to understand analytical methods. Data mining assists business analysts with finding patterns and relationships in the data—it does not tell them the value of the patterns to the organization. Furthermore, the patterns uncovered by data mining must be verified in the real world.

Data mining will not automatically discover solutions without guidance. Rather than setting the vague goal "Help improve the response to a direct mail solicitation," analysts might use data mining to find the characteristics of people who (1) respond to the solicitation, or (2) respond *and* make a large purchase. The patterns data mining finds for those two goals may be different. Also, data mining does not replace skilled business analysts or managers but rather gives them a powerful new tool to improve the job they are doing. Any company that knows its business and its customers is already aware of many important, high-payoff patterns that its employees have observed over the years. What data mining can do is confirm such empirical observations and find new, subtle patterns that yield steady incremental improvement (plus the occasional breakthrough insight).

Remember that the predictive relationships or observed patterns found via data mining are just models, and are not necessarily causes of an action or behavior. For example, even if one observes that many professional athletes sport a certain hairstyle, this does not imply that if you adopt that hairstyle, you are likely to become a professional athlete!

What Is a Model?

When an engineer wants to study how a new airplane design will perform, she may decide to first build a model of the airplane for study in a wind tunnel. By studying the model, she can learn a lot about how the design will work under different conditions. Of course, the model is not reality itself but only an approximation. It's something we can examine and manipulate in order to learn more about the real world.

Data mining uses data to find predictive patterns that are expressed as models that predict behavior in the real world. The data miner hopes to predict who will respond to the next solicitation by using information in the database from past solicitations. The model uses the differences in the information of the two groups—those who responded in the past and those who didn't respond—to identify relationships that can help predict future responses.

Of course, having the right information, or predictor variables, is a key ingredient in building a useful model. The quality of the data is crucial as well. If many of the important variables are missing or corrupted, it will be impossible to build reliable or useful models. Sometimes, domain knowledge about the problem indicates which variables to consider for a model. If you're running a diaper service and want to expand your client base, it would be helpful to know if there's a baby or toddler in the house before you make a sales call.

Some of the models used in data mining are relatively familiar. *Linear regression models,* as the name implies, predict a value for the response (the variable we are trying to predict) based on a linear equation of the predictor variables. For example, you might predict the gas mileage of an automobile from its weight and the size of its engine:

$$\text{predicted } mpg = 48.04 - 7.93 \text{ weight} - 0.302 \text{ size}$$

where weight is measured in thousands of pounds and size in hundreds of cubic inches. When the response variable is a yes/no variable (like whether the person responds or not to the campaign), then we may use a *logistic regression* model. The logistic regression model predicts the *probability* that someone will respond, usually denoted by p. Since p can only take on values between 0 and 1, and our linear model might predict values outside this range, we employ a mathematical trick, called a *transformation,* to obtain a more meaningful model. In this particular case it is called a logistic regression, because we take the logarithm (a common mathematical function) of the ratio between the probability of someone responding, and the probability of the person not responding $(1 - p)$. The resulting variable, $\log(p/(1 - p))$, can take on values from minus infinity to plus infinity. For example, a model to predict whether a patient has diabetes or not based on his age and glucose levels might look like this:

$$\text{predicted } \log(p/(1 - p)) = -5.83 + 0.0342 \times \text{glucose} + 0.028 \times \text{age}$$

where p is the probability of having the disease, and glucose and age are the predictor variables. We can then convert the predicted value for $\log(p/(1 - p))$ back to a probability by reversing the transformation. Other models used in data mining are more exotic. Many are nonlinear, but all share the characteristic that a response variable is predicted from a set of predictor variables. The model *selection*— including choice of form for the model, the choice of the set of predictor variables to enter into it, and the estimation of the *parameters* (the unknown constants in the model)—is one of the primary challenges of the data mining effort.

RETURN TO THE DIRECT MAIL EXAMPLE

To better understand how data mining works, let's look at the context of the real problem previously introduced. A key question about this problem is whether data mining techniques could help PVA discover what types of people would be most likely to respond, so that direct mail could be more effectively targeted. Rather than attack the problem itself, PVA decided to take a rather novel approach and ask for experts in data mining to work on the problem for it, via a competition.

In 1998, the PVA made available some of its data to a data mining competition, called the KDD Cup. In this competition, sponsored by the Fourth International Conference on Knowledge Discovery and Data Mining, 21 data mining teams tried to build a model that would predict who should be sent the next solicitation.

The PVA Data

The first step in mining the data is to prepare them for model building. Surprisingly, this step takes more time and effort than all the other steps combined. You may need repeated iterations of the data-preparation and model-building steps as you learn something from the model that suggests you should modify the data. *These data-preparation steps may take anywhere from 50% to 90% of the time and effort of the entire data mining process.*

The data for the PVA competition came from a recently completed campaign. The responses consisted of two response variables— *whether* the customer responded, and (if there was a response) the *amount* of the contribution. In addition, there were 481 *predictor variables,* variables to be used in building a model for the response. These predictors included transactional data (e.g., the amount donated in each of the previous 24 months); individual data (e.g., the number of children and title, such as Mr., Ms., General, His Highness); psychographic data (e.g., interested in gardening, Bible, etc.); and neighborhood demographic data (e.g., percentage in the customer's zip code never married, median family income). Needless to say, there were significant amounts of missing data, plus an unknown amount of incorrect data.

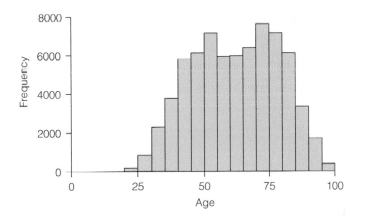

FIGURE 1 This histogram of the customer *ages* shows that the customer base is fairly old, with a median around 60 years old. There appears to be two groups in the data — one with a median near 50, and another with a median around 70.

Knowing how to start the modeling process is challenging with 481 potential predictor variables. Usually data miners like to explore the data first by plotting the variables separately in histograms (if they are *quantitative* variables) or bar charts (if they are *categorical* variables). Figures 1 and 2 show examples for two of the variables, age and homeowner status. Because *age* is quantitative, a histogram of the ages is shown. Here we see that the ages of these people are fairly high, with a median of around 60 years old. There seem to be two peaks, or modes—one near 50 and the other near 70. The *home-owner* variable, shown in Figure 2, is categorical, with values H for homeowner, N for not homeowner, and U for unknown. The major-ity of people in the database are homeowners, with equal splits among non-homeowners or unknowns.

Plotting one variable at a time gives some insights, but if we wanted to investigate whether owning a home was related to age, we would need to look at the two simultaneously, perhaps in side-by-side displays of the ages by each category of home (Figure 3).

This particular graphical display is called a boxplot. The center line within each box represents the 50th percentile (the median), and the two horizontal lines on either side of the median show the lower quartile Q_1 (the 25th percentile) and the upper quartile Q_3 (the 75th percentile). The box itself represents the middle 50% of the data

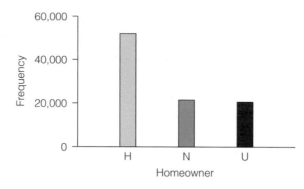

FIGURE 2 This bar chart of homeowner status shows that the majority of customers in the database are homeowners, with about one-quarter not owning their own homes and another quarter of unknown status

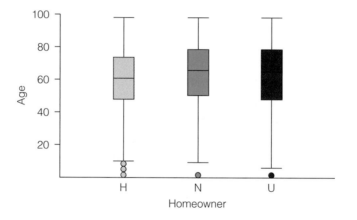

FIGURE 3 Side-by-side boxplots show that the age distribution for the three *homeowner* groups are very similar

(that is, from the 25th to the 75th percentiles). The length of the lines coming out above and below the boxes represents the extremes of the data unless the data point lies too far from the quartiles, in which case these points are marked as *outliers*—unexpectedly high or low values (all outliers are low values in this particular plot). Here we see that the age distribution looks similar over the three categories of home ownership. Now that we've investigated two of the 481 predictors, imagine repeating this for all 479 other predictors and more than 100,000 other pairs of predictors. Clearly, no analyst can hope to look at all these plots.

In order to begin, the analyst has to limit the number of variables to consider. This can be accomplished by relying on the analyst's domain knowledge, or using a data mining tool to help winnow the number of predictors to consider. Because past behavior is the best predictor of future behavior, it makes sense to focus attention on behavioral variables such as whether someone has donated in the past and if so how much. We still want to include some important descriptive information such as age and gender, factors that our experience tells us influence behavior.

However, the analyst may not want to rely on his or her own sense of which variables are important. Data mining tools are available that consider all the variables and return a rank ordering that indicates which ones are important. These tools can save a lot of time for the analyst when beginning a data mining project.

Building a Model

There are a large number of tools for building models, including both traditional statistical tools such as linear and logistic regression previously mentioned, but also more exotic-sounding tools including decision trees and neural networks. Rather than the traditional statistical approach of choosing the model based on the science of the data and the question, the data miner will typically try multiple tools in an attempt to find a useful model.

It is always important to think about the strategy you will use in building the model. For the PVA data, we are trying to maximize the net revenue (revenue less cost of the mailing). One of our response variables is in fact the amount of the contribution made by donors. So you might just want to predict that variable, a choice that the overwhelming majority of KDD Cup entrants chose. A zero contribution would mean that the prospect did not respond to the solicitation.

However, the three highest finishers (all very close in their results) tried a more sophisticated strategy. They recognized that there was another response variable in the data set—a flag that identified those customers who had responded. So the successful miners used this variable to first build a model that would predict the probability, or likelihood, of response. They then built a model using only those people who had made contributions to predict the amount of a

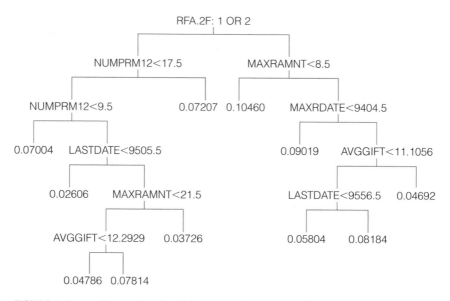

FIGURE 4 A tree diagram for classifying responders and nonresponders. The first split is on the variable *RFA2F*. Customers with values 1 or 2 on this variable fall to the right of the split; the others fall to the left. Each variable is split in a similar way. The final numbers at the ends of the branches show the proportion of the responders for that branch. For example, 10.46% of the customers who had values 1 or 2 on *RFA2F* and whose value on *MAXRAMNT* was less the 8.5 responded to the solicitation.

donation. Using the fact that it costs \$0.68 to send the solicitation, the modelers then combined these two models to predict the net profit or loss from each potential donor.

A modeling tool called a *classification tree* is often used to start the modeling process. Figure 4 shows part of a tree model run on the PVA data to predict whether a donor will respond to the next mail campaign. Table 1 describes the variables used in Figure 4.

The numbers at the end of the last branches show the probability of responding, given the values of the variables described in Table 1. For example, the first variable, RFA2F, is a measure of the frequency with which the donor has contributed in the past. If he or she gave either 1 or 2 gifts, move to the left, otherwise follow the branches to the right. In general, we move down to the left if the statement is true, down to the right if the answer is false. For example, a donor who gave 2 gifts in the past 12 months and had 22 promotions in the past 12 months would move left at the first branch (RFA2F = 1 or 2)

Variable	Description
RFA2F	Number of gifts received by PVA from the donor in the past 12 months
NUMPRM12	Number of promotions received by the donor from PVA in the last 12 months
MAXRAMNT	Dollar amount of the largest gift to date
MAXRDATE	Date associated with the largest gift to date
LASTDATE	Date associated with the most recent gift
AVGGIFT	Average dollar amount of gifts to date

TABLE 1 Variables selected by the tree algorithm

and then would move to the right branch where NUMPRM12 > 17.5. Such a donor would be predicted to respond to the next campaign with probability 0.07207.

This model may be used as the model itself, or a tree may be used just to screen the *potential* predictor variables for further exploration. Much of the choice depends on the time that the analyst has, and the level of expertise with both the data set and the various data mining tools that the analyst brings to the analysis. Next, the data miner may try to find the tree of best size. This might be accomplished by splitting up the data into training and test sets, fitting trees of *different sizes* to the training set, and choosing the size that did best in estimating the response in the test set. By different sizes, we mean different levels of depth of the tree. For example, in Figure 4, there are six levels of branches to the tree. Would more branches help predict more accurately?

Another approach would be for the data miner to start the modeling process by looking at plots of the 40 variables in more detail, deciding if transformations of the variables seemed appropriate before proceeding with another modeling technique, such as logistic regression. The myriad models available to the data miner and the vast number of combinations of predictors that can be used make the process of choosing a model a complex and fascinating challenge.

Results

To model the PVA data set, many strategies were employed. To test the models, teams were asked to predict which of 100,000 people in a test set should get the next mailing. These 100,000 had the same information as the original 100,000 for the training set, except, of course, it was not revealed whether they responded or how much they gave. The results were evaluated based on the predicted profit utilizing each team's proposed strategy.

If no strategy was used and the materials were sent to all 100,000 people, a net profit of $10,500 would be realized. The team with the best results on the test set increased that by 40%. They used a strategy of considering five versions of each predictor variable (for example, the variable itself and the logarithm of the variable were two of the versions). They then used an automated procedure to select the variables that provided the best prediction of whether someone responded or not using a logistic regression. Then, they focused only on those people predicted to respond, and used a similar procedure to produce a linear regression of *how much* each person would give. They then used a cutoff based on the predicted probability of responding multiplied by the projected donation to decide who should get the offer.

Other teams used a variety of methods, including tree models, neural networks, and other modeling tools. The fact that logistic and linear regression did best in this case should not be taken as proof that these methods always perform better. In fact, several studies have shown that for nearly every method you can find a data set for which that method will be the best. So, the data miner is best off using a variety of methods, remaining open to what the different models have to say, and using a model, or combination of models, that does best on a test set.

SUMMARY

We set out to improve the profitability of the PVA's direct mail campaigns. By better targeting to whom the solicitations should be sent, data mining resulted in not only as much as a 40% improvement in profit but a reduction of almost 45% in the number of mailings.

The data mining process requires examining, preparing, and visualizing the data before a predictive model can be built. Because there can be hundreds or thousands of potential predictor variables

(481 variables in the case of the PVA data), finding a relatively small set of predictors to use in a model can be challenging. Fortunately, the collection of methods and tools encompassed by data mining enables us to understand the relationships between these variables and build useful predictive models.

REFERENCES

Berry, M., and G. Linoff. (2004). *Mastering Data Mining Techniques*, 2nd ed. New York: Wiley.

Breiman, L., J. Freidman, R. Olshenand, and C. Stone. (1984). *Classification and Regression Trees.* New York: Chapman and Hall.

Cleveland, W. S. (1994). *The Elements of Graphing Data,* rev. ed. New York: Chapman and Hall.

Duda, R. O., P. E. Hart, and D. G. Stork. (2000). *Pattern Classification.* New York: Wiley-Interscience.

Edelstein, H. (2004). *Data Mining Technology Report.* Potomac, Md.: Two Crows.

———. (2004). *Introduction to Data Mining and Knowledge Discovery,* 4th ed. Potomac, Md.: Two Crows.

Hand, D. J., H. Mannila, and P. Smyth. (2001). *Principles of Data Mining.* Cambridge, Mass.: MIT Press.

Hastie, T., R. Tibshirani, and J. H. Friedman. (2001). *The Elements of Statistical Learning: Data Mining, Inference, and Prediction.* New York: Springer-Verlag.

Quinlan, J. R. (1992). *C4.5: Programs for Machine Learning.* San Mateo, Calif.: Morgan Kaufmann.

Wainer, H. (1997). *Visual Revelations.* New York: Springer-Verlag.

QUESTIONS

1. Transaction files are accumulated from credit card usage by banks, websites, mail-order clubs, catalog stores, and so forth. They also collect personal information such as number of children, income, age, and other demographic information. In order to predict the

buying inclinations of particular customers, these various data sources must be linked. Why is this difficult to do?

2. What is the difference between the OLAP approach to extracting information from a data file and the technique described in the essay as "data mining"?

3. What is the main difficulty in finding a good model when starting with 481 predictor variables?

4. What strategy was used in the model construction process by the winners of the competition in the PVA example?

5. In the classification tree (Figure 4), one tree ends at the probability 0.07004. What event has this probability?

6. Why is it advisable to split the available data into two or more separate data sets when building and evaluating a predictive model?

IMPROVING THE ACCURACY OF A NEWSPAPER

A Six Sigma Case Study of Business Process Improvement

RONALD D. SNEE

Performance Excellence

———— ❉ : : ❉ ————

Just before the turn of the 21st century, American companies such as GE, Honeywell, Motorola, DuPont, American Express, Ford, and many others, large and small, began reporting large improvements in business performance that produced billions of dollars in bottom-line savings. The strategy and methodology used to produce these improvements is called Six Sigma, a methodology based on statistical thinking and methods. Six Sigma continues today as the best-known approach to quality and business improvement. Although Motorola is credited with "creating" Six Sigma in 1987, the truth is that quality and business improvement efforts have been evolving steadily over the last hundred years. In this essay, we will see how Six Sigma was used to improve newspaper accuracy and then look briefly at how Six Sigma evolved by enhancing earlier process-improvement efforts.

A CASE STUDY

The easiest way to see how the Six Sigma methodology and tools work is through a real case study. The following discussion looks at one newspaper's efforts to reduce errors, and it is an elaboration of the "Newspaper Accuracy" case from Hoerl and Snee (2002). While this case is about reducing errors in newspaper publishing, it is illustrative of reducing errors in processes in general.

The five major phases of Six Sigma projects are (1) define, (2) measure, (3) analyze, (4) improve, and (5) control (DMAIC). These phases are discussed in the context of newspaper accuracy, with emphasis on the tools used in each phase, the purpose of each tool, and what was learned from the application of each tool. Some, but not all, Six Sigma tools are statistical in nature; the others focus primarily on the business process itself. Even though the tools in the second group are not considered statistical tools, they too are based on statistical concepts.

Define Phase

In the define phase a project is selected and the specific problem to be solved and process to be improved are identified. The resulting project and its objectives are summarized in what is called the project charter, a key tool of the project definition phase. A common cause of failure in improvement projects is for different people to have different understandings of what the project will do and accomplish. This can lead to disappointment and finger-pointing at the end of the projects. In the newspaper accuracy study the project leader established that error reduction was an important issue.

Some baseline data showed that while the newspaper's copy desk might catch and fix as many as 30–40 errors per day, a typical daily error rate was 20 errors. A goal set for this project was to reduce errors caught at the copy desk by 50% to fewer than 10 per day on average. The financial impact of correcting an error was established as $62.00 if caught at the copy desk, $88.00 if caught at the composing room, $768.00 if a page had to be redone, and $5000.00 if the presses had to be stopped and restarted. Of course the cost of an error that makes it all the way to the published paper is unknown, but is thought to be significant. Clearly, the earlier an error is caught, the better. Preventing errors altogether is better still.

Before any data were collected, a formal definition for an "error" was created so that the data would be accurate and everyone would be talking about the same thing when errors were discussed. An error was defined as (1) any deviation from truth, accuracy, or widely accepted standards of English usage, or (2) a departure from accepted procedures that causes delay or requires reworking a story or a graphic. Since different types of errors have different consequences, it was also decided to divide the errors into nine categories: misspelled words, wrong number, wrong name, bad grammar, libel, word missing, duplicated word, fact wrong, and other. The 11-person team consisted of the six sigma expert, the editor, two copy editors, two graphics editors, one reporter, and four supervisors.

Measure Phase

The measure phase is intended to ensure that you are working on improving the right metric, one that is truly in need of improvement and that can be measured well. In the measure phase, based on the objectives of the project and the customer needs, the appropriate process outputs to be improved are selected. Acceptable performance is determined and baseline data is gathered to evaluate current performance. This work includes the evaluation of the performance of the measurement system, as well as the performance of the process being studied.

Some of the tools typically used during the measure phase include the process map, cause-and-effect diagram, measurement system analysis, capability analysis, and control chart analysis of the baseline data. The process map is a schematic diagram of the steps in the process that also indicates the process inputs and outputs. It provides a picture of the process and also enables the identification of non-value-added work. For example, warehousing finished goods adds no value from a customer point of view, but some level of warehousing is typically needed to effectively supply customers.

Important process input and output variables are usually identified during the process mapping work. The newspaper writing and editing five-step process map is shown in Figure 1. If further detail is needed for a few key steps, these steps can be mapped further by creating substeps. Note the "revision cycle" in Figure 1, which may be a

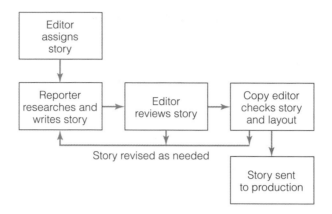

FIGURE 1 Newspaper writing and editing process map

source of non-value-added work. The size of the paper (number of pages), number of employees absent each day, and a major change in front-page story (yes, no) were identified by the team as being among the variables that could have an effect on errors.

A cause-and-effect diagram (sometimes called a fishbone diagram, for reasons obvious when you look at the diagram!) is a graph designed to show the process variables thought to have the greatest effect on the process output variables. The cause-and-effect diagram enables the team to study the relationships between the input variables and the output variables. In the case of the newspaper accuracy study, the key output variable is the number of errors. The cause-and-effect diagram is shown in Figure 2. Sometimes the input variables are also rated in terms of their relative impact on the process output variables, but that was not deemed necessary for this project.

A measurement system analysis is a study that focuses on the repeatability and the reproducibility of the measurements that will be made as part of the process improvement effort. In this case the measurement system analysis consisted of developing the errors collection scheme and validating it.

A process capability study was also conducted as part of the measure phase. A process capability study measures how well the process is capable of meeting the customer expectations or product specifications. Such studies are often done using control charts. A control chart is a plot of data over time with statistically determined limits

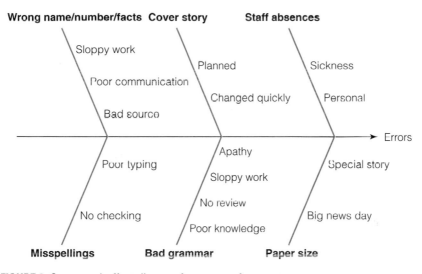

FIGURE 2 Cause-and-effect diagram for causes of newspaper errors

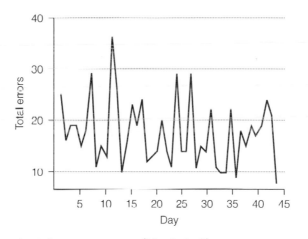

FIGURE 3 Run chart of newspaper errors (March–April)

that reflect typical variation. A run chart is similar to a control chart, but it does not have the statistically determined limits. Figure 3 shows a run chart of the errors data. From this chart, we can see that the number of errors appears to have been stable over time, with an average value of approximately 20 errors per day and daily variations from just below 10 to just below 40.

Analyze Phase

The analyze phase helps you avoid the "ready, fire, aim" approach by accurately diagnosing the root causes of problems prior to taking action. That is, we want to avoid the temptation to jump to solutions prior to properly identifying root causes. In the analyze phase you evaluate the baseline data to identify root causes of the observed variation and defects. Additional data are collected as needed. Two of the most commonly used improvement tools in the analyze phase are multivariable studies and failure modes and effects analysis. Multivariable studies are process studies in which we collect data on the key process and input variables, as well as the key outputs, without actually intervening in the process to introduce changes. The data are then analyzed using graphical and statistical tools (e.g., regression analysis, hypothesis testing, etc.) to identify the variables having the most significant impact on the output variables. Failure modes and effects analysis is a disciplined methodology for identifying potential defects and taking proactive steps to avoid them prior to actual occurrence.

When data on errors are evaluated, a Pareto analysis is often used to depict which categories of errors are causing the major problems. The Pareto chart (Figure 4) is just a bar graph where the bars are

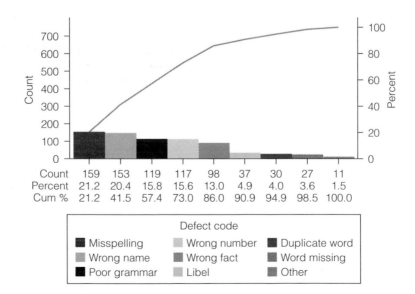

Count	159	153	119	117	98	37	30	27	11
Percent	21.2	20.4	15.8	15.6	13.0	4.9	4.0	3.6	1.5
Cum %	21.2	41.5	57.4	73.0	86.0	90.9	94.9	98.5	100.0

Defect code

■ Misspelling	▨ Wrong number	■ Duplicate word
▨ Wrong name	■ Wrong fact	▨ Word missing
■ Poor grammar	▨ Libel	▨ Other

FIGURE 4 Pareto chart for newspaper errors (March–April)

ordered either by number of occurrences or by severity. Knowledge of the important error categories can suggest root causes of errors. Often only a few categories account for the majority of the errors. This is the case for the newspaper errors study. In Figure 4 we see that the majority of the errors during the March–April time period were due to misspellings, wrong names, numbers, and facts, and poor grammar.

Attention was first focused on the categories producing the largest number of errors: misspellings, wrong names, numbers, and facts, and bad grammar. One root cause identified was that the reporters were not using a spell-checker. The attitude was, "I don't have time to spell-check. Besides, the copy editors will catch the errors anyway." The reporters were also not routinely checking their facts and their sources, which was a job requirement. The improve phase addresses how to deal with these root causes.

A multivariable study is conducted to identify variables that may be producing the errors. In the newspaper case, the variables studied were those identified in the measure phase—the size of the paper, number of employees absent, and major changes in the front-page story. While the size of the paper is controllable, the other two variables are not controllable. Day of the week (M, T, W, Th, F) and month of the year (March–December) differences were also studied. Day-of-week and month-to-month differences, if present, provide clues to other causes of errors. Work teams often perform differently on Mondays and Fridays than on the other days of the week. Analysis of the data indicated that the only variables that had an effect on errors were the size of the paper (more pages lead to more errors), and a change in the front-page story (new stories had to be created under tight schedules, increasing the error rate).

Improve Phase

In the improve phase you figure out how to change the process to address the root causes identified in the analyze phase, and thereby improve the performance of the process. Each process change is tested with a confirmatory study to verify that the predicted improvements in fact happen. Several rounds of improvements may be required to reach the desired level of performance. Note that the improve phase is the only phase in the process that actually makes improvements. The other phases are intended to properly set up (the define, measure, and analyze phases) and maintain (the control phase) the improvements from the improve phase.

In the newspaper case, it was reaffirmed that reporters had the responsibility to check the accuracy of their articles. Three job aids were also created: a "Spell-Check How-To," a list of "Ten Rules of Grammar," and the "Pyramid of Trust," which detailed the sources that could be trusted to produce accurate names, numbers, and facts. These new working methods were communicated in an "all hands" meeting in July. The importance of being careful when the front-page story changed with little notice, and publication of large-size editions of the newspaper, were also discussed. The interim goal of producing fewer than 10 errors per day reaching the copy desk was also reviewed and reaffirmed.

It was now time to test whether the changes were in fact having an effect. One month went by, and the data for the month of August were analyzed. The result was that total errors had not changed! Why were the errors still high? It was learned that the new procedures were not being used. Many employees did not feel that management was serious about the changes, and therefore did not take the changes seriously. The editor reiterated that the new procedures were to be used and that the management team was expected to lead this new way of working. Another "all hands" meeting was held to address the issue. One month later, when the latest data were analyzed, the total errors had dropped significantly (Figure 5). After one more month, the total errors had dropped by approximately 65%, as compared to the goal of a 50% reduction! The new procedures were clearly working. Of course, further improvement is always possible.

Control Phase

In the control phase you install a system to ensure that the improved performance of the process is sustained after the Six Sigma team has moved on to another project. The key tool of this phase is the control plan, which includes control charts and long-term capability studies.

Returning to our newspaper case, a control plan was put in place to hold the gains of the work done to date, and to keep the errors at

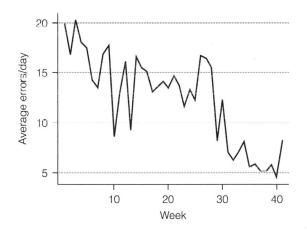

FIGURE 5 Weekly average newspaper errors (March–December)

the level obtained. The control plan specified the following measures to be monitored using control charts:

- Total errors
- Errors by category
- % Articles checked by the author
- % Articles spell-checked

The latter two measurements were found to be particularly useful in detecting when the reporters were not following procedures and thus errors could be a problem.

Checklists were also created, and roles and responsibilities—including backups—were defined to reduce hand-off problems between departments. This enabled people to view their work processes as part of an overall system. Work was also initiated to identify the sources of errors in graphics.

Results

This work resulted in errors being reduced 65%, producing quantifiable savings of more than $226,000 per year. The actual savings were even greater because of the fringe benefits not quantified, such as:

- Fewer missed deadlines, including the ability to deal effectively with extremely tight deadlines.

- Improved morale at the copy desk. Copy editors were freed up to make better use of their talents and training.
- Re-keying of names (rework) reduced.
- More efficient and less costly sources of information were found, resulting in reduced errors and less number-input time (less keying of data). News assistants were freed up to do more valuable work.

SIX SIGMA: A MAJOR STEP IN THE HISTORY OF STATISTICS IN QUALITY

So how did Six Sigma arrive on the scene? Where did it come from? Is Six Sigma "the answer," or will it also evolve and morph into something else? Our story begins around a hundred years ago with the work of Frederick Taylor, done shortly before and after the turn of the 20th century. Taylor concluded that the best way to improve productivity was to separate the doing of work from the planning and improving processes used to enhance the work. Thus he created the labor (do the work)–management (plan and improve the work) model for operating an organization, which exists today in most companies.

Since then, many statisticians have developed methods that have been successfully applied in process improvement efforts. These include

- The development of the *t* test by W. S. Gossett, writing under the pseudonym "Student." The *t* test is still used today to test the statistical significance of process improvements. (1908)
- A statistical approach to the design of experiments and analysis of variance techniques for analyzing data from experiments, developed by Sir Ronald Fisher. (1920s)
- The use of control charts to monitor the performance of manufacturing processes. Control charts were developed by Walter Shewhart. (1920s)
- The methods of acceptance sampling by Harold Dodge and Harry Romig. This methodology uses information resulting

from the inspection of a sample to decide if an entire shipment is acceptable or not. (1920s and 1930s)

- The adaptation of Fisher's design of experiment methods to industrial optimization by George Box and his coworkers. (1950s and 1960s)
- The development of exploratory data analysis (EDA) by John W. Tukey and his coworkers at Bell Laboratories. EDA emphasizes rigorous examination of data graphically prior to rushing to numerical calculations. (1950s and 1960s)

This work led to the widespread use of statistical process control (SPC) methods in the late 1960s and early 1970s, particularly in the auto industry. Through SPC, which combines a variety of tools such as control charts, statistical evaluation of process capability, and Pareto charts to identify sources of problems, it became widely recognized that product quality was directly related to the process that produced the product, and so the quality improvement efforts broadened to include process improvement as well as product improvement. Later, in the early 1980s, many companies adopted what became known as total quality management (TQM). TQM built on prior quality improvement methods and focused on management leadership, people and teamwork, process improvement and management, and serving customer needs.

In the mid-1980s, perhaps due to the influence of TQM, it became apparent that statistics needed to be viewed from a broader context in order to have significant impact on quality or business performance. In other words, if organizations could not change people's attitudes toward customers, teamwork, scientific approaches to improvement, and so on, statistical tools would not be effective. They either wouldn't be used or they would be used inappropriately—for example, to prove things people already knew, rather than to discover new things and make business improvements. This broader viewpoint of statistical tools has been referred to by different names but perhaps most frequently as "statistical thinking." This phrase is intended to imply that statistics should alter our way of thinking about the world around us, rather than just being a set of technical tools.

In 1996, the American Society for Quality defined statistical thinking as "a philosophy of action and learning based on three fundamental principles: all work occurs in a system of interconnected processes; variation exists in all processes; [and] understanding and reducing variation are keys to improvement."

The first point notes that we have the mental ability to view work critically and objectively; that is, we can step back from any activity and study it analytically. This skill is critical, because once people obtain it, they immediately see the necessity of data and statistical methods to study and improve their work. This point also implies that people need to understand how different processes are interconnected within an overall business system, so that they can resist the temptation to solve a problem in process A by simply transferring it to process B. For example, procurement may be tempted to solve its budget problems by buying cheaper raw materials, even though this might cause huge problems in manufacturing.

The second point ("variation exists in all processes") highlights the fact that variation is omnipresent in our world. Everything from how long it takes to get to work to how well people or machines perform is subject to variation from many causes. Once we recognize this fact, we see the need for statistical methods, because statistics is the science that allows us to draw conclusions in the presence of variation.

The third point ("understanding and reducing variability are keys to improvement") emphasizes the fact that in many cases, we don't just want to understand or study variation, we want to eliminate it! Insurance companies would like to eliminate variation between underwriters (i.e., have every underwriter perform as well as the best underwriter). Manufacturers would like to produce consistent products. Hospitals would like to provide consistent care to patients, rather than treating one well and another poorly. While there are certainly exceptions to this rule (e.g., no one wants to eat the same thing for dinner every night), it is the realization that we need to not only understand but also reduce unintended variation that enables statistical methods to be used proactively for improvement, rather than reactively to prove what is already known.

One of the challenges of using statistical thinking is getting the approach used throughout an organization. Six Sigma provided an overall structure and some roadmaps and tools to guide this effort. It has therefore become a popular means of actually implementing statistical thinking. It is important to keep in mind, however, that Six Sigma naturally evolved from TQM and other previous advances by, among other things, filling in four specific gaps. These are

1. Placing a clear focus on producing bottom-line results ($).
2. Integrating the human and technical aspects of improvement, that is, the "hard" and "soft" sides of improvement.
3. Providing an overall process of scientific inquiry (DMAIC) to help practitioners link and sequence the tools.
4. Providing a supportive infrastructure of formal roles and responsibilities, training systems, budgets, and project selection and review processes. This infrastructure provides "gas" to the initiative, and keeps it from getting stalled after the initial wave of excitement has worn off.

TO BE CONTINUED

The appropriate way to end this essay is to say that the story is "to be continued." The desire to grow and prosper is universal, and Six Sigma improvement efforts are currently thriving, with applications in manufacturing, research and development, sales and marketing, service processes, and many others. The ongoing need to improve will no doubt create a need to develop new and even better methods in the future. The one thing we know for sure is that processes experience unintended variation and always will; hence, statistical thinking and methods will continue to be an integral part of quality improvement methods.

ADDITIONAL READING

Box, G. E. P., W. G. Hunter, and J. S. Hunter. (1978). *Statistics for Experimenters*. New York: Wiley.

Breyfogle, F. W. (1999). *Implementing Six Sigma: Smarter Solutions Using Statistical Methods,* 277–80. New York: Wiley.

Creveling, C. M., J. L. Slutsky, and D. Antis, Jr. (2003). *Design for Six Sigma in Technology and Product Development.* Upper Saddle River, N.J.: Prentice Hall.

Deming, W. E. (1982). *Out of the Crisis.* Cambridge, Mass.: MIT Center for Advanced Engineering Study.

Hoerl, R. W., and R. D. Snee. (2002). *Statistical Thinking: Improving Business Performance.* Pacific Grove, Calif.: Duxbury Press.

Snee, R. D. (1990). "Statistical Thinking and Its Contribution to Total Quality." *The American Statistician* 44, no. 2:116–21.

Snee, R. D., and R. W. Hoerl. (2003). *Leading Six Sigma: A Step-by-Step Guide Based on Experience at GE and Other Six Sigma Companies.* Upper Saddle River, N.J.: Financial Times–Prentice Hall.

QUESTIONS

1. Shewhart invented the control chart in the 1920s to help process operators monitor and improve the performance of processes. The Six Sigma process described by Snee is a much more general improvement approach than simply using a control chart. Can you think of a reason why the control chart itself is considered a key element in the Six Sigma improvement process?

2. Reduction of errors in the newspaper case study was proposed to reduce production costs. What other reason is there for a newspaper to reduce errors, and how important do you think error reduction would be for this other reason?

3. Consider how the Six Sigma process (DMAIC) might be applied to improving the operation of a hot dog stand? Select an improvement opportunity, or need, and then identify at least one item or activity that might occur in each of the phases of DMAIC in identifying and implementing the process improvement.

4. In the fishbone diagram, there are items such as "cover story," "sloppy work," and "errors." What is the generic category for each of these items—that is, in fishbone diagrams for other processes, what categories of items do you need to record?

5. Why use graphical displays like Figures 1–5 in improving processes using the Six Sigma approach? What is the value of graphical displays?

6. What does a Pareto analysis do? Why is it useful? Who was Mr. Pareto?

7. The multivariable study described in the analyze phase mentions that some causal variables are controlled and some are not controlled. Why is it important to distinguish between controlled and uncontrolled variables in the Six Sigma improvement approach?

8. How is product quality different from process quality? Is product quality linked to process quality in any way? Explain with reference to the newspaper case study.

9. The essay makes frequent reference to "variation," such as, "Variation exists in all processes." Identify two or three examples of variation in the newspaper context of "variation."

10. "Understanding and reducing variation are keys to improvement." But how does variation reduction improve a company's profit? What is the link between "variation" and "profit"?

ASSURING PRODUCT RELIABILITY AND SAFETY

NECIP DOGANAKSOY

GE Global Research

GERALD J. HAHN

GE Global Research (retired)

WILLIAM Q. MEEKER

Iowa State University

———————— ⁂ : : ⁂ ————————

Are you buying a new automobile or laptop computer or CD player? What are the top three things that you wish from your purchase? We bet your list includes long-term, trouble-free operation. Most likely, you want your new product to operate flawlessly for many years to come. In our consumer-oriented society, your penchant for reliability sends a clear message to manufacturers that wish to delight customers, earn their repeat business, have them recommend the product to friends—and avoid lawsuits.

Reliability is quality over time. Safety, moreover, is closely related to reliability. We are all keenly aware of catastrophes that can result from poor reliability. The 1986 *Challenger* and 2003 *Columbia* space shuttle disasters and the Ford Explorer–Firestone tire failures of the 1990s are some well-publicized examples. At home, we are concerned with fire hazards created by anything ranging from a toaster malfunction to faulty wiring. Injury caused by possibly careless operation needs to be considered by manufacturers, and is especially critical for products that might be used "imaginatively"—a particular concern for the makers of children's products. Manufacturers don't intend for small pieces from toys and games to be eaten, but sometimes infants don't read the labels!

You know the price of a new car. You can evaluate its appearance by looking at it and its performance by test-driving it. Assessing its reliability, however, is not that easy. You frequently have to rely on consumer-review magazines or friends' recommendations. It is especially important that a product is reliable when first brought to market—a few early dissatisfied customers can give it a bad name.

But what does all of this have to do with statistics and statisticians? Plenty! Statisticians help designers make sure, before a new product is released, that it meets or exceeds customers' reliability expectations. And they often need to do so in a compressed time frame. For example, the manufacturer of a newly designed washing machine wants its product to operate failure-free for three, five, or even 10 years—but might have only six months to show this.

In this essay, we describe how statisticians help designers in the formidable task of ensuring high reliability in product design. We begin with an examination of the 1986 *Challenger* space shuttle disaster and how proper statistical analyses could have saved the day. We then use the washing machine example to illustrate up-front reliability validation— first for the motor, and then for the entire product. Similar concepts apply to more complex systems—such as medical scanners, aircraft engines, locomotives, and space vehicles—as we will briefly indicate. Finally, we show how statistical approaches to "just-in-time maintenance" help remove or reduce some of life's uncertainties.

RELIABILITY IN THE NEWS

We don't hear much about reliability when things go right—which, fortunately, is most of the time. It is, after all, not news that your plane landed safely. Unfortunately, the unexpected sometimes happens— especially in dealing with complex systems. Problems can frequently be avoided, however, or their impact reduced, if the right data are obtained in advance and analyzed properly.

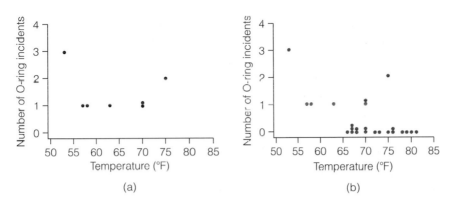

FIGURE 1 Plots of number of O-ring incidents per flight versus launch temperature on previous flights, (a) excluding and (b) including flights with no incidents

The 1986 *Challenger* space shuttle disaster provides a telling example, according to the report of the presidential commission that investigated this event (Dalal, Fowlkes, and Hoadley, 1989). The forecast for the launch date, January 28, was for temperatures in the low 30s, even though past launches had all been at temperatures above 50°F. The low predicted temperature was believed by some to be detrimental to the O-rings that were used to hold critical joints in place during launch.

A plot of the number of "unfavorable incidents" per flight versus launch temperature for previous flights, similar to the one shown in Figure 1a, was developed and presented to flight management. This plot provided no clear evidence of a relationship between temperature and O-ring "incidents." As a result, NASA proceeded with the launch the next day at a temperature of 31°F. Regrettably, the shuttle blew up and the entire crew was killed.

In the subsequent review it was determined that the critical analysis based on Figure 1a, substantiating the launch, was erroneous. This was because 17 additional past flights that had *no* O-ring incidents were incorrectly deemed to be irrelevant, and were excluded from the analysis. Figure 1b shows all the data, including the flights with no incidents. This plot suggests a strong association between temperature and number of O-ring incidents, with low temperatures being particularly risky.

Sadly, the 2003 *Columbia* shuttle disaster was also, at least in part, attributable to obtaining insufficient data and not paying enough attention to the data that were available. Some other well-publicized cases involving poor reliability of manufactured products or systems include

- The August 2003 power-grid shutdown, resulting in loss of power to tens of millions of homes in the eastern United States and Canada ("The Blackout of 2003: Across the Country; Disruptions Big and Small in a Swath of the Midwest," *New York Times,* August 15, 2003)
- Rollovers of Ford Explorer sport utility vehicles equipped with Bridgestone/Firestone tires, claiming nearly 90 lives and causing 250 injuries ("Tire Failures, SUV Rollovers Put Quality on Trial," *Quality Progress,* December 2000)
- Poor design of a refrigerator compressor, forcing GE to replace 1.1 million compressors ("Chilling Tale: GE Refrigerator Woes Illustrate the Hazards in Changing a Product," *Wall Street Journal,* May 7, 1990)

In each of these cases, investigations suggested that the right data analyses and prompt action could have eliminated or appreciably mitigated the problem. Many other reliability failures don't make the newspapers and don't have disastrous consequences, but nonetheless can still cause appreciable and unexpected inconvenience.

WASHING MACHINE EXAMPLE

Reliability Goals

A company's marketing department has concluded that sales for a newly designed washing machine could be increased by 50% if sufficient improvement in reliability can be gained over the current design. Such an improvement would result in increased sales of more than $1 billion over a five-year period—assuming the information gets out in time by advertising and word of mouth.

To achieve a 50% sales increase it was felt that failures for the new design must be fewer than

- 1 in 100 (or a reliability of 99%) during the first year of operation
- 1 in 50 during the first five years
- 1 in 10 during the first 10 years

The engineering department indicated that with an intensive effort, such improvements were feasible, and management signed on enthusiastically.

But how can we verify that the reliability objectives are being met prior to product release and large-scale production? The 10-year requirement will be the hardest to demonstrate, and we will focus on it in our discussion. Meeting this goal should also enhance the chances of meeting the other goals, and also reduce expenses on extended long-term warranty policies.

Reliability Model

A product's reliability is defined as the probability of it being failure-free for a specified period of time. As part of the design process, the engineering department developed a model for washing machine reliability. The goal is to have the system reliability, denoted by R_{system}, to be at least 0.90 after 10 years. The first step is to break down the system into independently operating subsystems, assemblies, or components. From engineering considerations and experience with previous designs, seven critical areas where failures could occur were identified. These were

1. The motor
2. The main bearing
3. The washer brake
4. The cycle/timing control
5. The water temperature control
6. The hot-water valve
7. The cold-water valve

We will refer to these as the system's "components," and use R_1, R_2, R_3, R_4, R_5, R_6, and R_7, respectively, to denote the 10-year reliabilities (i.e., proportion surviving without failure) for these seven components. The goal that at least 90% of the washing machines operate for 10 years (i.e., $R_{system} \geq 0.90$) requires that *all* seven components operate failure-free for 10 years in at least 90% of all washing machines. The machine fails as soon as the first one of these components fails. This is referred to as a "series system," and is sketched in Figure 2.

FIGURE 2 System reliability diagram with seven components in series

It is reasonable to assume in this example that components operate and fail independent of one another. Under this assumption, the proportion of washing machines that will survive 10 years can be obtained from the individual component reliabilities according to the following formula (from elementary probability):

$$R_{system} = R_1 \times R_2 \times R_3 \times R_4 \times R_5 \times R_6 \times R_7$$

Thus, to meet the goal of having 10-year system reliability $R_{system} \geq$ 0.90, the individual values R_1, R_2, R_3, R_4, R_5, R_6, and R_7 must each be appreciably larger than 0.90. If one of the components has a reliability of only 0.90, each of the other components needs to have a perfect reliability of 1.0—not an easy objective!

To develop a plan to meet the overall reliability goal, engineers often start with a "reliability budget," specifying target reliabilities for each component. Generally, these will not be the same for all components. This is because some components are more likely to fail than others; for example, the hot-water valve is expected to fail earlier than the cold-water valve.

Based on experience with similar components in previous models of the product, and through negotiation, the design engineers assigned the individual target values $R_1 = 0.97$, $R_2 = 0.99$, $R_3 = 0.99$, $R_4 = 0.985$, $R_5 = 0.985$, $R_6 = 0.98$, and $R_7 = 0.995$ to the 10-year reliabilities of the seven components. If these goals are met, the overall system reliability will be

$$R_{system} = 0.97 \times 0.99 \times 0.99 \times 0.985 \times 0.985 \times 0.98 \times 0.995 = 0.903$$

If one or more of the component reliabilities falls below its goal, then system reliability can be met only if one or more of the other components exceeds its target by a sufficient amount.

So how do we go about estimating the actual component reliabilities? So far, we have stated only what these need to be, not what they actually are. We will focus initially on one of the components—the motor.

Assessing Motor Reliability

For some components, reliability may be validated from previous experience. The motor, however, was completely redesigned to reduce noise and to improve reliability. Skilled engineers had worked hard to design a motor to meet the new goals, using top-quality materials and state-of-the-art design practices. They reviewed information on past motors to correct previous reliability problems. They also performed short, high-stress tests (known as highly accelerated life tests, or HALTs), subjecting components and a few prototype motors to carefully chosen temperature cycling, vibration, and over-voltage conditions, to discover, understand, and remove new failure modes. Units would still fail due to the wearing out of the bearing in the motor, but for the vast majority of machines this was not expected to happen during the first 10 years of life.

The engineering department was therefore confident of having developed a high-reliability motor. But had they? Would 97% of the motors last 10 years, as required? In light of the extensive changes in design, previous experience and engineering judgment provided only baseline estimates. To really find out, the engineers needed to conduct a statistically valid life test on a sample of new motors at conditions that simulate customer use.

The purpose of such testing is, first, to identify and eliminate any reliability problems that somehow were overlooked—and to do so as rapidly as possible. Second, it is to demonstrate with a high level of confidence that the desired reliability goals will be met. Moreover, all this has to be done in six months or less to meet product release goals!

GENERAL TESTING STRATEGY

How can we obtain the equivalent of 10 years of field experience in six months? We do so by running a sample of motors *continuously* at stresses that simulate their operation in washing machines, shutting down for only a brief cooldown between periods of continuous running (such on–off cycles cause temperature changes that excite certain failure modes). This allows us to run 24 cycles per day, simulating 3.5 years of field operation in each month of testing—presuming a use rate of four washes per week. All of this assumes that failures are the result of the actual running of the motor, independent of the elapsed time. This assumption seemed reasonable based on engineering considerations.

Statistics enters in two important ways. First, statistical concepts are used to develop a specific test plan to obtain the needed data. Then, statistical methods are used to analyze the results to assess whether the reliability goals are being met, and to help identify areas of improvement if they are not.

THE TEST PLAN

A test of 66 motors for six months (equivalent to 21 years of field operation) was proposed. Statistical evaluations showed this to be an appropriate sample size. A larger sample would have provided greater precision, but would require more time and money. Statistics helps us balance precision of the information obtained against cost.

The motors to be used for the test need to reflect, as closely as possible, the variability expected in large-scale production. Thus, the 66 motors were built at three different times, using multiple lots of materials for each of the key assembled parts. Also, good records of motors' histories and their test performance were maintained.

RESULTS AFTER ONE MONTH

After one month of testing (728 test cycles, equivalent to 3.5 years of customer use) four of the 66 motors had failed—to everybody's surprise and consternation. This resulted in an estimated failure probability of 6% (i.e., $(4/66) \times 100$), or an estimated reliability of 94% in 3.5 years—falling far short of the goal of 97% reliability in 10 years. This result illustrates the fact that it is often easier to show inability to meet a reliability goal than to demonstrate high reliability! Physical evaluation indicated that all failures were due to a manufacturing defect that, if present, would assert itself early in life. In fact, all four failures occurred during the first week of testing. Fortunately, once aware of the problem, it was easy for the manufacturing department to fix it and ensure that it would not occur on future product.

RESULTS AFTER THREE MONTHS

After three months of testing (2184 test cycles, equivalent to 10.5 years of customer use) four more failures occurred. One bearing failure took place at 1989 cycles. In addition, and unexpectedly, there were three failures (at 932, 1397, and 1558 cycles, respectively) due to the malfunction of a plastic part. Clearly, this further failure mode needed to be eliminated in order to meet the reliability goal. The design team felt confident in its ability to do so by changing the geometry of the plastic part. Thus, it was decided to continue the test with the existing units, and a separate program was initiated to validate the design fix for the plastic part.

Assuming that both the manufacturing defect and the plastic part failure fixes are completely successful, and that no additional failure modes are introduced in the process, the only relevant failure out of 59 test units is the one bearing failure. This results in an estimated 10.5 year motor failure probability of 1.7% (i.e., 1 in 59), or an estimated reliability of 98.3%, exceeding the 10-year reliability goal of 97%.

This estimate, however, is based upon testing a relatively small sample of motors and is therefore subject to much statistical uncertainty. A statistical confidence bound takes this uncertainty into consideration. A lower 95% confidence bound on reliability was calculated to be 92.2%. This means, roughly speaking, that we can claim with 95% confidence[1] that the 10.5-year motor reliability is at least 92.2%.

The calculated lower confidence bound of 92.2% is appreciably less than the 10-year goal of 97%. Thus, although the results at this time suggest that the motor will likely meet its reliability goal, assuming successful fixes of the two identified problems, we have not yet demonstrated this with the desired degree of statistical confidence.

Continuing the units on test will further reduce uncertainty—that is, narrow the gap between the estimated reliability and the lower confidence bound.

---

[1] The term *95% confidence* means that, before the data are collected, the *procedure* used to compute the *lower confidence bound* has a 95% chance of resulting in a bound that is exceeded by the actual unknown reliability. Once the data have been generated, the lower confidence bound either exceeds the actual reliability or does not—we just do not know which is the case. See Hahn and Meeker (1991) for further discussion of confidence levels and procedures to compute one-sided confidence bounds and two-sided confidence intervals.

RESULTS AFTER SIX MONTHS

After six months of testing (4368 test cycles, equivalent to 21 years of customer use), a total of seven bearing failures had been observed. These occurred at 1989, 2807, 3502, 3612, 3889, 4006, and 4341 cycles. There were also four additional plastic-part failures. The bearing failures are shown in Figure 3 in what is known as a Weibull distribution probability plot. The intermediate textbooks listed at the end of this essay describe statistical tools for making such a plot.

The Weibull distribution, like the better-known normal distribution, is a statistical model that, when appropriate, allows one to smooth observed data. This distribution has been found—on both theoretical and empirical grounds—to provide a reasonable fit to time to failure data in many applications.

The computer-fitted line in Figure 3 provides reliability estimates assuming bearing failure is the primary cause of motor failure (i.e., assuming successful elimination of the other two observed failure modes). Approximate upper 95% confidence bounds on the failure probabilities over time are shown by the dotted curve. These estimate

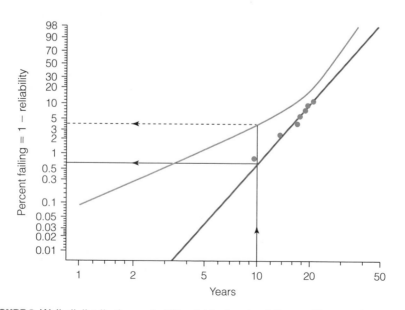

FIGURE 3 Weibull distribution probability plot for bearing failures with approximate upper 95% confidence bound (gold curve) and 10-year failure probability estimate

the worst reliability that one could reasonably expect in light of statistical sampling variability. From Figure 3, the estimate for 10-year reliability is now 99.4% (failure probability of 0.006, or 0.6%), with a 95% lower confidence bound of 96% (upper confidence bound on failure probability of 0.04, or 4%). The fact that the plotted points in Figure 3 scatter around a straight line, moreover, suggests that a Weibull distribution provides a good representation for the motor bearing failure times, at least within the range of the data. (Even though Figure 3 shows only the seven bearing failures, the line was fitted taking all data into consideration.)

Because the 95% lower confidence bound on 10-year reliability of 96% still falls below the goal of 97%, we just missed demonstrating with 95% assuredness that the new motor design meets its target reliability, even after elimination of two failure modes. However, it can be shown that we can make the desired demonstration with 92% confidence—and it was agreed that this was good enough.

EPILOGUE

The six-month test results were sufficiently favorable to allow management to give the go-ahead for full-scale production of the motor. Ten of the surviving motors were selected randomly to continue testing for another 4000 cycles to monitor their reliability further and to obtain more precise estimates.

Also, each of the failed motors, as well as the unfailed ones that were removed from test, were taken apart and subjected to careful engineering analyses to obtain added information to help improve reliability of future products.

Assessing Washing Machine Reliability

Evaluations similar to those conducted for the motor were performed for the other key components of the washing machine, especially those for which a change was made from the previous design.

Assurance that each of the parts meets its reliability goal is what mathematicians call a necessary, but not sufficient, condition for ensuring that the system reliability goals are met. It is always possible that when the parts are used together in a system, they interact, or in some way operate differently from when they are run separately, contradicting the assumption of independence made in the development of the reliability model. Thus, the reliability of the assembled washing machine needs to be validated further at various levels. Again, statistical planning and analysis play an important role.

IN-HOUSE TESTING

Twenty washing machines, randomly selected from prototype production, were life tested by the manufacturer in environments that were typical of field operation. These tests included running a full load of soiled towels mixed with sand through the machine's full functional operation, at the hottest water setting. In addition, some machines were wrapped in a plastic bag to accelerate possible effects of moisture. The details of this program resemble those for testing the motor, and are therefore not discussed further here.

EARLY FIELD TESTING

In-house testing, irrespective of how well constructed, can never fully represent all the diverse uses and environments that might be experienced in field operation. One of our favorite examples is that of short circuits that occurred in ovens due to cockroaches accessing the circuit board. This failure mode had not been envisaged by the design engineers, and therefore was not simulated by the in-house test.

A two-part program of field testing prior to general product release (sometimes referred to as "beta testing") was, therefore, conducted. In the first part, 100 washing machines were given to company employees, mostly with large families, for home use. In addition, 60 additional machines were distributed among three laundromats, where they were subjected to heavy and frequent use. The performance of both groups was carefully monitored. The results available by the time of product release were scanty. However, they gave added support to the previous findings and provided no further evidence of reliability problems.

RELIABILITY MONITORING AFTER PRODUCT RELEASE

The preceding measures taken prior to product release help ensure that the product put on the market meets customers' high reliability expectations. Hopefully, everything is smooth sailing thereafter. But, as the former baseball celebrity Yogi Berra would remind us, "it ain't over until it's over." Changes occurring after initial product release, perhaps due to cost-cutting measures by the manufacturer or by suppliers, might erode reliability. Therefore, all material or manufacturing process changes were to be reported and evaluated. In some cases, added testing was required to ensure that the changes did not have a detrimental impact on reliability.

For further protection against subtle changes in reliability, five units were selected at random from each week's production and put on a rapid-cycling life test for periods ranging from one week to three months. The results were monitored daily and compared with those of the earlier life tests to confirm that there had not been any deterioration in reliability—and to provide a statistically based early warning signal if there had.

Finally, a system was created to feed back information about the performance of the product experienced by customers. This provides the most pertinent, and also the most voluminous, data about product reliability. Such information will help improve future designs— but it is of little solace to customers who have already purchased an unreliable product. In such situations, the customer bears the inconvenience, while the manufacturer is responsible for the associated warranty costs (to say nothing about blemishes on reputation)—an unsatisfactory situation for both.

Recently, more enlightened companies have recognized that it is essential to build high reliability into the initial product design, as illustrated by our example. Unfortunately, this process is not yet fully perfected, and some reliability problems occasionally still slip through. Thus, a statistically based system designed to capture information about product malfunctions in field operation and to address these as rapidly as possible provides useful insurance. If problems are uncovered, the resulting action might range, depending on problem severity, from a change in the design of a future product to a recall of previously shipped units. Such scrutiny also might provide information about any rarely occurring safety problems—such as a fire hazard under extreme-use conditions—that had not been identified by previous testing, perhaps because of their infrequent occurrence.

Last but not least, the performance in the field, together with that of earlier testing, may form the basis of advertising claims. Such claims need be justified by valid statistical data and analyses to meet the scrutiny of industrial and governmental regulating agencies, as well as legal challenges from competitors.

THE BOTTOM LINE

As a result of the care taken in product design, and the comprehensive testing described here, the new washing machine met its reliability goals. As expected, this resulted in a significant increase in sales—although the specific dollar impact, and the savings due to the reduction in warranty costs, was difficult to quantify precisely. However, the business leadership was convinced that these were in the millions of dollars.

FROM WASHING MACHINES TO SPACE SYSTEMS

We have used the relatively simple example of a washing machine as an illustration. Similar concepts apply to ensuring high reliability for more complex systems, such as automobiles, locomotives, power-generation equipment, medical-scanning devices, and space systems. In most of these cases, it is vital to achieve an extremely high level of reliability in light of the potential risks of human injury. These and other products call for more complex statistical evaluations than those for the washing machine. The suggested readings at the end of this essay provide details.

Accelerated Testing

In the washing machine example, one can simulate 10 years in the field in three months by accelerating the use rate. For many other products that operate continuously, such as refrigerators and power-generation equipment, this is not possible. So what do we do then?

The answer often lies in testing under more severe environments, such as high temperature or humidity, than encountered in normal operation. The purpose is to accelerate the physical or chemical degradation process that causes certain failures, such as the weakening of an adhesive bond. Or it might involve increasing the stress, such as the voltage or pressure, to which the unit is exposed. In both cases, physical understanding of the failure mechanism is required to identify an appropriate accelerating variable(s) and to provide a mathematical model to relate the accelerating variable to time to failure.

Statistical concepts are used to develop an accelerated life test, conducted at a number of conditions of the accelerating variable(s), to gain the maximum amount of information for the money spent. Then, after data on times to failure of the failed units (and survival times on the unfailed ones) are obtained, statistical methods are used to fit the relationship between the accelerating variable(s) and time to failure using the assumed model. The fitted relationship is then used to predict reliability at the operating conditions. In addition, confidence bounds are calculated to quantify the associated statistical uncertainty (Hahn, Meeker, and Doganaksoy, 2003).

Helping Design Robust Products

Products need to be insensitive to both variability in manufacturing conditions and to fluctuations in the operating environment. The *Challenger* space shuttle tragedy would have been avoided if the O-rings had been designed to function over the wide range of temperatures experienced at launch. Statisticians help engineers design such "robust" products. This includes planning and analyzing experimental programs to assess the impact of component design on system performance in different environments.

More Complex System Models

For many products, added reliability protection is gained by building redundancy into critical areas of the design. Some typical examples are

- Alternative systems for landing a spacecraft
- Two headlights on an automobile to reduce the risk of an accident if one light fails
- Redundant array of inexpensive disks (RAID)—memory systems that employ two or more identical disk drives in mission-critical computer systems

Such systems fail only if all of the redundant parts fail. Probability theory can be used to evaluate the resulting reliability.

Just-in-Time Maintenance

All who have flown on airplanes can attest to the inconvenience of unscheduled delays due to equipment malfunctions. Couldn't these faults have been detected and fixed earlier, perhaps at routine maintenance? Indeed, the havoc created by the failure of a part can often be avoided by replacing the part before it is likely to fail.

Statisticians work with their engineering counterparts to make this hope a reality. To improve aircraft engine reliability and reduce unscheduled flight delays, they obtain data to determine how often an engine should receive routine maintenance—and what parts to replace.

Remote monitoring and diagnostics, moreover, allow one to detect impending faults proactively. This involves placing measurement devices into aircraft, locomotives, medical-scanning devices, power-generating equipment, and even automobiles to "take the pulse" of the system continuously—and feed back information to provide advance warning of likely failures. Such warnings range from a siren at a remote control desk to a red light on your automobile dashboard. At best, such early warnings prompt action to avoid failures altogether. At worst, they mitigate the impact of failures by, for example, rushing a spare part or a replacement locomotive to the scene (Pool, 2001). Indeed, we foresee the day when monitoring devices embedded in people provide early warnings of impending heart attacks and automatically dial an emergency number—even before any symptoms are experienced.

Statistical methods advance such efforts by, for example, utilizing past data to establish warning thresholds that maximize failure detection rates—while maintaining acceptable false alarm rates.

CONCLUSION

Product and system reliability is a major concern to consumers and manufacturers alike. The inability to meet reliability goals can lead to everything from minor inconveniences (e.g., no toast for breakfast) to human tragedies. Ensuring high reliability in the design of manufactured products is, first and foremost, an engineering challenge. But statistics and statisticians play an important role. They ensure that meaningful data are obtained quickly in a cost-efficient manner, and that the results are translated into useful information for incisive decision making during product design.

In this area, as in many others, statisticians do not work in a vacuum. They need to be intimately knowledgeable about the product with which they are dealing and fully sensitive to the business environment. They need to be articulate team players who can work successfully with engineers, scientists, managers, and sometimes even lawyers and accountants.

The eventual failure of most products, as of all humans, is inevitable. A tire or an aircraft engine will fail sooner or later. Statisticians strive to have such failures happen in "old age," and in a safe and predictable manner. And when failures do occur, statisticians help minimize their detrimental impact.

ACKNOWLEDGMENTS

We would like to thank Frank Forbes, James Loman, and the editors and student readers of this essay for their highly useful comments and suggestions.

REFERENCES

Dalal, S. R., E. B. Fowlkes, and B. Hoadley. (1989). "Risk Analysis of the Space-Shuttle: Pre-Challenger Prediction of Failure." *Journal of the American Statistical Association* 84 (December): 945–57.

Hahn, G. J., and W .Q. Meeker. (1991). *Statistical Intervals: A Guide for Practitioners.* New York: Wiley-Interscience.

Hahn, G. J., W. Q. Meeker, and N. Doganaksoy. (2003). "Speedier Reliability Analysis." *Quality Progress* (June): 58–64.

Pool, R. (2001). "If It Ain't Broke, Fix It." *MIT Enterprise Technology Review* (September), https://www.techreview.com/.

ADDITIONAL READING

Elementary

Doganaksoy, N., G. J. Hahn, and W. Q. Meeker. (2000). "Reliability (or Product Life) Data Analysis: A Case Study." *Quality Progress* (June): 115–21.

Hahn, G. J., N. Doganaksoy, and W. Q. Meeker. (1999). "Reliability Improvement: Issues and Tools." *Quality Progress* (May): 133–39.

Meeker, W. Q., et al. (1999). "Reliability Concepts and Data Analysis." Chap. 48 in *Juran's Quality Handbook,* 5th ed. New York: McGraw-Hill.

Intermediate (More Detailed)

Meeker, W. Q., and L. A. Escobar. (1998). *Statistical Methods for Reliability Data.* New York: Wiley.

Nelson, W. (1982). *Applied Life Data Analysis.* New York: Wiley.

———. (1990). *Accelerated Testing: Statistical Models, Test Plans, and Data Analyses.* New York: Wiley.

QUESTIONS

1. This chapter mentions various "well-publicized cases involving poor reliability of manufactured products." Suggest some other examples, possibly from your own experience, of poor reliability (that may or may not have been well publicized). What might the manufacturer of the product have done to avert this problem or reduce its impact?

2. Consider how much you rely on others to ensure your own safety. What do you think (or hope) are things that manufacturers of products and providers of services do to provide high reliability and make "things go right"? What role does statistics play in these efforts?

3. How would you define reliability for a CD player? Why might its assessment sometimes be ambiguous, and how can we help avoid such ambiguity? Suggest a program to help ensure high reliability for a newly designed CD player.

4. What costs does a manufacturer experience when products have poor reliability in field operation? How can such losses be quantified for management? What are some obstacles that you might encounter in convincing managers focused on short-term profits to invest in proactive reliability improvement?

5. What mistake did the *Challenger* engineers make in examining the reliability of the O-rings in various temperature environments? Express this in general terms not specific to the *Challenger* incident.

6. In the washing machine example, suppose that slowing of the motor, suggesting a forthcoming motor failure, also increases the chances that the cycle/timing control will fail. Is this taken into account in the calculation of R_{system}? Explain.

7. What shape would you expect a histogram of time to failure to have assuming that failures are due mainly to a product wearing out? Why might the Weibull distribution provide a good fit to this histogram? (Check this out by looking up the Weibull distribution on the Internet or the reading list for this essay.)

8. This essay suggests that for some products, especially ones that are in frequent operation, increasing the use rate of the product will not provide the needed information about long-term reliability. As a result, one needs to test "under more severe environments" to generate timely reliability information. What are some examples of such accelerated testing and why should it be conducted at multiple accelerated conditions?

9. In the washing machine example, units were continued on test beyond three months to "further reduce uncertainty   that is, narrow the gap between estimated reliability and the lower confidence bound." Why would continuing the test narrow the gap?

10. Some systems, such as the washing machine described in this essay, require all its components to work for the system to operate successfully, as sketched in Figure 2. Other products have built-in redundancies to allow the machine to operate as long as at least one component in each redundant group is working. Relate these two situations to the flow of electricity in series and parallel circuits.

11. In what ways is field testing different from the "in house" testing that a manufacturer conducts during product development?

12. Why is the "robust design" of products important? Give some examples of robust product design.

13. This essay states that "we foresee the day when monitoring devices embedded in people provide early warnings of impending heart attacks . . . even before any symptoms are experienced." How might statistics play a role in making this a reality?

14. Assume that you are a battery (or incandescent lamp) manufacturer. You believe that your product has a longer average life than that of your three main competitors. You wish to make an advertising claim to that effect. Suggest how statistics might help you in validating such a claim and in defending it against legal challenges by your competitors.

15. How do statistics and statisticians help in assuring high product reliability and safety? What personal characteristics should a statistician working in this field possess to be successful?

RANDOMNESS IN THE
STOCK MARKET

RICHARD J. CLEARY

Bentley College

NOREAN RADKE SHARPE

Bubson College

———————— ⁂ : : ⁂ ————————

On May 7, 2003, the website http://www.msnbc.com reported the following information about stock market activity for that day: "Stocks were slightly lower Wednesday as Wall Street took a breather after nearly two months of gains after high-tech bellwether Cisco Systems issued a gloomy quarterly sales outlook." This is a typical stock market report; each day the change in the market is followed by apparent explanations for the change. Investors listening to these reports may believe that the stock market is a predictable system.

Such reports never read: "Stocks were slightly lower Wednesday because the stock market is a random process. Some days it goes up; some days it goes down. Today it went down. Over the long term it seems to go up more than it goes down so it has been a good investment historically." This statement is just as good an explanation as the first one, but it is not deemed newsworthy.

We can use statistics to determine how much variation in a system is associated with other variables, and how much is genuinely random. Financial analysts explore complex models in an attempt to predict the future behavior of the market. Investors, on the other hand, need to find a balance between the analyst who proposes precise reasons for each daily movement in the market and the anarchy implied by a completely random system. In addition to neglecting the troubling concept of randomness, daily stock market reports have the advantage of hindsight. Developing a plausible theory for market behavior in retrospect is interesting, but making prospective, or predictive, forecasts is much more difficult. To paraphrase many investment commercials, does past performance indicate future results? In addition, are recent trends that we see in a stock or stock index useful as forecasting tools? Does it make a difference if we consider daily, weekly, or monthly prices? How much of the variation is genuinely random?

Dabbling in the stock market is a hobby for millions of Americans. People pay for advice on buying stocks and mutual funds, and they develop their own theories about which stocks are good investments. In the following example we describe a number of methods that an investor might use to make short-term forecasts, even in the presence of random variation. In addition, we demonstrate the impact of historical prices on future prices and discuss approaches commonly used for long-term forecasting.

SHORT-TERM FORECASTING: A CASE STUDY

We examine the historical stock price for a leading company, Intel Corporation, to demonstrate the degree to which stock prices are random and the impact of recent performance on short-term forecasts. Intel Corporation, located in Santa Clara, California, was founded by three engineers in 1968 to develop technology for silicon-based electronic chips and is currently listed on the NASDAQ as INTC.

Intel experienced a growth in net revenues from $8.8 billion to $28.8 billion between 1993 and 2002,[1] but sales were relatively flat from the late 1990s into 2002. However, Intel remains the world's largest maker of chips, with approximately 80% of the microprocessor market. Industry analysts suggest that the semiconductor industry, in general, has been affected by volatility in sales of personal computers and other electronic and computing devices.

[1] Intel's 2002 annual report, http://www.intel.com.

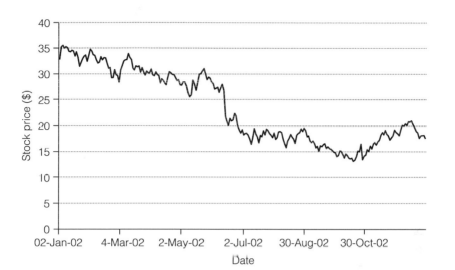

FIGURE 1 Daily stock price for Intel (January 2002–December 2002)

Figure 1 is a time series plot of the daily stock price for Intel between January 2, 2002, and December 12, 2002. Using the values of Intel stock over the past year, how might we predict Intel's stock price in the short term, using the information contained in these daily returns?

Clearly our forecasts for Intel stock will depend on the time frame chosen as a reference. Note that the stock price appears to decrease slightly over the first six months of 2002, but remains fairly stable over the last six months of the year. Looking only at the last quarter of the year, October through December, might give an impression of a stock on the rebound. Different investors might study Figure 1 and reach different conclusions about where Intel's stock price is heading.

Statisticians and financial analysts have developed a variety of tools to use in forecasting future values of time-dependent data such as stock prices. Because of the apparent recent monthly stability and the continued daily volatility in price, we could use a simple technique called a *moving average,* in which we average the stock price over a consistent number of recent trading days and use this value as a forecast. For example, if we use the prior 30 days of stock price to compute the average, then the first closing price in the data series is dropped and the next closing price in the series is used to compute the next 30-day average. This process is continued until a series of

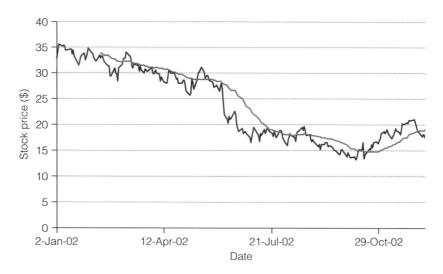

FIGURE 2 Daily stock price with moving average

30-day moving averages has been computed and the most recent average can be used to forecast the next daily stock price. This approach and other more sophisticated techniques are referred to as *smoothing* methods, because these methods smooth out the variation, or noise, in a series. Smoothing methods are easy to implement using modern software and are often proposed to novice investors.[2]

A plot of the 30-day moving average for Intel's closing price appears in Figure 2. This "smoothed" graph provides a good visual summary of the stock's recent history. Despite their ease of use, moving-average models are not particularly useful for decision making because the natural forecast is constant—simply the average of some number of most recent closing prices. Thus for forecasting, we focus on two other commonly used techniques: the *simple linear regression* (or *trend*) model and the *autoregressive* model.

A basic statistical tool for computing trends over time is *simple linear regression,* which determines the equation of a line that "best fits" the data. The term *best fit* is defined to mean that the line chosen minimizes the sum of the squared errors between the observed stock

---

[2] *Business Week,* June 2, 2003, pp. 104–6. This article shows 200-day and 50-day moving averages and suggests making stock purchases based on the relationship between the moving average and the current price, or on the ratio of the 200-day to 50-day moving averages.

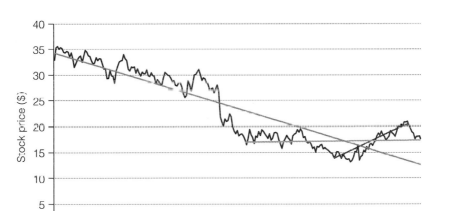

FIGURE 3 Linear models for stock price using different time frames

prices and the corresponding predicted points on the line. The result is an equation that estimates the value of the stock price at any particular time, where we measure time by a count of the trading days over the range of the data used. Figure 3 shows the resulting best fit lines using three different subsets of the original data. The result of a simple linear regression model can be expressed algebraically as well as graphically. Here the equation has the form ($Price_t = b_o + b_1 \times t$), where $Price_t$ represents the price after t trading days. The term b_o is a constant while b_1 is the *coefficient* of the time variable.

Note that when we use our entire set (241 trading days), we get a line that slopes downward. The corresponding equation has a coefficient of -0.09, which suggests that a share of Intel stock has lost $0.09 per trading day over the course of the year. Predictions based on this model will show a similar decline, with the stock price eventually falling to zero. (To avoid such drastic conclusions, analysts typically focus on proportional change rather than absolute change by using logarithms of stock prices instead of the prices themselves. Our focus on the short term reduces that need here, so we stick with the original data for simplicity.)

Now consider regression lines using the prior six months and three months of stock data as pictured in Figure 3. Not surprisingly, the six-month trend yields a coefficient of 0.01, much closer to zero,

	Forecast (one day)	Actual (one day)	Difference	Difference (%)	MAPE (%)
MA(30)	18.93	17.54	1.39	−7.92	10.02
Linear model (12 mos.)	12.64	17.54	4.9	27.94	12.90
Linear model (6 mos.)	17.36	17.54	0.18	1.03	9.55
Linear model (3 mos.)	20.71	17.54	−3.17	−18.07	5.34
Autoregressive	18.15	17.54	−0.61	−3.48	3.12

TABLE 1 Comparison of "next day" forecasts

suggesting a nearly constant stock price. The three-month trend suggests that the stock is gaining in value at a rate of $0.13 per day—and will continue to do so in the foreseeable future. These linear equations with time as the predictor variable have some benefits; they do a good job of summarizing the recent behavior of a series of values, and they are easy to interpret and implement.

Despite the attractive simplicity of linear models, analysts often prefer *autoregressive* models for short-term forecasting. This set of models assumes that adjacent error terms are related, and therefore, historical stock prices influence future prices. Thus autoregressive models are more adaptive, because they respond to successive changes in stock prices. The model for Intel uses the most recent value of stock price as the independent variable and the current stock price as the dependent variable. Expressed as an equation, we summarize this model as $Price_t = b_0 + b_1 \times Price_{t-1}$. The explanatory variable is now simply time-lagged values of the closing stock price, and thus is often referred to as a "lagged" variable.

In Table 1, we compare the forecasts for our three linear models (using the three different time frames) and the autoregressive model with the actual Intel stock price on December 13, 2002, which is the first trading day after the data shown in Figure 1. These forecasts, along with the actual stock price and percentage error, are shown in Table 1. The value in the last column represents mean absolute percentage error (MAPE), which is computed as the average of the absolute value of the percentage difference between the forecasted values and the actual values over the entire time frame of the model. This measure is typically used to compare forecasting models, because it is an average error over the historical period used to develop the model.

Of course, the fact that we are trying to look only one day ahead is a short-term forecast. We could also compare how the models do looking further into the future. Yet the presence of day-to-day variability suggests that long-term predictions would be even less reliable. Could we recommend any of these models to an investor hoping to make a profit in the short term?

Let's consider the information the regression models provide more closely. Because the linear model using the past 12 months of data reflects a downward trend, it *under*forecasts, and because the linear model using only the prior three months of data reflects an upward trend, it *over*forecasts. The linear model over six months had the most accurate forecast for this particular day. However, the low MAPE for the autoregressive model indicates that it might be the most appropriate to use for short-term forecasting of daily stock prices.

IMPLICATION OF A RANDOM WALK IN STOCK MARKET

If the price of Intel stock were entirely random, with no deterministic component, we could model it with a random walk: $Y_t = Y_{t-1} + e_t$, where e_t represents random error. In a simple version of the random walk, we assume that the errors follow the *standard normal distribution* (a bell shaped curve with mean 0 and standard deviation 1). This would mean that on any particular day the stock would be as likely to go up as it would be to go down; that average daily change would be 0; and that the stock would only rarely change in price by more than about $3.00 per day.

Note how similar this random walk model is to the first-order autoregressive (AR) model discussed in the prior section. If the coefficient is 1 and the constant is 0 in the AR model, then these two models are equivalent. In fact, the equation for the AR model in Table 1 has an estimated coefficient of 0.99 and an estimated constant of 0.19—very close to the form of a random walk! Thus we might conclude that daily stock prices at Intel over this time period resemble a random walk.

Let's examine the AR model ($Y_t = 0.19 + 0.99 \, Y_{t-1}$) more closely. Note that because the constant term is positive, this seems to indicate a growth in closing price of about $0.19 per day. However, the coefficient on the previous day's price is slightly less than 1, and this indicates a decline in value. The price of the stock determines which component dominates. If today's closing price is $30.00, the model forecasts tomorrow's price as $0.19 + 0.99 \times 30 = 29.89$, a modest decline. If today's price is only $10.00, our forecast for tomorrow is $0.19 + 0.99 \times 10 = 10.09$, an increase. Note that if the coefficient on the previous close happened to be exactly equal to 1, our model would predict a slow upward drift in Intel's closing price of about $0.19 per day. In fact, this sort of model is an excellent fit for how the stock market has behaved historically. Market indices such as the Dow Jones Industrial Average have shown growth over the long term but a great deal of random variation in the short term.

While the random walk view of the stock market is well-known,[3] investors have continued to try to forecast stock prices because the potential profit is enormous. If we know that a stock is about to go up in price, while other investors are uncertain, we can cash in by effectively always being able to "buy low" and "sell high." During the tech stock boom in the late 1990s, "day traders" thought they could capitalize by making quick trades in the market through recognition of stocks that were about to go up. Many succeeded in making a profit in the short term, while the entire stock market was increasing in value, but the practice became much less widespread when the market started to fall.

Most individuals find it *counterintuitive* that a series of cumulative random numbers can indeed not only resemble the behavior of a particular stock or the S&P 500 index but can even seem to simulate a trend, as well as cyclical behavior. Figure 4 shows a series of cumulative random numbers that started with Intel's price on our first day.

Each random number simulates the daily change in Intel closing price, assuming a standard normal error distribution. Note that this random walk resembles Intel's stock behavior during 2002. Over a specified period of time, daily fluctuations in most stocks can be shown to be similar to a random walk. Most investors *under*estimate the random variability and *over*estimate deterministic factors in the

---

[3] G. Malkiel Burton. (2000). *A Random Walk Down Wall Street*, 7th ed. New York: Norton. The first edition was published in 1973.

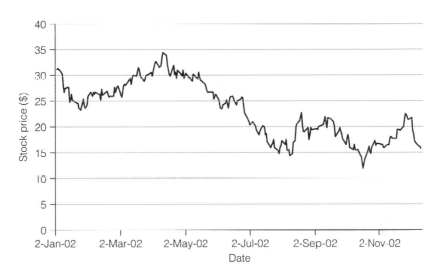

FIGURE 4 Series of cumulative random numbers

short term. Is forecasting the long-term trend of a stock using weekly or monthly prices (which have averaged out more random noise) a different question?

LONG-TERM HISTORICAL PERSPECTIVE

Despite the fact that volatility in stock prices can be shown to resemble a random walk, many investors continue to devote time to forecasting the stock market. Certainly, it makes a difference whether the objective is to obtain a short-term forecast or to generate a long-term prediction. Suppose instead of examining Intel daily stock prices for the past year, we consider Intel monthly stock prices for the last 17 years (see Figure 5). Note the difference between the series in Figure 5 and the stock prices over the past year in Figure 1. (The data to the right of the vertical line in Figure 5 denote the stock data shown in Figure 1.) Here, it is evident that the decline in stock prices is fairly recent and that in fact, there was a long-term increasing trend present between 1986 and 2000.

If the long term behavior is not entirely random, what explains the sustained growth and the subsequent decline in Intel stock price? Long-term trends can be affected by internal variables such as product quality, advertising expenditure, and competitive strategy. The

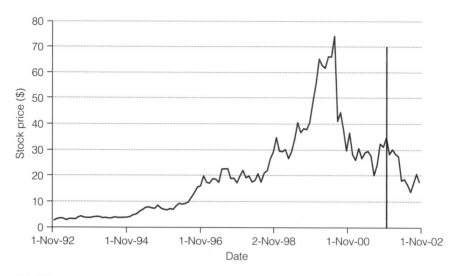

FIGURE 5 Intel monthly stock price (1986–2002)

internal decisions of a corporation clearly play a role in the long-term success (or failure) of a corporation.

In the absence of internal corporate variables (particularly if they are proprietary), forecasters use external economic and industry indicators. For example, Intel's stock value can be shown to be related to measures such as the U.S. gross domestic product (GDP), a measure of the market value of goods and services produced by labor and capital in a given country, and the index of leading economic indicators (LEI), an aggregate measure of economic activity. Both of these indicators have shown slowed growth over the past few years, similar to the slowed growth in many corporations in the semiconductor industry. Finally, the decline in Intel stock prices could also reflect slowed growth in sales (decreased demand) and a built-up inventory. To test this theory, industry indicators can be used in combination with economic indicators to forecast long-term trends of semiconductor stocks.

While the preceding external explanatory variables can capture a portion of the variation in stock prices over the long term, the difficulty remains that the cycles observed in stock prices are challenging to forecast. The main reason is that their length (duration) and height (depth) cannot be forecasted reliably quantitatively. In fact, cycles appear to be primarily the result of cumulative random errors, where the errors (or white noise) are independent and normally distributed.

Thus, although long-term trends in economic variables can be modeled with the hope of making good investment decisions, the underlying volatility still contains a random component. Although "the trend is your friend," there is always a level of risk associated with investing in the stock market.

FINANCIAL MODELS AND MANAGERS

Trying to successfully model the stock price of a company based on external economic measures requires a certain level of statistical sophistication. For example, the selection of variables (i.e., which external variables are important); the diagnosis and treatment of collinearity (i.e., the correlation of multiple economic indicators with each other); and finally, the consideration of how many lagged variables to use are nontrivial issues. Many of these techniques are currently employed by professional investment managers, with the objective of selecting the most promising stocks for their portfolios.

In fact, stock analysts critiquing our essay might say, "Yes, I'll admit that one particular stock may be nearly random. But building a portfolio is a much more predictable process. When we can choose a variety of stocks, we can build a winning investment scenario." Would they be right?

Apparently not. A recent study[4] of 237 investment strategy recommendations by financial newsletters between 1980 and 1992 revealed that less than 25% of newsletters have average returns higher than a passive portfolio. In general, few newsletters can do better than the S&P 500 index[5] and there is a lack of evidence that the newsletters can forecast direction of the stocks. Stated differently, a set of randomly chosen stocks typically equals or outperforms the advice of the majority of investment newsletters.

Professionals who manage investment funds do not fare better. The percentage of professional managers who in the moderate term

----

[4] J. Graham and C. Harvey. (1996). "Market Timing and Volatility Implied in Investment Newsletters' Asset Allocation Recommendations." *Journal of Financial Economics* 42, no. 3:397–421.

[5] The S&P 500 index is a group of 500 stocks selected by Standard and Poor's and often used to represent or summarize the behavior of the stock market as a whole.

(three years) or the long term (10 years) have outperformed the S&P 500 is consistently below 50%, and the median returns of managers are consistently below the median return of the S&P 500 index over the same time period.[6] This lack of success, or volatile behavior in the stock market, indicates that individuals who randomly select their stocks would have as much success (and failure) as the professional managers.

SUMMARY

So, we return to our original questions: Is randomness inherent in stock fluctuations? If so, is the random variation in stock prices handled differently for short-term and long-term forecasting—using daily versus monthly stock prices? Thus far we have suggested that short-term volatility in the market appears to be predominantly random, while long-term trends may be predictable and influenced by external factors in the economy. Retrospective, as opposed to prospective, modeling is far more feasible. Multiple smoothing techniques can be used to reveal underlying historical trends but are inappropriate for long-term forecasting. If moving averages or monthly prices are used, more of the random variation is *smoothed* out of the series, resulting in a more predominant trend component. To model the trend component, a variety of advanced forecasting tools are available for skilled financial analysts. For novice investors, however, an appreciation for the inherent randomness in the system is important if they want to be successful.

Explaining the behavior of the stock of an individual company, a portfolio of stocks, or the entire market through an index like the S&P 500 is an interesting and important problem. Through careful analysis of sophisticated statistical models, professionals in finance hope to gain a slight edge by which they can make a bigger profit than the "typical" investor. Simple models and simple advice, however, are unlikely to be more valuable to the small investor than an understanding that much of the market is inherently random and difficult to predict.

---

[6] Makridakis, S., Wheelwright, S., and Hyndman, R. (1998). *Forecasting,* 3rd ed. New York: Wiley.

ACKNOWLEDGMENTS

The authors acknowledge the help of former statistics students Geeta Athalye, Ann Iannuzzi, Brenda Lord, and Andrew Miller (Babson College) and the fall 2003 sections of Quantitative Research Methods (Bentley College) for assistance in preparation of this essay.

QUESTIONS

1. What reason might an investor have for wanting to graph a moving average of a stock price series for a company of interest?

2. Why was the moving average considered inadequate for short-term forecasting, and what feature of alternative methods (autoregressive models and simple linear regression) makes these methods more useful?

3. If stock prices really did follow a random walk model, what would be the best predictor of the next day's stock price?

4. The autoregressive model estimated from the Intel data indicated that the stock would tend to rise if the price was below $19.00 and tend to fall if the price was above $19.00. Verify this property of the model and suggest how it might be used by an investor.

5. Is it surprising that the random walk model mimics the Intel stock price variability in 2002? Discuss.

6. The quotation "the trend is your friend" is comforting given that markets tend to rise in the long term. What about short-term trends—are they useful, too? Discuss.

7. Either by using a computer program or by tossing a fair coin 100 times, simulate the movements of a stock price that follows a random walk model. For simplicity, assume that the daily changes are always $+1$ or -1. Do you observe an apparent trend in the stock price? Would such a trend be useful for predicting the future price changes?

8. Look on the Internet for stock investment advice regarding Intel Corporation. What forecasting "tips" are provided? How would your impression of these tips change after reading this essay?

9. What other economic or industry indicators might influence Intel stock price over the long term? How might you use these

to develop a long-term forecasting model? Do you think it would be appropriate to lag these indicators in the model?

10. Explain the importance of selecting the appropriate time frame for a forecasting model.

11. Do you think that the autoregressive model is as sensitive to the choice of time frame as the linear model? Why or why not?

12. Why are autoregressive models appropriate to use in forecasting stock prices?

13. For which of the following situations would an autoregressive model be appropriate?

    a. Forecasting tomorrow's temperature based on today's temperature

    b. Forecasting the number of runs scored in the next Boston Red Sox game based on the runs scored in the more recent game

    c. Forecasting a student's GPA next semester based on prior semesters

    d. Forecasting next year's total snowfall for a particular region based on prior year's snowfall

14. This essay demonstrates the inherent random variation in the stock market. Other essays have made the same argument for gambling. Consider the similarities between investing in the stock market and casino gambling, or other games of chance. (For example, consider the difference between the outcome of a single roll of the dice and the expected long-run average return on a game of chance.)

15. Develop smoothing models (e.g., moving averages using different time periods and exponential smoothing) for the Intel stock price data set using different periods of averaging and weighting constants, respectively. How does the accuracy of these models in forecasting Intel stock price compare to the accuracy of the models developed here? Over the short term? Over the long term?

16. Develop a multiple regression model using values of GDP and LEI to predict Intel monthly stock prices over the time frame of the data set (1986–2002).

ADVERTISING AS AN
ENGINEERING SCIENCE

WILLIAM KAHN

Capital One Services, Inc.

LEONARD ROSEMAN

Capital One Services, Inc.

———— ※ : : ※ ————

Messaging is an essential part of life. Whether it is primitive ("Get out of my way, or else!") or subtle ("Wanna dance?") its purpose is to convey information and change behavior. The best messages are tailored for the intended recipient. When messages are not effectively tailored they can be annoying, wasteful, and will fail to achieve their purpose.

This essay discusses the application of the scientific method to a particular form of messaging called advertising. Advertising is messaging designed to drive people's economic behavior. Many types of messaging fall into the realm of advertising, including communications designed to impact our retail, political, financial, and personal activities.

Just as advertising encompasses many different sorts of activities, so does science. One of the early stages of scientific inquiry is recognizing that there is something interesting to be studied—something worthy of deeper consideration. This leads directly to natural science activities such as cataloging, naming, and classifying. Later stages of scientific effort focus on causal understanding and manipulation.

These stages require experimental and engineering activities such as experimentation and modeling. Advertising, while receiving significant attention for thousands of years, has just recently begun to enter its experimental stage.

We discuss here how statistical tools, especially those of experimental design and multivariate analysis, can help us understand how best to serve a particular group of consumers who asked for communication via e-mail. These people enrolled in the November 2002 Turkey Mail program from Butterball (a brand of turkey products from ConAgra Foods) because they were interested in communications about how to buy, prepare, cook, and eat turkey around Thanksgiving.

This research was successful at two levels. First, and specifically, we learned which registrants responded to what kinds of messages. But second, and more generally, we proved to a skeptical advertising community that modern statistical techniques can quickly and efficiently identify how to tailor e-mail advertising for recipients, thus making the advertising significantly more beneficial for advertisers and consumers.

ADVERTISING: THE INDUSTRY

For some projects, such as developing and launching a NASA space probe or a new computer chip, for every billion dollars invested in the enterprise, thousands of highly trained scientists and engineers are involved in ensuring every detail of the work is optimized to the highest level of performance. After all, a billion dollars is a lot of money, and we need to make sure we get maximum benefit from this investment. For other projects, such as developing and launching a sequence of advertising campaigns, it is common that a billion dollars can be invested with only a handful of scientific or engineering professionals being involved—or none at all. Much advertising activity is not only not optimized to the highest level of performance, it is barely even sensible. Inconceivable but true: we know of a major food distributor that spent more than a billion dollars in advertising in Mexico in 2001 and didn't even keep track of the details of where the spending occurred let alone perform a statistical analysis showing what messages worked where and for which recipients.

Despite this lack of rigor, advertising is, nonetheless, a big industry. In 2001 about $400 billion was spent in the United States alone
(for comparison, NASA spent less than $15 billion): $260 billion
in mass advertising (radio, TV, print, outdoor signs, etc.) and
$139 billion in direct advertising (direct mail, call center, e-mail,
etc.). With spending of this magnitude, representing 4% of the U.S.
gross national product, one would expect that the best possible combination of skills is used to optimize the return on this huge financial
investment. But this does not appear to be the case. In particular, the
statistical profession is currently making only small contributions to
this industry. In the future we expect it to have significantly greater
impact.

The research described in this essay represents an application of
modern experimental design to advertising. We hope it will stimulate
others to go further. Much of the value of an engineering approach to
advertising comes from the systematic application of core principles
and not just from technical sophistication.

THE SCIENTIFIC METHOD

Science encompasses an extremely broad class of methods serving a
similarly broad class of human needs. A core process in scientific
inquiry is the following four-part cycle, shown in Figure 1.

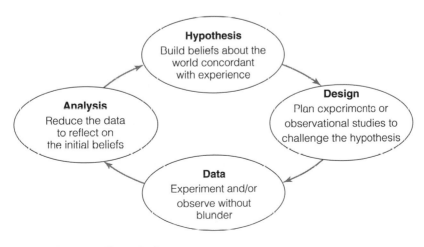

FIGURE 1 The scientific method

The scientific method starts with a hypothesis about the phenomenon in question, typically based on previous experience. It then structures experiments or observations that provide objective data to support or refute this hypothesis. Appropriate data are gathered from these experiments in an objective manner, without introducing the biases of the experimenter. These data are then summarized and analyzed via appropriate statistical methods. Finally, this analysis provides support for or against the original hypothesis. This process is intended to be objective, that is, unbiased and completely open to all the evidence. This differs significantly from other environments, such as legal arguments in courtroom trials, where evidence that supports your case is emphasized and evidence that does not support your case is minimized. Our point here is not to claim moral superiority for science but simply to note the uniqueness of the approach.

With careful designs, data gathering, and analysis, additional questions and hypotheses beyond the original hypothesis are often raised. New designs need to be developed to evaluate these new hypotheses. In this manner, the scientific method renews itself, continuously building knowledge as it proceeds through this cycle over and over again. Virtually all major scientific developments in recorded human history have resulted from this iterative process of discovery.

Let's see how the scientific method was applied to advertising in the 2002 Turkey Mail campaign.

TURKEY MAIL

With the advent of the Internet, an increasing number of mass marketers have been trying to learn how to have individualized interactive communications with their customers. Specifically, Butterball, which sells a billion dollars a year in poultry and poultry-related products, wanted their website to be viewed as the premier source of information about poultry products, preparation, and usage. The Butterball Internet site therefore has many useful tips for selecting poultry, safe handling, cooking, recipes, Thanksgiving lore, holiday planning, and personal help. Consumer interest in turkeys obviously peaks around Thanksgiving, and the Butterball advertising team wanted to send out a series of e-mails providing information and links to their website during this period.

As of November 2002, 57,584 individuals had registered at the site and requested to receive e-mails. In order to increase usage of the Internet site, as well as to increase sales, the Butterball brand managers faced two questions: What messages should be sent to the registrants and when should they be delivered?

Applying the scientific method to advertising, we performed the following steps.

Background

For each registrant we had certain basic information including date of registration, e-mail address, and self-reported zip code, age, household size, sex, and marital status. Each registrant explicitly provided permission to receive e-mail from Butterball.

The ultimate financial metric is closely related to the amount of Butterball products sold. However, tracking the impact of e-mail messaging through to store purchases was not of primary immediate interest. As a major, multidecade brand, it is important to Butterball to constantly increase the awareness of Thanksgiving, of turkey generally, and of Butterball in particular. One immediate measure of individual engagement is the customer visiting the Butterball website by "clicking through" on a hyperlink in the e-mail. Brand managers believed that stimulating registrants to revisit a website was a sensible proxy measure for helping maintain the long-term future of the brand.

Hypothesis

As Thanksgiving approaches, people's thoughts increasingly turn to the upcoming holiday. There are many possible messages and sequences of messages that could be e-mailed to registrants. The team used their experience to brainstorm factors that should have an impact on effectiveness of the e-mail messages. Factors of potential interest included number of weeks in advance, day of week, subject line, graphical content, textual content, and reading level. Within each of these factors, many levels, or settings, of the factors would be possible. For example, weeks in advance (prior to Thanksgiving) could sensibly be set at values anywhere between 0 and perhaps 6.

Design

In most advertising environments, the specific message and timing is decided by a priori belief, and then sent to everyone. Unfortunately, this provides no way to learn what might be better because we have no way of knowing how people would have responded to other messages. In a slightly more ambitious environment, all factors but one, perhaps the subject line, would be the same. Then a random half of the e-mails would get one subject line, and the other half a second subject line. Data on people's responses to each subject line would support learning about the impact of this factor. We would remain ignorant, however, about the impact of all other factors and combinations of factors.

Fortunately, this is where statistics comes in. Using designed experiments it is possible to vary many factors at once and afterward develop a nearly perfect understanding of the impact of each factor, or combination of factors, of interest. Factors discussed in our brainstorming sessions included subject line, personalization, graphical content, reading level of the text, delivery day, e-mail length, and number of links. Many of these factors require changes to the body of the e-mail, which at the time was technically hard to do. However, weeks in advance, subject line and which day of the week the e-mail was delivered were easy to vary, and these factors were chosen for our experiment. (In subsequent experiments on other products, we have also varied tone, value proposition, number of links, discount level, and graphical content.)

Each registrant to the Internet site received four e-mail messages, one three weeks before Thanksgiving, one two weeks before, one a week before, and one that week (zero weeks before).

Each of the first three messages sent contained one of three subject-line themes: Planning, Elegance, or Festive. Some registrants received the same subject line all three weeks; others received a different subject line all three weeks. All 27 combinations were used. The fourth and final message always had the Planning subject line. The first three e-mail messages sent each had three web links to the three focus points: Planning, Elegance, and Festive. For every e-mail a recipient might click on any of the Planning, Elegance, or Festive links.

Messages could be sent out on any day of the week, but we were particularly interested in learning if the end of the week was better than the middle. The first three e-mails were sent out either on Tuesday or Friday, and the fourth was sent out either on Wednesday or Friday.

For the first three e-mails, each registrant could be assigned one of six (2×3) possible day–subject combinations. The fourth e-mail had a single subject sent on one of two days. So there are a total of $((2 \times 3)^3) \times 2 = 432$ day–subject line e-mail combinations.

A lot of work is always required to go from a statistician's experimental design "on paper" to a properly conducted, real experiment. In our case all the e-mails had to be created, approved by multiple parties, and proofread multiple times. The computer programming for the messaging had to checked and rechecked. Different versions (text, HTML, etc.) of each message had to be built so that they could be read by any e-mail program.

Finally, we randomly assigned the 57,584 registrants to the 432 e-mail messaging streams subject to all the previously described constraints. This random assignment is the spiritual and intellectual heart of experimental design, as it is the randomization that allows scientifically definitive, inductive statements to come out of the analysis.

Data

The tool we used to launch the e-mails built a log file of events associated with every e-mail sent. We observed every e-mail bounce, e-mail open, and website click through. Automatically, addresses discovered to be invalid were removed from further messaging. As is true in all research, funny, anomalous behavior was observed. E-mails that bounced were also opened. E-mails were bounced multiple times. E-mails that were never opened were clicked through. We spent some time satisfying ourselves that each of these behaviors could indeed be explained and were not the result of coding errors. Now that we had all this data, what could we learn from them?

Analysis

Data are normally analyzed for two reasons—to provide support for or against beliefs about the world and to help generate new beliefs. The new beliefs fuel the scientific method. However, before the data are rigorously analyzed, it is usually wise to carefully examine them for special events: errors, mistakes, or unusual occurrences. First, basic summaries and plots are made, and individual values are examined for plausibility and sensibility. Next, plots of two or more variables are constructed, and common sense brought to bear on the relationships seen. Only then should formal models be used to extract information, while statistically controlling for the impact of the other factors in the analysis. A tremendous benefit of statistically designed experiments is that they allow us to experiment with many factors at once, while still enabling us to understand the unique impact of each one.

Creating sensible statistical models requires training, experience, and a substantive understanding of the process being modeled—there are no effective shortcuts or automated tools. After building these statistical models, the results need to be presented to others, in part for their review and in part to give practitioners the guidance they need to run the process better the next time.

While data analysis is often done using advanced mathematics to ensure correctness, the presentation must be concrete and simple enough for the audience to understand. Because the presentation is often in the form of basic tables and graphs, many nonstatisticians believe that the field of statistics consists of simply summarizing data in tables and graphs. Nothing could be further from the truth. Summary tables and graphs show the results; they are not the analysis. The statistical analysis reveals which tabulations and graphs are important, and these are the ones shown.

The statistical analysis we performed for this research, after a thorough data-cleaning process, was a model predicting whether or not an e-mail would be clicked through, that is, whether or not the people receiving the e-mail would click on a hyperlink to the Butterball site. Other events, such as "e-mail bounce" and "e-mail open," were of interest, though are not discussed here for brevity's sake. Since the primary goal of the study was to learn how to engage the registrants, our primary measure of success was whether the e-mail recipients were interested enough to click through to the website.

In the experiment we controlled seven factors: day of week (two levels) for each of the four weeks and subject line (three levels) for each of the first three weeks, thus producing a $((2 \times 3)^3) \times 2$ full factorial design. It turns out that these factors interacted with each other, meaning the effect of one variable depended on the level of another

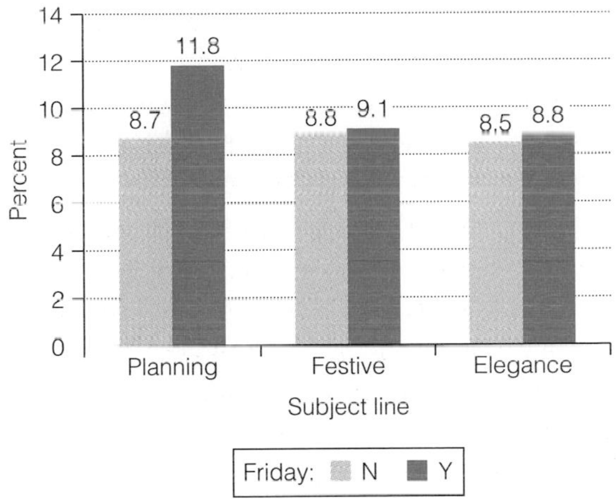

FIGURE 2 Click-through rate: the Friday effect

variable. For example, it may be that the best subject line depended on which day of the week it was sent. The results below are based on the 167,890 e-mails sent, of which 15,019 (8.9%) were clicked on through to the website. Let's look at some specific examples.

SUBJECT BY DAY OF WEEK

It did turn out that the effect of day of the week on click-through rate was not the same for each subject line. As can be seen in Figure 2, the difference between the click-through rate on Friday versus Tuesday for Subject 1 (Planning) was much greater than for Subject 2 (Elegance), or Subject 3 (Festive). Note that we are discussing here the difference between differences—that is, an interaction effect. Note also that the day of the week factor is coded in these graphs as either Friday-yes, or Friday-no, so we will call this the "Friday effect." Only the first three weeks are shown here in order to create a fair comparison—Planning was the only message sent out on the fourth week.

From a purely empirical point of view it may be sufficient to know that planning messages are best delivered on Friday, but it may also be of interest to know *why* planning depended on day of the week but the other subject lines did not have this dependence. Our initial hypothesis is that this interaction is due to "planning"-oriented people being less time stressed on Friday than on a weekday than are "festive"- or "elegance"-oriented people. But, at this point, this is

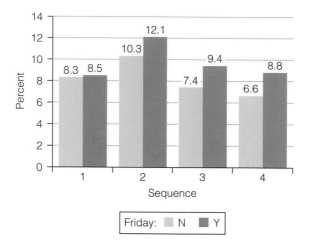

FIGURE 3 Percentage click-through rate: effect of day of week

merely a newly generated hypothesis, an example of how additional questions arise from the analysis and lead to the development of new testable hypotheses for the next round of experimentation.

SEQUENCE BY DAY OF WEEK

Averaging across the three subject lines, we see that the second e-mail sent had the highest overall response rate. We also see in Figure 3 that the Friday effect exists for all weeks, but just barely for the first week.

Again, empirically knowing that two weeks in advance is best may be all that is required, though understanding *why* may be worth investigating in further experimentation.

SEQUENCE BY SUBJECT

Planning did not have higher impact than Festive or Elegance on the first e-mail, but did have higher impact on the second and third e-mail drops (Figure 4).

This may suggest that planning Thanksgiving, at least for this sub-population, doesn't begin in earnest until two weeks before the holiday.

Effects of the Covariates

We have dozens of variables describing each registrant, which were provided through the website at registration. However, the completeness of this data is highly variable since it is dependent upon

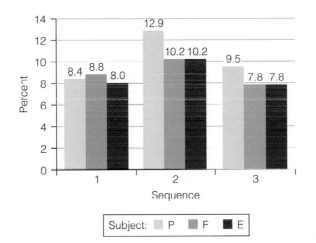

FIGURE 4 Percentage click-through rate: effect of subject

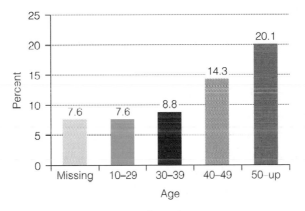

FIGURE 5 Percentage click-through rate: effect of age

when they registered (the website registration process changed over time) and whether they were willing to give us information about themselves. Statistical techniques were therefore needed to make sense out of this erratic data. Below are some of the conclusions reached by the team, based on statistical analysis. We do not list all conclusions in the interest of brevity.

AGE

Older registrants click through at a much higher rate than younger registrants (Figure 5).

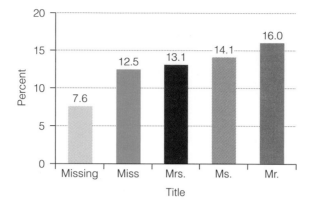

FIGURE 6 Percentage click-through rate: effect of title

The trend with age is intriguing, with the self-identified oldest group having a click-through rate almost triple that of the youngest group. Several sociological/psychological ideas are suggested by these results, which will be tested in future campaigns. Why do you think we have seen such a large age effect?

TITLE

Registrants who do not supply a title click through at a low rate, while self-identified men click through at a high rate (Figure 6).

Men are more interested in these e-mails than women, and those in the "Miss" category are least interested among those who gave a title. No one predicted this pattern before the campaign, but afterward everybody said, "Of course. Only very interested men would register, and single women don't care about cooking turkey." This kind of retrospective commentary is common in all experiments, and no doubt serves some psychological need. But, it is important to understand rigorously what was learned through the scientific process. It is therefore helpful to record researchers' expectations prior to conducting the experiment. In this way, we document all the relevant hypotheses and can demonstrate the new knowledge that was generated.

Tailoring the Message to the Customer

The analysis, so far, has allowed us to identify the best messaging across all the registrants. It also allowed us to identify the best prospects for future messages. But we have not yet determined

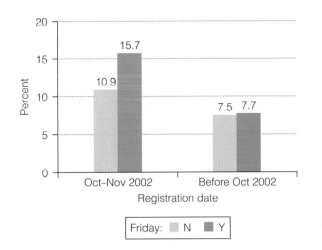

FIGURE 7 Percentage click-through rate: effect of registration date

whether the best message for each person is the same. Is it possible that some messages are best for some people, and other messages are best for other people? In advertising, this would allow specific personalization—each message customized to each individual. This would be scientifically driven advertising and would have high economic, and social, value.

DAY BY REGISTRATION DATE

Recent registrants had a bigger day effect, preferring Fridays more than earlier registrants (Figure 7).

That recent registrants had a overall higher click-through rate is not surprising—they have shown recent interest. That they click 44% more often when e-mailed on a Friday was surprising. Of course, we were explicitly looking for this effect, so seeing these results only validated one of our existing hypotheses, but the size of the effect was more than we expected.

SUBJECT BY REGISTRATION DATE

The difference between early (before October) and recent (October or November) registrants was greatest for Subject 1 (Planning), as shown in Figure 8.

Planning was more important to recent registrants than anyone realized. In the future we would need to more effectively serve the planning needs of recent registrants.

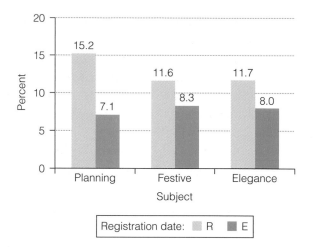

FIGURE 8 Percentage click-through rate: effect of registration date and subject

Additional Analyses

Additional analyses were performed. We have highlighted only these particular analyses because in a single, fully integrated logistic regression, these were the biggest effects, after controlling for everything else that was also having an effect. Given the large sample size, all the effects we have mentioned are highly statistically significant. For the factors, that is, those variables we explicitly controlled and randomized over, this multivariate regression is not needed; simple one- or two-way analyses are correct by themselves. But for the covariates, those variables that are observed but not controlled, it is this simultaneous modeling that allows us to identify which effects it is appropriate to show in simple analyses. Technically, the multivariate regression allows us to understand the "confounding pattern" of the design and hence which effects are really multiple variables working together.

Putting It All Together

In summary, the Planning message delivered on Friday had the highest click-through rate, and this was particularly pronounced for recent registrants. Older people and men were more likely to click, and this response peaked two weeks before the holiday. Additional focus on serving the planning needs of registrants is suggested, as well as explicitly advertising this service to men and older people.

In many traditional engineering environments one is strongly driven to perform a cost–benefit optimization. In our e-mail advertising we would like to know what the value is to registrants of receiving e-mails that they click on, and the cost to registrants of receiving e-mails that they asked for but do not click on. Understanding such personal, societal, operational, and brand management economics would support formal optimization of this brand management investment. Anecdotally, we know that a large proportion of registrants are recent immigrants to the United States trying to learn about becoming Americans through learning about this most American of holidays. Making estimates of the value of this e-mail service is clearly difficult, but probably would be worth at least some thought.

Sending certain messages to some people and other messages to other people, all based upon formal statistical studies, is a contribution to the engineering science component of advertising. Other areas of marketing are aspiring to this effective use of technology as well.

Path Forward

With the current targeting model we can identify better messaging to send to each registrant in 2003. Of course, 2003 is not 2002. While we do expect the 2002 model to apply, there will be differences. Thus, in 2003 we will direct some large fraction of all messaging to the ones identified as best from the 2002 model. But we will also explore new variations in an effort to be even better prepared for 2004. The cycle of the scientific method goes on.

SUMMARY

Advertising, along with other aspects of marketing, is a tremendously large and important industry ripe for the application of modern experimental design. As statistical techniques make inroads into the field, messaging will improve materially. We expect that advertising of all types will become increasingly targeted and personalized—a more valuable, informative, educational service, and less wasteful, intrusive, and annoying. Advertising technology is not yet sophisticated, but existing methods are more than adequate to transform the field into an engineering science. Statisticians must patiently, but emphatically, bring their special tools, insights, and approaches to bear on this large sector of the economy and guide it into the 21st century.

ACKNOWLEDGMENTS

The Butterball brand managers proactively expanded the sophistication of advertising in allowing us to help them learn how best to serve their Turkey Mail registrants. Creating new approaches to pressing commercial problems requires an unusually bold and confident management style. We admire the brand managers' drive to continuous improvement and appreciate the confidence they had in us as we worked to learn how to improve advertising generally even as we learned how to serve their customers specifically.

REFERENCES

Box, G. E. P., W. G. Hunter, and J. S. Hunter. (1978). *Statistics for Experimenters*. New York: Wiley.

Department of Commerce. (2002). *Survey of Current Business* 82, no. 12 (December): D-5.

Nail, J., with C. Charron, E. Schmitt, and Sadaf Roshan. (2002). "Forrester Report, Mastering Advertising Measurement." TechStrategy Report. September. Http://www.forrester.com.

QUESTIONS

1. What words do the authors use to describe the stages of the scientific method?
2. What tool of the scientific method does the current advertising industry lack?
3. What population is being studied in the Turkey Mail study?
4. What was the primary short-term objective of the Turkey Mail study?
5. Explain why there are 432 possible combinations of features to include in the study e-mails to registrants.

6. What was the reason for the random assignment of the 57,584 registrants to the 432 message combinations?
7. From the results of the study, give an example of two factors interacting in determining the click-through rate.
8. A covariate is a variable that is not included in the design of the experiment (i.e., is uncontrolled) but is measured for each subject. What covariates are discussed in the study?
9. Do you think the "benefit" of the study was worth the "cost" to Butterball? (There is no exact answer here, but offer your views.)

PART FIVE

HOBBIES AND RECREATION

※ : : ※

Baseball Decision Making by the Numbers

Hal Stern

Predicting the Quality and Prices of Bordeaux Wines

Orley Ashenfelter

BASEBALL DECISION MAKING BY THE NUMBERS

HAL STERN

University of California, Irvine

———— ⁜ ∷ ⁜ ————

During an October 1996 play-off game, the San Diego Padres came to bat in the top of the eighth inning trailing the St. Louis Cardinals by a single run. The first two batters reached base, and then Tony Gwynn, one of the greatest hitters of all time, came to bat with runners on first and second base and nobody out. This situation is an example, one of many such examples, of a point during a baseball game that requires a baseball manager to decide among a number of competing strategies. In this particular case, San Diego manager Bruce Bochy's choices included (1) asking Tony Gwynn to sacrifice (make an out, one of three allowed in the inning, while advancing the runners), and (2) allowing Tony Gwynn to bat as usual. The sacrifice offers the benefit of increasing San Diego's chances of tying the game but at the cost of making a big inning less likely. Hitting as usual allows for the possibility of a big inning if Gwynn gets a hit but risks a double play (a play leading to two of the three outs allowed in the inning), which would damage San Diego's chances of tying the game. Revisiting this decision made by a baseball manager in a key play-off game in 1996 allows us to see how an understanding of the key statistical concepts of uncertainty and variability, along with consideration of all relevant data, might change the way we think about making such decisions. We will return to that October day to reveal what happened after considering in more detail how statistical thinking can be used in baseball.

DECISION MAKING IN BASEBALL

Individuals and businesses make decisions all the time; these range from a car owner's decision regarding the deductible to choose on his or her auto insurance policy to a large corporation's decision whether or not to build a new manufacturing plant. A number of factors are considered in any such decision problem, including the possible actions or decisions, the outcomes that may result from those actions, the cost or benefit associated with the various outcomes, and the likelihood of the various outcomes under each possible action. Decisions made during baseball games appear often to be guided by a long-term, collective intelligence known as "the book." For example, playing it by the book might suggest that one choose to sacrifice in the situation just described because the improved probability of achieving a tie is the most important result to achieve. There is nothing wrong with applying information of this sort to guide our decisions; after all it is built on the experiences of a large number of people. However, decisions made in this way are prone to biases because individuals may tend to recall the negative outcomes more often than the positive outcomes when one deviates from the traditional strategy. In the baseball context, sportswriters are sure to point out negative consequences of nontraditional plays but are not likely to question decisions made by the book.

The field of statistics has been characterized as the science concerned with decision making in the face of uncertain outcomes. In particular, thinking statistically requires that we consider the availability of any data that might help us to identify the costs or benefits of the various outcomes and to estimate the likelihood of the different outcomes. It is natural to wonder then whether a data-based evaluation of the evidence would support the strategies favored by collective wisdom in baseball and other sports. As it happens, the start of the 21st century has seen baseball itself debate the merits of a more quantitative approach. A number of younger baseball executives have embraced the use of quantitative methods, even if it occasionally leads to nonstandard decisions. For example, the Boston Red Sox hired baseball writer Bill James, long known for his statistical analyses, as an adviser prior to the 2003 season.

BUILDING A MODEL FOR BASEBALL

One of the primary arguments against a formal quantitative approach to baseball decision making is that each baseball situation is different, involving different teams and players. Under such a view, which is not without merit, one would argue that it is impossible to collect enough data to inform about a particular current situation. The field of statistics teaches us, however, that one can often learn a great deal by considering average or typical behavior.

Combining information across different situations that involve different players and teams is facilitated by making use of a model or representation of baseball that captures the main features of the game without being 100% accurate. We often make use of models to understand how large systems work. In the case of baseball we can think of a model as being similar to baseball simulation games such as Strat-o-matic, developed in the 1960s, and their modern-day computer relatives.

SUMMARY OF THE GAME OF BASEBALL

Before proceeding to construct a model of baseball, we provide a thumbnail summary of the game of baseball. If you are a baseball fan (and you likely are if you've chosen to read this essay), then the following information is already familiar to you. If you are not familiar with baseball (perhaps you are reading this as an assignment!), then hopefully this description will be sufficient to make the rest of the essay more meaningful. Figure 1 provides a rough approximation to the infield portion of a baseball field (not drawn to scale) and shows the location of first base, second base, third base, and home plate. A baseball game between two teams consists of nine innings. In each inning both teams have a turn to bat. During a team's turn, the players on that team take turns coming to home plate in a specified order to serve as the batter. Each batter either reaches base successfully (becoming a runner) or the batter (or one of the existing runners if there

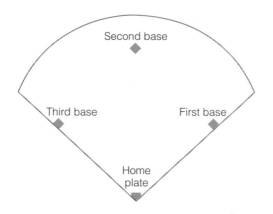

FIGURE 1 The infield portion of a baseball field, with the locations of the bases identified

are any) is put out. Depending on how far the ball is hit, the successful batter may reach first base, second base, third base, or even, in the event of a home run, return to home plate. Any runners on base may advance when the current batter hits the ball. Each time a batter-runner returns to home plate without being put out, the batting team scores a run. After the batting team records three outs its turn is over. The team that scores the most runs during the nine innings wins the game.

A STATISTICAL MODEL FOR BASEBALL

We consider a statistical model that views each half-inning of the baseball game as a random process in time, also known as a *stochastic process*. The particular model that we use to represent the progress of a baseball team during one of its innings is a *Markov chain* model. The basic idea for our model is to consider the situation on the baseball field when each batter comes to bat as one of a limited number of possible states or situations. For example, the first batter in each inning comes to bat with no runners on base and nobody out. Later in the inning a batter may come to bat with runners on first base and second base with nobody out (the situation facing Tony Gwynn in our San Diego–St. Louis game). To characterize the situation facing a batter we concern ourselves with the bases occupied (first, second, third, or combinations of those) and the number of outs (0, 1, 2).

There are eight possible scenarios regarding the bases occupied, which we denote by indicating the numerals corresponding to the bases that are occupied by runners: 0, 1, 2, 3, 1-2, 1-3, 2-3, 1-2-3. In addition there are three possible scenarios regarding the number of outs, which we denote by the corresponding numeral: 0, 1, 2. Combining the number of outs and the bases that are occupied makes for $3 \times 8 = 24$ possible situations that a batter can expect to see. In what follows, a situation is described by listing the bases occupied and the number of outs. For example (0, 0) indicates no runners on base and zero outs, the situation facing the first batter in each inning. When Tony Gwynn came to bat during the play-off game described earlier, the situation was (1-2, 0).

To complete the model we must determine all of the possible outcomes for a batter, and, most important, how those outcomes will affect the situation on the baseball field. If a batter hits a home run, then all runners and the batter score, and the next batter will come to bat with no runners on base and the same number of outs as before the home run. If the batter strikes out, then all runners will stay exactly where they are and there will be one additional out when the next batter comes to bat. All told there are many things that can happen during one baseball batter's turn at bat. Each possible outcome results in a change or transition from one of the 24 possible situations to another (or to the three-out situation that signifies the end of the inning). Our Markov chain model is completely specified if we can give the probability of going from a given situation on the field to any other situation through the actions of a single batter.

HOW MANY SINGLES? HOW MANY DOUBLES?

There are a few methods that can be used to determine the probabilities of going from one of the 24 possible situations to another. One natural approach is to collect data from actual baseball games. In fact, a Canadian researcher, George Lindsey, did just that during the late 1950s (actually, he asked his father to collect the data). Lindsey recorded the situation when a batter came to the plate more than 27,000 times. He also recorded the situation after the batter's turn was complete. In addition to the starting and ending situation, he recorded the number of runs that scored from the time the batter came to the plate until the end of the team's turn at bat. For now, we focus on the data concerning the starting and ending situation. The number of runs scored is useful information; we will discuss it further later in the essay. Lindsey's 27,000 observations can be used to count the number of times that a transition was made from each state to every other situation. One limitation of relying on actual game data, though, is that they are highly variable and consequently we need a large number of observations to generate reliable estimates of the probability for a transition from one situation to another. Lindsey's data seem like a lot but are not enough for some of the rarer situations. The incredible amount of information available on the Internet means that it would now be possible to obtain the needed data, but one still needs to worry about whether the nature of the game is changing over the time for which the data are being collected. We use an alternative approach that uses estimates of the probability of simpler events—for example, the probability that a batter hits a home run, makes an out, walks, and so forth—to construct estimates for the probabilities that we want.

Table 1 provides the probabilities of various baseball-batter outcomes based on data from 1989. Results are provided separately for the two leagues, American and National, because the leagues have slightly different rules. Approximately 2% of batters' attempts result in home runs. This means, for example, that the probability of a move from the situation (2, 0), indicating a runner on second base with zero outs, to the situation (0, 0) is 0.02, because no other outcome will allow such a transition to occur. In the same way we can use the information in Table 1 and knowledge about the rules of baseball to find the probability of a move from each situation to

Event	American League	National League
Out	0.672	0.685
Walk	0.092	0.091
Single	0.171	0.159
Double	0.040	0.040
Triple	0.005	0.006
Home run	0.020	0.019

TABLE 1 Probabilities of various baseball outcomes (based on 1989 data)

every other situation. In some cases we must supply information not available in Table 1. For example, a single (a hit that allows the batter to reach first base) may be a "short" single, which advances all runners one base; a "long" single, which advances all runners two bases, or a "medium" single, which advances runners one base except that it advances a runner from second base to home plate (scoring a run). We make educated guesses about the relative frequency of these different types. In addition, we can decide how much effort to apply toward creating a realistic model of baseball. The model used here includes other baseball events such as errors (misplays by the defensive team on routine plays), sacrifice flies (fly-ball outs that allow runners to advance), and double plays (two outs on a single batted ball), in addition to the events of Table 1. This increases the ability of the model to better mimic actual baseball games. Estimates of the frequency of such events are required; the estimates used here were chosen so that the number of such events suggested by the model matches the number observed in games. Note that we intentionally omit various events that represent strategic decisions (e.g., sacrifice bunts, intentional walks, stolen bases) because these are exactly the kinds of things we'd like to address with the model.

Our discussion so far has implicitly assumed that the probability of each event is the same regardless of who is batting, which teams are playing, and which bases are occupied. The most troubling aspect is the assumption that the probabilities are the same for all players—there are substantial differences among hitters. For example, some hitters hit home runs much more frequently than implied by Table 1, while others rarely hit any. We will return to this important point later.

LEARNING ABOUT BASEBALL USING THE MODEL

The Markov chain model of baseball, like the earlier mentioned baseball simulation games, can be used to learn about the game of baseball. In particular one can ask about the number of runs that a team can expect to score from a given situation—for example, runners on first and second base with nobody out—or about the likelihood of scoring any runs at all from that situation. Quantities like these can be derived using formal mathematics or by computer simulation. The mathematical approach is more efficient, so that is what we use to provide the numbers in the remainder of the essay. To understand the intuition behind the mathematical formulas, we describe how one might use a computer simulation of baseball to calculate the quantities of interest. Quite simply, to learn about how a baseball inning is expected to proceed from a given situation—for example, the situation (1-2, 0) that is serving as our primary example—we use our model to play an inning from the given starting situation. We use the transition probabilities that we calculated from Table 1 to determine a random outcome for the next hitter. Then we update the situation (that is, we advance any base runners and record any outs), and then use the same transition probabilities to determine the outcome for the next hitter. We repeat until the third out ends the inning. For each inning simulated in this way, we record the number of runs scored. We then repeat the process over and over from the same starting point. Then, collecting the information from many such simulated innings allows us to develop various summaries that might be of interest. For example, we can estimate the probability of scoring at least one run by looking at the proportion of simulated innings in which one or more runs scores. We can estimate the average number of runs scored over the simulations; this is known as the expected number of runs scored.

Table 2 gives the probability of scoring at least one run and the expected number of runs scored for each of the 24 situations calculated using the probabilities in Table 1. The numbers in Table 2 are calculated using mathematical formulas for Markov chains rather than simulation (for more details, see the Additional Reading). Notice that the expected number of runs scored in each half-inning is 0.44 for the National League (NL) and 0.49 for the American League (AL) (found by looking at the entry for (0, 0) in Table 2). If these numbers are multiplied by nine they come close to reproducing the average number of runs scored by teams in 1989: 3.96 in the NL and 4.29 in the AL. The probability of scoring any runs in an inning is approximately 0.25 for each league; this number decreases to about 0.15 if the first batter makes an out.

There is an interesting historical comparison that can be made at this point. Lindsey's data from the late 1950s allow one to compute the same quantities that we display in Table 2. The results obtained

SITUATION		AMERICAN LEAGUE		NATIONAL LEAGUE	
Bases		Expected	Probability	Expected	Probability
Occupied	Outs	Runs	of Scoring	Runs	of Scoring
0	0	0.49	0.26	0.44	0.24
	1	0.27	0.16	0.24	0.15
	2	0.10	0.07	0.09	0.06
1	0	0.85	0.39	0.78	0.36
	1	0.52	0.26	0.48	0.25
	2	0.23	0.13	0.21	0.12
2	0	1.06	0.57	0.99	0.54
	1	0.69	0.42	0.64	0.40
	2	0.34	0.24	0.32	0.23
3	0	1.21	0.72	1.14	0.70
	1	0.82	0.55	0.77	0.54
	2	0.38	0.28	0.35	0.27
1-2	0	1.46	0.59	1.36	0.56
	1	1.00	0.45	0.93	0.43
	2	0.48	0.24	0.45	0.23
1-3	0	1.65	0.76	1.56	0.74
	1	1.10	0.61	1.04	0.59
	2	0.51	0.37	0.48	0.36
2-3	0	1.94	0.83	1.84	0.82
	1	1.50	0.74	1.42	0.72
	2	0.62	0.37	0.58	0.36
1-2-3	0	2.31	0.81	2.20	0.80
	1	1.62	0.67	1.54	0.66
	2	0.82	0.43	0.78	0.41

TABLE 2 Expected number of runs and probability of scoring any runs from a given situation

by Lindsey (not shown) are similar to those displayed in Table 2 but there are differences. There are two reasons that small differences are to be expected. First, Lindsey's data, like any sample, are subject to the variability inherent in the sampling process. A sample of a different 27,000 situations (perhaps involving different teams or a different season) would yield different results. A second cause for differences is the fact that the results in Table 2 are based on 1989 data, whereas Lindsey's results are based on data from the late 1950s.

There are slight changes in player ability, rules, and equipment that limit our ability to rely on data from another era.

APPLYING THE MODEL FOR PLAYER EVALUATION

There are a couple of obvious uses for information like that provided in Table 2. One possible use is to measure the performance of players. As an illustration, consider the common situation when one pitcher is replaced by another late in the game. For example, a team's starting pitcher may be replaced after allowing the bases to become full with zero outs in a particular inning (our situation [1-2-3, 0]). The relieving pitcher finishes the inning and allows those three runners to score but no others. How should we measure the relative performance of the two pitchers? Well, the current baseball rules tell us that the three runners are the responsibility of the pitcher who left them there. Thus the starting pitcher would get the blame for allowing those three runs and the relieving pitcher would get credit for not allowing any runs. But Table 2 tells us that in the NL an average team would score 2.2 runs when starting from the situation (1-2-3, 0). Thus the starting pitcher can be given blame for the expected 2.2 runs and the relieving pitcher would be blamed for allowing 0.8 runs more than expected. This seems more equitable than the resolution under the existing rules. In fact, one might go further and use our Markov chain model, along with some supplemental information, to compute the probability of winning the game from a given situation. Then, players could be given credit for their contribution to the team's probability of winning and losing; after all, that is what the game is about.

APPLYING THE MODEL FOR DETERMINING STRATEGY

The results in Table 2 can also inform us about the value of different strategies. Consider a NL team in the situation (1, 0). That team can expect to score 0.78 runs on average and would score one or more runs with probability 0.36. A successful sacrifice bunt, a play in which the batter intentionally makes out while allowing the runner to advance, leaves the team in the situation (2, 1). From that situation the probability of scoring at least one run is 0.40, and the expected number of runs is 0.64. The successful sacrifice bunt improves our probability of scoring but decreases the expected number of runs. Unfortunately, not all sacrifice bunts are successful. A failed sacrifice bunt will likely leave us in the situation (1, 1), for which the probability of scoring is 0.25 and the expected number of runs is 0.48. There are other possible outcomes for an attempted sacrifice, too: the batter may reach base safely without any outs being recorded or the batter and runner may both be put out in a double play. We don't worry about these other possible outcomes here, as they just make things more complicated. The actual decision facing the manager is whether to give up the current situation, (1, 0), for an uncertain future depending on the outcome of the sacrifice attempt. To make that decision we must estimate the probability of a successful bunt. If we assume that 80% of sacrifices are successful, then choosing to sacrifice will gives us 0.61 expected runs (80% of the time the sacrifice is successful yielding 0.64 expected runs and 20% of the time the unsuccessful sacrifice yields 0.48 expected runs) and probability of scoring 0.37. Then, we can see that the decision as to whether to sacrifice or not involves a trade off; we expect fewer runs if we sacrifice (0.61 versus 0.78) but we increase the chance of scoring any runs (0.37 versus 0.36). The small increase in probability of scoring hardly seems worth the loss in expected runs scored, especially if we are early in the game. The assumed success rate also seems a bit high. This leads us to seriously question the effectiveness of the sacrifice bunt in this situation.

The decision about whether to sacrifice is more interesting for the situation that we started the essay with. Suppose the batting team has runners on first and second with zero outs (situation [1-2, 0]). Then, the expected number of runs in the inning is 1.36 and the probability of scoring is 0.56. A successful sacrifice bunt leaves us in the situation (2-3, 1) and an unsuccessful sacrifice bunt leaves us in the situation (1-2, 1). If the probability of a successful sacrifice bunt is 0.80 (80% of bunts are successful), then calculations like those just described indicate that the expected number of runs in the inning is 1.32 and the probability of scoring is 0.66. Now we see a large increase in the probability of scoring for a fairly modest decrease in the expected number of runs. Here the sacrifice seems like a reasonable strategy, at least if we think the success rate is sufficiently high.

ALL PLAYERS ARE NOT CREATED EQUAL

One dictionary defines a "model" as a cheap representation of something. In our case the Markov chain model of baseball is cheap compared with actually collecting enormous amounts of data or running experiments on a baseball field. But you get what you pay for; a cheap model also implies the potential for inaccuracy. Here our Markov chain model assumes that the probability of a single, double, triple, walk, home run, or out is the same for every player. That is clearly untrue. The variability among players is great. In 2002, Barry Bonds's probability of a home run was four times the probability presented in Table 1, and his probability of a walk was three times the probability presented in Table 1. On the other extreme are players who rarely hit home runs or those who walk infrequently. It would be possible to improve upon our model by allowing each player in the lineup to have a different set of probabilities for moving us from one state to the other. Though the mathematical formulas are more complicated, the simulation approach that we described could still be used to determine the effectiveness of different strategies for different players. It may be worth using a weak hitter to sacrifice in a particular situation even when the probability calculations based on our simple model suggest it is a poor idea in general. The limitations of our model mean that application of our methods won't necessarily provide an unambiguous answer for the baseball manager; our methods do, however, provide a great deal of valuable information.

BACK TO ST. LOUIS IN OCTOBER 1996

When we left the St. Louis–San Diego baseball game, San Diego trailed by one run in the eighth inning and Tony Gwynn was coming to bat, with runners on first and second base. The previous section suggests that *on average* a sacrifice bunt will increase the probability of scoring at least one run but will decrease the expected number of runs scored. Manager Bruce Bochy asked Gwynn to sacrifice. This is not an unreasonable decision, but there are a couple of reasons to question it, given the caveats of the previous paragraph. First, recall the model used here assumes all hitters are equivalent, whereas San Diego had one of the best hitters of all time coming to bat. Second, the model also ignores the fact that San Diego would be relying on weaker hitters in the next inning (putting San Diego at a disadvantage even if the sacrifice was successful and they achieved a tie). Sure enough, the decision was effective in the short term—the sacrifice attempt was successful in advancing the runners and San Diego scored one run to tie the game—but failed in the global sense: San Diego lost the game when St. Louis scored in the bottom half of the eighth inning and San Diego failed to score in the ninth inning.

It would be presumptuous to claim that Bruce Bochy made the wrong decision. After all, the Markov chain model of baseball ignores an important factor, variation among players, that managers can and do use in making their decisions. What the Markov chain model does do is allow us to quantify the trade-offs involved in making decisions about baseball strategy.

SUMMARY

Decision making in all fields requires an understanding of the possible consequences of each action and the likelihood of those consequences. Probability plays a key role in quantifying the advantages and disadvantages of different strategies and leading us to optimal decisions. In addition, the field of statistics tells us how to collect the data that will inform our decisions. In the case of baseball, a thorough analysis leads to interesting insights regarding how the game is played and how players might be evaluated.

ADDITIONAL READING

Albert, J., and J. Bennett. (2001). *Curve Ball: Baseball, Statistics, and the Role of Chance in the Game.* New York: Copernicus.

Bennett, J. (1998). *Statistics in Sport.* London: Arnold.

Bukiet, B., E. R. Harold, and J. L. Palacios. (1997). "A Markov Chain Approach to Baseball." *Operations Research* 45, no. 1:14–23.

Hooke, R. (1988). "Statistics, Sports, and Some Other Things." In *Statistics: A Guide to the Unknown,* 3rd ed., edited by J. M. Tanur, F. Mosteller, W. H. Kruskal, E. L. Lehmann, R. F. Link, R. S. Pieters, and G. R. Rising. Belmont, Calif.: Duxbury Press.

Karlin, S., and H. M. Taylor. (1975). *A First Course in Stochastic Processes,* 2nd ed. New York: Academic Press.

Ladany, S. P., and R. E. Machol. (1977). *Optimal Strategies in Sports.* Amsterdam: Elsevier Science.

Lewis, M. (2003). *Moneyball: The Art of Winning an Unfair Game.* New York: Norton.

Lindsey, G. R. (1963). "An Investigation of Strategies in Baseball." *Operations Research* 11, no. 4:477–501.

Stern, H. S. (1997). "Baseball by the Numbers." *Chance* 10, no. 1:38–41.

QUESTIONS

1. Why does the general public tend to perceive deviations from traditional strategies in baseball as more likely to be unsuccessful than the actual historical data show?

2. The Markov chain model is clearly erroneous for various reasons—for example, it does not allow for different players to have different success rates at bat. Is it possible that the model is still useful? Discuss in the context of the essay.

3. Use the Markov chain model and the information in Table 1 to compute the probability of a transition from (1, 2) to (0, 2) for each league. (As described in the essay, the first position is the bases with a runner, and the second is the number of outs in the inning so far.) Hint: Which events could produce such a transition?

4. In the section "Applying the Model for Player Evaluation" an incident concerning a change of pitcher is described. Explain how the attributed runs against is computed to be 2.2 for the retiring pitcher and 0.8 for the new pitcher.

5. How can the model be used to evaluate the performance of batters?

6. One common defensive strategy is the intentional walk. If the batting team has the situation (2, 1), then the defensive team may intentionally walk the next hitter to create the situation (1-2, 1). The justification is that this makes it possible for a double play to occur and end the inning. Based on the information in Table 2, do you believe this is an effective strategy?

PREDICTING THE QUALITY AND PRICES OF BORDEAUX WINES

ORLEY ASHENFELTER

Princeton University

───────────── ❧ ∶ ∶ ❧ ─────────────

THE ECONOMIC CHALLENGE OF WINE PRICES

If you drink a bottle of root beer today, it will taste about the same whether it was made this year, last year, or two years ago. If you save a bottle unopened in a refrigerator, it won't taste much different 10 years down the road than it would if you were to drink it today. Root beer won't go bad, and it won't improve with age. If you tried to sell a 10-year-old bottle of root beer, you could expect to get about what you'd paid for it. For certain, you'd be foolish to buy a hundred cases, expecting to hold them for 10 years and then sell them for 20 times your purchase price. As an investment vehicle, root beer is a nonstarter.

Good red wines from the Bordeaux region of France aren't like root beer. This may be an obvious fact about how they taste, but it is also true about how they behave economically. For one thing, wine has good years and bad years. Wine from a good year tastes a lot better, and sells for a lot more, than wine from a bad year. An even bigger difference is that red Bordeaux wines get better with age. A new bottle has an unpleasant, puckery taste, but if you age the wine under proper conditions, its quality, and its price, will improve dramatically as it ages.

VINTAGE	Latour	Lafite	Cheval Blanc	Pichon Lalande	Cos d'Estournel	Montrose	Average
CHATEAUX (VINEYARDS)							
1960	464	494	486				**479**
1961	5432	4335	3534	1579	1170	1125	**4884**
1962	1064	889	821	281	521	456	**977**
1963	471	340			251		**406**
1964	1114	649	1125	410	315	350	**882**
1965	424	190		258			**307**
1966	1537	1274	1260	734	546	482	**1406**
1967	530	374	441	243	213	236	**452**
1968	365	223	274				**294**
1969	319	251		152	123	84	**285**
Average	**1935**	**1504**	**1436**	**649**	**553**	**530**	

TABLE 1 London auction prices for select mature red Bordeaux wines, 1990–91 (per dozen bottles in $US)

Note: A blank in the table indicates that a wine has not appeared in the market recently. Lower quality wines are usually the first to leave the market. The row and column "averages" were computed using a statistical method that estimates the missing values.

These two key differences—good and bad years, and improvement with age—make wines much more interesting, statistically and economically, than root beer will ever be. With root beer, if you know the brand, you have pretty much all the information you need to predict the price. Red Bordeaux is a different story altogether: just look at the wine prices in Table 1. Each column is for a different "house brand," or chateau. The chateaux were named long ago by the owner of the land where the grapes are grown and made into wine. Each row is for a different vintage, the year when the grapes were grown and harvested. The six chateaux in the table are generally regarded as among the best in the Bordeaux region. The 10 years were chosen from a long time back because it takes a long time for the wine to age and its value to be agreed on by experts.

As you look over the table you can see that there have been quite large differences in quality and hence price from year to year. Until recently, these quality differences have been considered a great mystery. In this essay I show that the factors that affect fluctuations in wine vintage quality can be explained in a simple quantitative way. In short, I show that a simple statistical analysis predicts the quality of a vintage, and hence its price, from the weather during its growing season. Along the way, we will learn how the aging of wine affects its price, and we will learn under what circumstances it pays to buy wines before they are at their best for drinking. Since this procedure for predicting wine quality has now been in use for more than a decade, I also provide an appraisal of its successes (and failures), and a discussion of the role this information has played in the evolution of the wine trade.

THE TWO MAIN INFLUENCES ON PRICE

From Table 1 you can see that if you know the average price of the vintage (shown in the last column) and the average chateau price (shown in the last row) you know a lot about the price of each wine. For example, by examining the last column of Table 1 it is clear that 1961 was the best year in this decade and that it was followed by 1966, and then 1962 and 1964 in quality (and price). There would be no dispute about this ranking from wine lovers anywhere in the world. Likewise, in the bottom row the average prices by chateau indicate that Latour is the most outstanding chateau in the group. Finding the 1961 Latour entry in the table reveals that, indeed, this is the best wine of the decade in this group. In fact, a more advanced statistical analysis reveals that information on chateau and vintage alone explain more than 90% of the variation in the prices. In short, there is not much room for other factors to play a big role in price determination.

The columns are ordered to show the ranking of the chateaux based on their prices: Latour, Lafite, Cheval Blanc, Pichon-Lalande, Cos d'Estournel, Montrose. In fact, as Edmund Penning-Rowsell points out in his classic book *The Wines of Bordeaux* (London: International Wine and Food Society, 1969), the famous 1855 classification of the chateaux of Bordeaux into quality grades was based on a similar assessment by price alone. Surprisingly, the 1855 classification ranks these chateaux in only a slightly different order: Lafite, Latour, Pichon-Lalande, Cos d'Estournel, Montrose. (Cheval Blanc was not ranked in 1855.) Likewise, a ranking of the quality of the vintages based on price alone would be 1961, 1966, 1962, 1964, and 1967. The remaining vintages (1960, 1963, 1965, 1968, and 1969) would be ranked inferior to these five, and perhaps because of this fact, many of the wines from these inferior vintages are no longer sold in the secondary market.

The rest of this essay presents a summary of statistical analysis of these wine prices. To simplify the description of the method, I assume in what follows that we have replaced all missing values with statistical estimates, so that every vintage year now has prices for all six chateaux, and for each chateau there is a wine price for all 10 vintages. (The estimates came from a regression of the logarithm of price on indicator variables for vintages and chateaux, but I omit the details.) The main goal of the analysis described here is to predict the price that a wine will have at maturity, and to make the prediction using only the information that is available when the new wine first comes on the market. To do that, we need a way to evaluate the effect on price of the two main influences, the chateau and the vintage. (1) *Chateau:* The effect of chateau is easier to assess, because the other influences balance. Every chateau produces wine in every year, so the effects of good and bad years are present equally for all the different houses. The average price, averaged over a long stretch of years, gives a good indication of the effect of the chateau. (2) *Vintage:* For the effect of vintage, there is an additional complication, due not to chateau but to age. The potentially "contaminating" effects of the houses balance and so wash out of the averages: for each vintage, we can compute the average price, averaging over all houses. Then, because each average incorporates contributions from the same set of six houses, the vintage averages will not be biased by differential effects of the houses.[1]

Now we need to consider the effect of age. Think about comparing the prices for this year's vintage and for a vintage of 10 years ago. In most years, the 10-year-old wine will have a higher price than the new wine, in part because of the cost of storing it for 10 years and in part because someone had to forego consumption in order to save it. As we shall see, the people who store these wines are compensated for doing so by those who want to consume older, tastier wines. But just

---

[1] Although the general idea is correct, there are certain technical requirements (e.g., additivity of effects) that must be met if the averages are to be totally free of bias due to chateau.

because the price of the older wine is higher does not mean that its year was better. There are two influences at work here, age and vintage. To reveal the effect of vintage by itself, you have to estimate and remove the effect of age—a process that statisticians call "adjusting for age." (An everyday parallel may help: At maturity, some people are taller than others, but just because a 10-year-old is taller now than a seven-year-old, you're not justified in concluding that when they both reach adulthood, the 10-year-old will still be the taller one. On the other hand, comparing each person's height to the average height for a person of the same age might let you predict with good accuracy which one will be taller at maturity. In effect, you *adjust* for age and compare the adjusted ages of the two people to decide which one will eventually be taller. To assess the effect of vintage, we need to make a similar sort of adjustment for age.)

ADJUSTING FOR AGE

To make a suitable adjustment, we want to eliminate differences between chateaux and differences due to good and bad years, so we can look at the effect of age separately from other influences. To do this, we create an *index*, much like the consumer price index (CPI), which is used to measure the effect of inflation. Just as the CPI is based on a fixed "market basket" of goods and services, the wine index used here is based on a set of wines from a fixed selection of chateaux. Figure 1

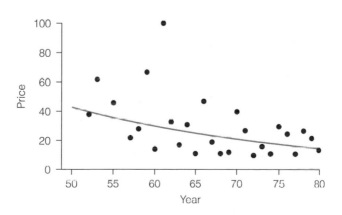

FIGURE 1 Red Bordeaux wines, price index relative to 1961 vintage, by year

shows a plot of the average price of this market basket of wines, suitably scaled to make the price for 1961 equal to 100.

The points correspond to individual vintages (years). You can think of the pattern in the plot as having two parts: an overall trend, represented by the curve, and variability above and below the curve. The curve represents the average or typical effect of age on price. The variability is due to unspecified factors other than age that make some wines worth more or less than expected from the general pattern. Points far above the curve correspond to very good vintages; points below the curve correspond to poor vintages; and the curve itself tells the value one could expect if the vintage that year had been of average quality.

Why is the relationship curved? Because over time, changes in the price of wines tend to be *constant in percentage terms.* If the price goes up 5% per year, for example, a graph of price versus age will go up slowly at first, then more and more steeply. (Growth curves for world population show a rough version of this same behavior.)

Curves are somewhat harder to work with than straight lines. (For example, there are many families of curves that might have the right sort of shape; different families have different equations; and the constants in the equations are often somewhat tricky to find from the data. Lines are simpler: all lines have equations of the same form, and once you know the slope and intercept, you can find the equation of the line.) Fortunately, for changes that are roughly constant in percentage terms, you can usually get a much straighter pattern if you plot the logarithm of the price instead of the price itself. That's what Figure 2 shows. The graph lets us describe the effect of age by fitting a line and finding its slope. Once you have the line, you can use it to predict future prices.

Now we know how to adjust for age. This adjustment lets us figure out the effect of good and bad years, after a suitable adjustment. Using historical data, and looking back, we can adjust prices for age and assign an age-corrected value for each vintage year. This is essential progress toward our goal, but it still leaves the main challenge ahead of us: Can we tell, at the time that a wine is new, whether its vintage is a good one, and if so, how? The answer: Yes, we can tell, based on the weather for that year.

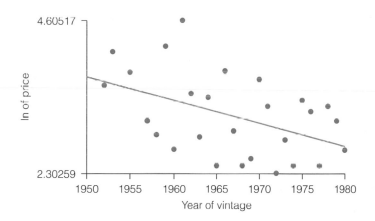

FIGURE 2 Red Bordeaux wines, price relative to 1961 vintage year, logarithmic scale

THE EFFECT OF WEATHER

If you're a wine grape, and you get to choose your weather, you should choose a year with above-average winter rainfall, above-average temperatures during the summer months, and below-average rainfall in September and October. How do we know this? From statistical analysis: we first adjust wine prices for age in order to assign adjusted values to vintage years, and then look for patterns that relate adjusted prices to the weather. These relationships are summarized in a formula that predicts adjusted prices from the weather. Finally, we compare predicted prices with the actual prices to see how much of the price variability has been accounted for using the weather.

It is well known that the quality of any fruit, in general, depends on the weather during the growing season that produced the fruit. What is not so widely understood is that although in some localities the weather will be quite predictable, in others it will vary dramatically from one year to the next. In California, for example, it never rains in the summer, and it is always warm in the summer. There is a simple reason for this. In California a high-pressure weather system settles each summer over the California coast and produces a warm, dry growing season for the grapes planted there. In Bordeaux this sometimes happens—but sometimes it does not. Summers in Bordeaux can be hot and dry, hot and wet, cool and dry, and, most unpleasant of all, cool and wet. In general, high quality vintages for Bordeaux wines correspond to the years in which August and

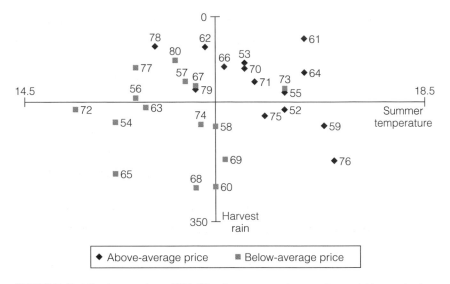

FIGURE 3 Red Bordeaux wines, 1952–80: price, summer temperature, and harvest rain

September are dry, the growing season is warm, and the previous winter has been wet. Except in places where irrigation is common to make up for low winter rainfalls, this finding will not surprise winemakers anywhere in the world.

Figure 3 establishes that it is hot, dry summers that produce the vintages in which the mature wines obtain the higher prices. This figure displays for each vintage the summer temperature from low to high as you move from left to right, and the harvest rain from low to high as you move from top to bottom. Vintages that sell for an above-average price are displayed in purple, and vintages that sell for a below-average price are displayed in green. The two axes in Figure 3 are placed to correspond to historically typical values of average temperature and rainfall, so that the four quadrants correspond to the four combinations of above- or below-average temperature and above- or below-average rainfall. The axis for rainfall is reversed from normal because higher rainfall, unlike a higher temperature for the growing season, leads to lower quality wines. Thus, the hot, dry growing seasons are in the northeast quadrant of the diagram.

If the weather is the key determinant of wine quality, then the dark points should be in the northeast quadrant of the diagram and the light points should be in the southwest quadrant of the diagram, and the other two quadrants should have a mixture of dark and light points. It is apparent that this is generally the case. (The correlation between price and position on the graph is not perfect, of course, because the graph shows only two of the features that affect price.) Nevertheless, even anomalies, like the 1973 vintage, tend to corroborate the fact that the weather determines the quality of the wines, because although the wines of this vintage, which are of somewhat above-average quality, have always sold at relatively low prices, insiders know that they are often bargains (and indeed I have bought and consumed a lot of them!).

Ideally, weather's effect on wine quality and price could be tested with a controlled experiment. However, weather could only be controlled in a laboratory setting, which would be prohibitively expensive. Moreover, in other respects the lab would almost surely be too different from actual vineyards for the results to be trustworthy. This inability to create a controlled experiment is a dilemma economists often face. Instead, researchers must rely on so-called natural experiments. A natural experiment is a set of circumstances that occur naturally (or at least is external to our control) and exhibit sufficient variation in the variables being tested so that their true effects can be extracted. The case of weather in Bordeaux presents a nice natural experiment. The weather differs sufficiently from year to year and the quality of the grapes is recorded sufficiently (through wine prices) to measure weather's true effects on quality.

To capture the effect of weather on prices, I analyzed data on weather and prices from three decades from the early 1950s to 1980.[2] The data set needed to be large in order to protect against mistaking chancelike variation for a systematic pattern, and the vintages had to be old enough for their quality to be established. The analysis indicated that four variables account for 80% of the differences in average price of Bordeaux wine vintages: the age of the vintage, the average temperature over the growing season (April–September), the amount of rain in September and August, and the amount of rain in the months preceding the vintage (October–March). Age alone accounts for only slightly more than 20% of the variability in price, suggesting

---

[2] All analyses use as data the vintages of 1952–80, excluding the 1954 and 1956 vintages, which are now rarely sold.

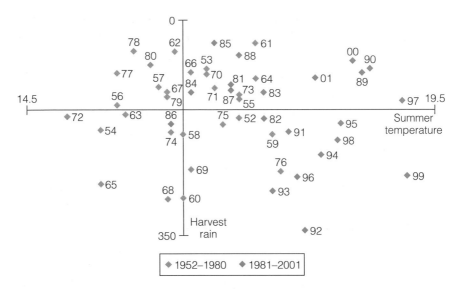

FIGURE 4 Red Bordeaux wines, 1952–2001: price, summer temperature, and harvest rain

that the weather is the main factor that determines the quality of a vintage.

Using a statistical method called regression analysis, the four variables were combined to give an equation for predicting the average price of a vintage based on its age and the weather:

$$\log(\text{Price}) = -12.15 + 1.17\,\text{WinterRain(m.)} + 0.62\,\text{SummerTemp}\,(^{0}\text{C}) - 3.86\,\text{HarvestRain(m)} + 0.24\,\text{Age(yrs.)}$$

Using this model, it is possible to predict the relative price at which the new vintage should be sold as soon as the growing season is complete. In fact, this has been done for several years, with results published in the newsletter *Liquid Assets: The International Guide to Fine Wines.* The basic idea for these predictions is displayed in Figure 4. This figure adds to Figure 3 by including the data for the vintages 1981–2001, but keeps the axes in the same place based on the historical normal rainfall and temperature data.

LOOKING FOR OUTSTANDING VINTAGES

Two things are immediately apparent from Figure 4. First, all but one of these recent vintages (1986) was produced by a growing season that was warmer than what is historically "normal." On the other hand, the average rainfall during the harvest in the later period shows no difference from "normal." Indeed, the prevalence of such warm weather in the summer in the last two decades no doubt accounts, in part, for the deeply held conviction of many Europeans that global warming is already upon us. This unusual run of extraordinary weather has almost certainly resulted in a huge quantity of excellent red Bordeaux wines. As some have observed, global warming is not entirely a bad thing, at least for some in the agricultural community.

Second, the weather that created the vintages of 1989, 1990, and 2000 appears to be quite exceptional by any standard. Indeed, the question must be asked, Is it appropriate to predict that the wines of these vintages will be of outstanding quality when the combination of temperature and rainfall that produced them is so far outside the normal range?

The statistical issue here is an important one that goes by the name *extrapolation*. Here's a simple example: In the United States, children between the ages of one and six grow an average of about three inches per year. This relationship gives a simple formula that does a good job of predicting average height from age. The formula works well for ages in the range from one to six (interpolation), but it would be foolish to expect the same formula to work for adults (extrapolation).

Before making the predictions for 1989 or 1990 I asked the distinguished Stanford statistician Lincoln Moses for advice. Moses suggested two informal tests. (1) *Would the last major extrapolation have been correct?* The idea here is to use the past to indirectly test the ability of the relationship to stretch beyond the available data. In fact, the last major extrapolation for which all uncertainty had been resolved was the vintage of 1961, which had the lowest August–September rainfall in Bordeaux history. Just as the unusual weather predicted, the market (see Table 1), and most wine lovers, have come to consider this an outstanding vintage. (2) *Was the warmth of the 1989 and 1990 growing seasons in Bordeaux greater than the normal warmth in other places where similar grapes are grown?* The idea here is to determine whether the temperature in Bordeaux is abnormal by comparison with grape growing regions that may be even warmer. In fact, the temperature in 1989 or 1990 in Bordeaux was no higher than the average temperature in the Barossa Valley of South Australia or the Napa Valley in California, places where high-quality red wines are made from similar grape types.

Based on these two informal tests, I decided in 1991 to predict that both the 1989 and 1990 vintages in Bordeaux were likely to be outstanding. Ironically, many professional wine writers did not concur with this prediction at the time. In the years that have followed minds have been changed; there is now virtually unanimous agreement that 1989 and 1990 are two of the outstanding vintages of the last 50 years.

Finally, Figure 4 indicates that the 2000 vintage is in a league similar to the outstanding vintages of 1989 and 1990. And what does the wine press say about this vintage? It is not hard to find out, as these wines have been advertised for sale over the last two years using the fantastic praise heaped upon them. For example, Robert Parker, widely considered the most influential taster says, "2000 is the greatest vintage Bordeaux has ever produced. Remarkably consistent from top to bottom, there has never been a year where so many exceptional wines were produced." And yet we learned this without tasting a single drop of wine!

MARKET EFFICIENCY: ARE THERE ACTUALLY ANY GOOD BUYS?

An important principle in economics known as the *efficient market hypothesis* says that prices adjust to take all available information into account. According to this hypothesis, there are no bargains and no bad buys, because the price always adjusts. An old joke has it that an economist who truly believes the efficient market hypothesis won't bother to pick up a dollar bill off the sidewalk: if it had any value, someone else would have grabbed it already; if it's still there, it must not be worth anything.

For Bordeaux wines, the efficient market hypothesis says there should be no big bargains or bad buys based on the weather, because by the time the wine is available to buy, the past year's weather is known, and the price has already made an appropriate adjustment. Is this in fact true? Although many markets are known to be efficient in this way, the efficiency of a market is not automatic, and the efficiency of the Bordeaux market, though plausible as one hypothesis, is not a forgone conclusion.

In short, efficiency is an empirical question, and we can use the price data to evaluate the hypothesis: Were the relative prices of the vintages when they were first sold at market good forecasts of the relative prices of the wines when they matured, and if so, were these forecasts as good as the predictions made using the data on weather alone?

Table 2 reveals the answer to both questions. This table shows selling prices for wines from the 12 vintage years 1961 through 1972 (columns), sold in the years 1971 through 1989 (rows). In order to eliminate the effect of inflation and to put wines of different vintages on a similar scale, each price is given as a fraction of the price of a "benchmark portfolio" of wines.[3] In Table 2, the entry of 1.68 in the row 1971 and column 1961 means that in the year 1971, the wines of 1961 sold at 1.68 times the benchmark price. At the bottom of the same column, we see that by 1989, the wines of the 1961 vintage were selling at 2.09 times the benchmark price.

The increase from 1.68 to 2.09 for the 1961 vintage was an exception to the general pattern, however. The next year's vintage, 1962, was more typical, selling for 0.79 times the benchmark in 1971 and slowly falling to 0.61 by 1989. By looking at the rest of the columns in the table, you can see that almost all the other vintages followed this same pattern. Thus, most of the older vintages began their lives in the auction markets at prices that are substantially above what they will ultimately fetch. Historically, then, initial price has been a poor predictor of ultimate value.

How good a prediction is possible using the weather? The bottom row of Table 2 shows predictions made by the statistical model.[4] In all vintage years but 1961, the 1989 price was lower than the starting price, and for all but one of those vintages, the predicted price was also lower; the model "knew" what the auction market did not—the wines were overpriced. For many of these vintages, the prediction based on the weather was remarkably close to the 1989 prices.

What about the efficient market hypothesis? The data in the table tell us that early market prices are not a good predictor of what the

[3] That benchmark is the average price of the wines from the superior vintages of 1961, 1962, 1964, and 1966.

[4] This model was created using an entirely different set of data. Using different data sets eliminates the possibility that the model does well just because it is predicting facts that it already knows.

VINTAGE

Year of Sale	Benchmark Portfolio	1961	1962	1963	1964	1965	1966	1967	1968	1969	1970	1971	1972
1971	£54	1.68	0.79	0.41	0.76		0.79						
1972	£97	1.58	0.76	0.26	0.70	0.27	0.96	0.77		0.75			
1973	£119	1.62	0.71	0.28	0.74	0.24	0.93	0.62	0.28	0.70	0.83		
1974	£85	1.31	0.77	0.39	0.84		1.08	0.78	0.30	0.70	0.88		0.30
1975	£76	1.65	0.77	0.29	0.78	0.35	0.60	0.57	0.31	0.41	0.84	0.61	0.44
1976	£109	1.67	0.83	0.30	0.66	0.29	0.65	0.51	0.23	0.36	0.69	0.54	
1977	£165	1.67	0.83	0.26	0.63	0.26	0.87	0.50	0.23	0.36	0.70	0.51	0.32
1978	£215	1.67	0.76	0.26	0.65	0.18	0.91	0.45	0.25	0.31	0.70	0.53	0.25
1979	£274	1.61	0.73	0.20	0.66	0.23	1.00	0.49	0.24	0.29	0.71	0.50	0.23
1981	£296	1.75	0.62	0.22	0.70	0.04	0.93	0.47	0.25	0.29	0.82	0.52	0.22
1982	£420	1.80	0.71	0.15	0.60	0.18	0.89	0.39	0.17	0.24	0.77	0.55	0.19
1983	£586	1.77	0.53	0.10	0.59	0.18	1.11	0.36	0.18	0.21	0.91	0.48	0.20
1985	£952	2.19	0.53	0.12	0.50	0.21	0.78	0.30	0.11	0.14	0.68	0.46	0.13
1986	£888	2.10	0.56	0.25	0.54	0.17	0.80	0.30	0.15	0.19	0.65	0.46	0.14
1987	£901	2.11	0.56		0.53		0.80	0.32	0.19	0.20	0.64	0.49	0.18
1988	£854	2.01	0.56	0.21	0.61	0.14	0.82	0.34	0.23	0.20	0.67	0.58	0.17
1989	£1,048	2.09	0.61	0.28	0.53	0.19	0.77	0.27	0.24	0.18	0.66	0.43	0.15
Predicted price		1.74	0.72	0.29	0.76	0.16	0.78	0.49	0.21	0.29	0.60	0.53	0.014

TABLE 2 Price per case of a portfolio of Bordeaux chateaux relative to the price of the portfolio for the vintages of 1961, 1962, 1964, and 1966

prices will be once the wines mature and their quality is established. This is true even though the information about the weather is known when the wines are first offered at auction.

It is apparent from Table 2 and from the preceding example that most vintages are overpriced when the wines are first offered on the auction market, and that this state of affairs often persists for 10 years or more following the year of the vintage. Remarkably, the overpricing of the vintages is especially apparent for those vintages that, from the weather, we would predict to be the poorest. This suggests that in large measure the ability of the weather to predict the quality of the wines is either unknown or ignored by the early purchasers and sellers of the wines.

One interesting way to see the inefficiency in the wine market is to compare the prices of the vintages of 1962, 1964, 1967, and 1969 in calendar year 1972. As the weather data in Figure 3 indicate, and the prediction in the bottom row of Table 2 confirms, in 1972 one should have expected that the 1962 and 1964 vintages would sell for considerably more than the vintages of both 1967 and 1969. In fact, in 1972 these four vintages fetched nearly identical prices, in sharp contrast to what the weather would have indicated. However, by around 1979 the prices of the 1969s and 1967s had fallen to around what would have been predicted by the weather.

One of the most fascinating surprises with respect to vintage pricing is provided by the 1982 vintage. As the data in Figure 4 indicate, this was a warm year, although not especially a dry one, and the wines should be good. However, what has happened to these wines is quite remarkable, with their prices virtually taking on a life of their own. Today, the wines from this vintage are priced at double and triple the prices of the superior 1989 and 1990 vintages, and there is no indication—yet—that there will be any correction soon. That there is something odd about the vintage is attested to by the current state of the auction markets, where huge quantities of 1982s are put up for sale at current prices, but where no one wants to part with their 1983s or 1985s. This suggests that the auction market may have evolved to include two classes of buyers, those who buy for the wine, and those simply looking for a status symbol. The sellers of the 1982s are the former, and the buyers are the latter.

The 1982 vintage phenomenon raises serious problems for those who would like to invest in wine but not consume it. How does one predict that a vintage will become a status symbol? For those who consume their wines, anomalies like 1982 cause no problem—they simply buy the wines based on fundamentals, and then drink the wines that do not appreciate in price and sell those that do. Wine investors do not, sadly, have the more pleasant half of these two options.

CONCLUSION

There is much variability in prices of mature Bordeaux wines, but as shown, much of it can be explained. First, virtually all of the variability can be explained once the vintage of the wine and its producer is known. Second, a straightforward statistical analysis of the weather in which a vintage is grown, and consideration of its age, can be used to describe much of variability in prices across vintages. Finally, the historical reputation of the chateau that produced the wine explains much of the remaining variability in prices. But why does the market for immature red Bordeaux wines appear to be so inefficient when the market for mature wines appears to be so efficient? There may be several related explanations, but one in particular stands out. For growers to maintain the same income from year to year *the price of the young wines must be inversely related to the quantity produced, and independent of the quality.* In other words, the fewer grapes they produce, the more money they need per grape to maintain the same income, and quality must be kept independent of the price or their income would be subject to volatility. Although the actual pricing of young Bordeaux wines falls short of this ideal, it is clearly closer to it than would occur if purchasers used the information available from the weather for determining the quality of the wines. The producers do attempt to raise prices when crops are small, despite the evidence that the quantity of the wines (determined by the weather in the spring) is generally unrelated to the quality of the wines (determined mainly by weather in the remainder of the year). Moreover, it is common for the proprietors to claim that each vintage is a good one, independent of the weather that produced it. Indeed, there is no obvious incentive for an individual proprietor to ever claim anything else!

A more fundamental question arises about the motives of the early purchasers of the wines. Why have they ignored the evidence that the weather during a grape growing season is a fundamental and easily measured determinant of the quality of the mature wines? And will they continue to do so as the evidence for the predictability of the quality of new vintages accumulates?

ADDITIONAL READING

For a description of the analysis with more of the statistical detail, including computer output for the regression analysis, see http://www. liquidasset.com/orley.htm.

QUESTIONS

1. What are the main determinants of Bordeaux quality, according to the essay?
2. What causes the missing values in Table 1, and how does this suggest the relationship that you would expect between the average of the nonmissing values for a vintage and the imputed average for that same vintage?
3. What is the effect of age on Bordeaux wines, if one is to ignore vintage and chateau?
4. Is there any evidence for global warming from the data shown for Bordeaux? What about long-term trends in rainfall?
5. What was price in pounds (£) of the benchmark portfolio in 1971? In 1989? What is your annual compounded rate of return over this period if you had bought the benchmark portfolio in 1971 and sold it in 1989?
6. Table 2 shows that the model was quite accurate in predicting 1989 prices, relative to the benchmark portfolio, for vintages 1961–72. Imagine you are looking at these predictions for 1989 in 1971. How would you use this information to make a wise investment in wine to be liquidated in 1989?

CREDITS

INDEX

Italic page numbers indicate material in tables or figures.

absconding of bees, 120
abundance index, 113, 116
abundance of tiger prey, monitoring, 105–18
accelerated life tests, 345, 350, 351, 352–53
accelerating variables, 353
acceptance sampling, 332–33
account signatures
    defining, 295–98
    design of, 298–99
    flagging accounts, 302–4
    scoring calls, 301–2
    starting a baseline, 299
    updating, 299–301
acquired immune deficiency syndrome (AIDS),
    227, 251
"action level" for radon remediation
    decision analysis, 159, 160, 162, 165, 166
    EPA recommendation, 152, 156
actual voters, 21
adjusting for age, wines, 411–12
administration records, 43, 46
advanced-generation lines, 249
advertisers, 374
advertising
    claims, 351
    defined, 373
    as engineering science, 373–89
    industry dollars, 374–75
    scientifically-driven, 375, 385, 387
    technology of, 387
    Turkey Mail program, 374, 376–87, 388
African Americans
    allele frequencies, 264, 265
    in health care surveys, 93
    prevalence of HIV, 233, 234
    school choice results, 78, 84
African hybrid bee colonies, 133
Africanized bee invasion, prediction of, 119–34
    arrival time, 121–28, 132–33
    density, 128–33
Africanized honey bees (AHB), 119–21
Agency for Healthcare Research and Quality, 90
age variable, 315, 315, 383–84
aging of wine, 407–9, 410–11
agouti mouse
    genetic mapping, 285–88, 285
    strain, 274, 274, 276
agricultural productivity, 120–21, 249–50
AHB. See Africanized honey bees (AHB)

AIDS. See acquired immune deficiency
    syndrome
aircraft, 340, 354
albino mouse
    genetic mapping, 280, 285–88, 286, 287
    strain, 272, 273–75, 274, 276
allele counts, 261–62, 263, 265
allele pairs, 287
alleles, 245–46, 257, 273–76
allelic independence, 267
Amalgamated States of America, 36
American League, 398, 398, 400, 401
American Society for Quality, 334
American Transit 4-A satellite, 185
Amur River, 107, 107
Amur tiger, 105–8, 106, 109, 117
analysts, business, 310–12, 317, 319
analyze phase of Six Sigma projects, 328–329
analyzing data
    in advertising experiment, 379–82, 386
    data mining, 309–13
    HIV infection clinical trials, 239–40
    in scientific method, 375–76, 375
    short-term memory experiment, 219–24
    Young Men's Survey, 232–34
Androutsopoulos, I., 144
animal models, 251. See also tiger prey
    abundance, monitoring of annual living
    area average radon concentration, 152–53,
    155, 158, 159, 162
anthrax
    antibiotics for, 198, 199, 201, 203–4, 207
    outbreak model, 197–209
    spores, 197, 198, 207
    types of, 198
anthrax outbreaks
    in Russia, 201–3
    in United States, 197–99, 203–5, 206–7
antiviral drugs, 252
apportionment, 35–67
    Census 2000, 42–48
    defined, 36
    example, 36–39
    implementation, 48–60
    and U.S. Constitution, 39–41
apportionment population, 46–48, 46–47
Arizona, Africanized bee invasion, 125–26
arrival time predictions of AHB, 121–28,
    132–33

Articles of Confederation, 39
Asia, 107
Asians and HIV, *233*
"as nearly as possible," 48–57, 62
aspirin in heart attack study, 14–15
assignment in census counting, 43
association and causation, 12–16
astronauts, 183–84
asymptotic confidence interval procedure, 206
athlete injuries, 225
Atlantic coastal areas, 121, 123
atmospheric behavior and predictability,
      171–73, 177, 180
Australia, Barossa Valley wines, 417
autistic adults, 225
automated fraud detection, 295, 304
automobile industry, 333
automobiles, 352, 354
autoregressive model, 362, 364–65
average, 159

bacillus anthracis bacteria, 198
back-calculation method, 207
bag of words for spam filtering, 139–43
Bali, Indonesian island, 107
Bangkok, Thailand, HIV clinical trials, 237–38,
      237, 240–41, *240*
bar charts, 315, *316*
bargains, wines, 415, 418
bar graph, 328
Barossa Valley, South Australia, 417
Bartlett, M. S., 129
baseball decision making, 393–406
baseball field, 395–96, *396*
baseball simulation games, 395, 400
baseline data of newspaper accuracy case, 324
base types of genomes, 257
Bayesian hierarchical modeling, *156,* 157, 206
bee parasite (varroa mites), 126–28, *127, 132*
bees
      European honey bees (EHB), 119–21,
            127, 133
      Indian, 127
      See also Africanized bee invasion
behavioral information database, 308, 314
Bell Laboratories, 214, 333
benchmark portfolio of wines, 419, *420*
Berra, Yogi, 351
best fit line, 362–63
beta testing, 350
between-subject differences, 217
bias
      in clinical trials, 238
      in radon surveys, 153, 154, 157–58
      right truncation of incubation period, 202–3
      in school choice self-selection, 71, 74
      selection, Young Men's Survey, 229–33
      time order, 225
binary representation of e-mail, 140, 143

Binghamton, New York, 175, *176*
binomial distribution, 23, 204
binomial probabilities, 282
biological invasions. *See* Africanized bee
            invasion; Anthrax outbreaks; HIV
            dynamics and genetic differences;
            varroa mites
biological weapons, 198
biomedical research, 288
bioterrorism, 197, 207, 245
birth-death model of AHB, 129–32
bisexual men, 230
bivariate distribution, 191
Bjerknes, Vilhelm, 172
black mouse strain, *272,* 273–75, 280
blacks. *See* African Americans
blindness, 243–44
Block, Felix, 213
blood oxygenation, 214
Bochy, Bruce, 393, 404–5
Bohr, Niels, 311
Bonds, Barry, 404
boosted decision trees, 144
Bordeaux region, France, 407–9, 413
Bordeaux wines, predicting quality and prices,
      407–23
      adjusting for age, 411–12
      market efficiency, 418–22
      outstanding vintages, 416–18
      price charts and graphs, *408, 411, 413, 414,*
            416, 420
      prices, 407–11, 418–22
      weather effects, 413–16
Boston Red Sox, 394–95
bottle-feeding babies, 234–35
bovine spongiform encephalopathy
            epidemic, 207
Box, George, 333
boxplot, 315–16, *316*
brain
      activity, 213–14, 215–16, 220, 223–24
      function, 218
      movement, 216
      research (*see* short-term memory experiment)
      structural image, 211, *222*
brand management investment, 387
breastfeeding and HIV transmission,
      234–35
Brentwood Post Office anthrax outbreak, 198,
      199, 203–4, *203*
Briard dogs, 243–44
Brown Simpson, Nicole, 256
Bureau of Economic Analysis, 96
Burris, John, 29, *30*
Bush, George W., 31, *32,* 69
business analysts, 310–12, 317, 319
business process improvement.
      See Six Sigma
Business Week, 362n2

Butterball company, 374, 376–77, 388
  demographic data on, 377, 382, 84
  website, 376–78
"by the book," 394

calibration relationship, 191–92, *191*
California
  baseball team, 393, 396, 404–5
  door-to-door survey, 229
  House representatives, 50–51, *50*
  radon concentration, *163,* 164
  semiconductor industry, 360
  wine country, 413, 417
call-by-call updating, 300–301
call characteristics, 294–98, *294,* 301
call data, 294–98
call scores, 301–3
Canada, radon threshold, 160, 165
candidate preference polling, 19–20, 32
capability studies, 326, 330
capture-recapture method, 110
Carney, John, 30–31, *31*
Carper, Tom (governor), 27, *28,* 29
Carreras, X., 144
carrying capacity, 129, 130, *130*
Caspian Sea, 107
catalogs of space debris, 185, 186, 187
categorical variables, 315, 383–84
causal effects, 72
causation and association, 12–16
cause-and-effect diagram, 326, *327*
cause-and-effect relationships, 14
CDC. *See* Centers for Disease Control
cell of aley, 188 89
cell phone fraud detection, 293 306.
    See also account signatures
censored observations, 239
census of U.S. population, 40,
    42–48, 57
Center for Applied Demography and Survey
    Research, 28
Centers for Disease Control (CDC), 227, 228,
    230, 241
centiMorgan (cM), *284,* 286
CERISE program, 185
Challenger space shuttle disaster, 339, 340,
    341, 353
chance variation, 11, 12, 15, 17
chaos/chaotic atmospheric behavior, 172–73,
    174, 180
chateaux (vineyards), *408,* 409–10, 422
checklists, 331
Cheval Blanc chateau, *408,* 409
children's products, 340
China, 107, 108
chromosomes
  described, 245–46
  DNA profiling, 256
  mice, 272–73

non-sister pairs, 281
  parental, 278
  sex, 273
  sisters, 277–79, *278*
Cisco Systems, 359
classification tree, 318–19
Cleveland, Ohio, 70
click-through rate (to website), 380–85,
    381–86
climate forecasts, 178
climatological distributions, 178–79
clinical trials of HIV infection, 234–40
cloud cover, 176, *176*
cluster sampling, 234
coat color of mice, 272, 274, 285, *285,* 288
Cobb, George, 11–16
cockroaches, 350
coefficient of time variable, 363
cognitive psychology, 214–16
coin toss, 7–8, 23
cold habitat, 107, 108
Coleman, James, 71
collection methods for space debris, 186
College Station, Texas, 126
collinearity, 369
collision probability, 186, 187, 192
Columbia space shuttle disaster, 193, 339,
    341–42
commercial spam filters, 137, 141
common-source outbreak, 199–200, 201
complete probability distribution, 296
complex screening, 22
complex trait dissection, 247
compliance in clinical trials, 235, 236
components of system, 343 44, 345, 349
computational efficiency, 300
computers
  apportionment computation, 54
  baseball simulations, 395, 400
  computer-fitted line, Weibull distribution,
    348, *348*
  computer system redundancy, 353
  models of atmospheric behavior, 172
  MR images, 220
  weather forecasts, 174
ConAgra Foods, 374
concussion injuries, 225
conditional probability, 16
conditional probability distribution, 296, 298
confidence bounds on product reliability, 347,
    347n1, 348, 349, 353
confidence interval
  animal density, 116
  anthrax outbreak, 205–6
  HIV infection reduction, 240
  for profile match probabilities, 266
confidential interviews, 232
constant in percentage terms, 412
Constitutional Convention, 39

consumer price index (CPI), 411
consumers, tailoring messages to, 374, 376,
    384–85. *See also* customer entries
control charts, 326–27, 330, 331, 332
control group, 236–38
controlled intervention, 236–38
control of experiment conditions, 218
control phase of Six Sigma projects, 330–31
control plan, 330–31
conventional counting methods, 43–44
copy editors, newspaper, 325, 329, 330, 332
Corpus Christi, Texas, 128
correction factor, counting animal abundance,
    114–16
correlation of allele frequencies, 265
Cos d'Estournel chateau, *408, 409*
cosine of day of year, 175
cost-benefit optimization, 387
counties in radon surveys, 156–58
counting census results, 42–44. *See also*
    apportionment
counting methods of animal abundance,
    111–12
courtroom statistics
    in DNA profile case, 256–68
    pattern of deaths, 3–18
covariates, effect of, 382–84, 386
Creutzfeldt Jakob disease, variant, 207
Crick, Francis, 245, 272
criminal conviction, 268
crosses of mouse strains, 273–75
cross-sectional data, 94–95
cross-training, 220–21
cumulative call score, 302–3
cumulative random errors/numbers, 366,
    367, 368
cumulative score thresholds, 303
curves, 412
customer behavior, 308, 310
customer expectations, 339, 340, 350
customer record, 307–8, 314
customers. *See* consumers
cutaneous anthrax, 198
cycles in stock market, 368

daily maximum temperature, 174–75
daily stock prices, 361–67, *361, 362*
Darwin, Charles, 255
Daschle, Thomas (senator), 198
data, in scientific method, 375–76, *375*
data analysis
    in advertising experiment, 379–82, 386
    tools, 310
data analyst choices, 83–84
data analysts, data mining, 310–12, 317, 319
Database and Information System
    Characterising Objects in Space (DISCOS)
    catalog, 186
databases, 308, 309

data collection, 310
data mining, 308
    competition, 314
    techniques, 307–22
data monitoring committee, 239
data-preparation step, 314
data warehouse, 308, 309
day of week as factor of interest, 377, 378
day-of-week differences, 329
Dayton, Ohio, 76–78
"day traders," 366
dead heat, 27–29
death penalty, 10, 16–17
deaths
    from lung cancer and radon exposure,
        150–51, 160
    People v. *Simpson,* 256
    United States v. *Kristen Gilbert,* 4, 6–7, *6, 14*
decennial census, 40–42
decision analysis, dollars and lives in radon
    exposure, 159–65
decision making in baseball, 394–95, 404–5
decision trees
    for data mining, 317–18
    for radon measurement and remediation,
        161–65, *161, 163*
deer, 108. *See also* red deer; sika deer
define phase of Six Sigma projects, 324–25
Delaware 2000 elections
    governor, 29–30, *30*
    lieutenant governor, 30–31, *31*
    presidential, 31, *32*
    Senate, 27–29, *28*
    voter turnout, 32
democracy, 39
demographic data
    data mining, 308, 314
    HIV survey, 232
    influence on obesity, 250
    preelection polls, 25
density dependent probabilities, 131
density of animals, 113–17, *113*
density predictions of AHB, 128–33
descriptive approach, 310
design, in scientific method, 375–76, *375*
design engineers, 340, 343, 345, 347, 350
designing experiments
    advertising, 375, 378–79, 380, 387
    clinical trial methods *vs.* experimental
        design, 236
    history of statistical approach, 332–33
    short-term memory, 212–19, 224–25
design process, reliability of, 340, 343
detection of fraud. *See* cell phone fraud
    detection
deterministic atmospheric prediction, 171–73
dew points, 176, *176*
diabetes, 273, 313
differential net undercounting, 43

diploid, 246
direct democracy, 39
direct mail solicitations, 309, 314, 320
DMAIC, Six Sigma, 324, 335
dmoz open directory, 145
DNA
    data, 261–63
    fingerprinting, 255–69
    profile probabilities, 260–61, 263–64, 263
    profiles, 255, 256–59, 268
    sequence, 257, 273
Dodge, Harold, 332
domain knowledge, 310–12, 317, 319
domain theory, 310
dominant alleles, 274, 285
double-blind study, 238
double plays, 399
Dow Jones Industrial Average, 366
Drosophila genetics, 284
drug regimen for HIV infection, 234–35
D7S9 marker, 257–58, 258
duplication, 232–33
dynamical weather forecasts/models, 172–73,
    174–77

early warnings of product failure, 354
economic indicators, 368, 369
EDA. See exploratory data analysis
editor, newspaper, 325, 330
education reform, 69
effects of interest, isolating, 214–16, 218
effects of selection bias in school choice, 71, 74
efficacy results, 239–40, 240
efficient market hypothesis, 418–22
EHB. See European honey bees
Einstein, Albert, 35
election polls, 19–34
elections
    voter turnout, 21, 32
    See also Delaware 2000 elections
eligible voters, 19–20, 21–22
El Niño, 178
e-mail
    advertising messages, 377–82
    and spam filters, 135–47
ensemble forecasting, 173
enumeration methods, 45
EPA. See U.S. Environmental Protection Agency
epidemics, 207
epinephrine, 3–4
equations of motion, 186–87
equilibrium size distribution, 130–31, 131
error rates
    control of, in genetics, 283, 288
    and target size of study, 239
error reduction in newspapers, 323–32
errors, statistical
    in counting, 42–43
    cumulative random, 368

independent, 368
mean absolute percentage error (MAPE),
    364–65, 364
measurement, 42
model assumptions, 205–6
non-sampling, 26–27
normally distributed, 368
sampling error, 22–23, 205–6
sum of squared errors (SSE), 52–54, 53,
    61–62
systematic, 154, 157–58
See also margin of error
errors in baseball, 399
errors in newspapers, 324–25, 327, 328,
    329–31, 331
estimating
    animal abundance, 109–11, 116–17
    census counting, 43
    effects of noncompliance, 79–81
    national health-care parameters, 92–93
    sampling variation, 233
Eta model, 175–76
ethics of research, 236
ethnic groups
    Asians and HIV, 233
    Hispanics, 93, 233
    and school choice results, 78, 84
    See also African Americans
European honey bees (EHB), 119–21, 127, 133
European Space Agency, 186
events of interest, weather forecasting, 174
expected number of runs scored, 400, 401, 403
expected value, 159
experimental design
    in advertising, 375, 378–79, 380, 387
    methods of, vs. clinical trial methods, 236
experiments. See randomized experiments;
        short-term memory experiment
expert testimony, 10–11
exploratory data analysis (EDA), 333
exponential distribution, 190, 200
exponentially weighted moving averaging, 300
extended forecasts, 172
extrapolation, 417

factors of interest, 377, 378, 380
failure detection rates, 354
failure modes and effects analysis, 328
failure of products, 355
false alarm rates, 138, 139, 354. See also false
        positive
false alarms, 221–23, 301, 303
false negative, 136, 138, 139, 143, 144, 303
false positive, 136, 138, 139, 143, 144
FBI. See Federal Bureau of Investigation
Federal Bureau of Investigation (FBI), DNA
        data, 257, 261, 262, 263
feedback information, 351
field testing, 350

financial analysts, 359–60, 361, 369–70
financial newsletters, 369
fingerprints, 255
fire hazards, 339, 351
Firestone tire failures, 339, 342
first filial generation, 274
fishbone diagram, 326, *327*
Fisher, Sir Ronald, 332
flagging accounts, 302–4
Florida anthrax outbreak, 197–99, 203–6, *203*
flux rate, 188–92, *188*
follow-up rate, randomized
    experiment, 82
food distributors, 374
Ford Explorer-Firestone tire failures, 339, 342
forecasting the weather, 171–81
forecast probabilities, 179
forecast uncertainty, 173, 180
forensic science, 255, 268
forested environments of tigers, *106,* 108,
    109, 117
Fourth International Conference on Knowledge
    Discovery and Data Mining, 314
France
    Bordeaux region, 407–9, 413
    French satellite, 185
    HIV infection clinical trials, 237–38, *237,*
        239, 240, *240*
fraud detection system. *See* cell phone fraud
    detection
fraud signatures, 301–2
fraudulent activities
    spam, 135
    See also cell phone fraud detection
frequency distribution, 178–79
functional MRI (fMRI), 214

Galton, Francis, 255
gametes, 246
gamma distribution, 124, *124, 125*
garbage in space, 183
gastrointestinal anthrax, 198, 201
Gaussian distribution, 179
gay bars, 230
Gc marker, 257–58, *258, 259*
Gehlbach, Stephen, 4–10, 11, 12
gene association studies, 251
generalizability of a study, 236
genes, 245, 272–73
gene therapy
    for blindness, 243–44
    for HIV infection, 252
genetic background information, 245–46,
    272–79
genetic distance, 249, 279–82, *284, 286*
genetic linkage maps, 246, 249, 279, 283
genetic mapping, 272, 279–88
genetic markers, 245, 251, 257, 282–84,
    285–87, 288

genetic research, 244–45
genetics, statistical, 243–54
    agricultural productivity, 249–50
    HIV dynamics and genetics, 251–52
    linkage analysis, 247–48
    Mendelian genetics, 245–46
    obesity and genetics, 250–51
genome, 245, 272–73
genome scan, 286, *287*
genomics, 244–45
genotype, 245, 257–58
genotype counts, 261, *262*
genotypic probabilities at marker, 267
geologic province, 157–58
geometric mean (GM), 155, *155, 156,* 157, 158
geometric standard deviation (GSD), 155, *155*
Gilbert, Kristen, 3–7, 9–13, 15–17
global warming, 417
Goldman, Ronald, 256
goodness-of-fit test, 276
Gore, Albert, 31, *32*
Gossett, W. S., 332
Graham, Paul, 136
grand jury, 5–10, 12
graphs, 380
grid, weather forecasting, 174
gross national product, 89, 96, 375
group quarters, census counting, 44–45
Guatemala, 121, *122*
Guerrero Province, Mexico, 125
Gwynn, Tony, 393, 396, 397, 404
GYPA marker, 257–58, *258,* 259, 264

Hahn, G. J., 347n1
Hamilton Post Office anthrax outbreak, 198,
    199, 203–4, *203*
Hardy-Weinberg law, 267
Hart Senate Office Building, 198
Haystack Observatory, 188
HBGG marker, 257–58, *258, 259,* 264
health. *See* medical entries; public health entries
health care expenditures, 89–91, 98–99
health care surveys, 89–101
health insurance coverage, 91–92, 94,
    96–97, 99
health policy, 89, 91, 96
health problems
    epidemics, 207
    obesity, 250–51
    radon exposure, 150–51, *151,* 154–59
    See also anthrax outbreaks; HIV research
heart attacks and aspirin study, 14–15
heart rate, 221
heating-degree days, 158
heritable variation, 247
heteroscedasticity, 114
heterozygotes, 246, 251, 273–75
highly accelerated life tests (HALTs), 345, 350,
    351, 352–53

High School Achievement: Public, Catholic and Private Schools Compared, 71
High School and Beyond survey, 71
Hildalgo County, Texas, 125
Hispanics, 93, *233*
histograms
  data mining, 315, *315*
  incubation periods, 201
  of transit time of AHB invasion, 123, *124*
historical record of weather, 172, 177
HIV/AIDS epidemic, 207
HIV dynamics and genetic differences, 251–52
HIV Network for Prevention Trials (HIVNET), 235
HIV research, 227–42
  HIVNET clinical trial, 234–40
  Young Men's Survey (YMS), 228, 229–34, 241
Hoerl, R. W., 324
Hoffer, Thomas, 71
homeowners
  data mining variable, 315, *316*
  decision on radon exposure, 161–65, *161, 163*
homoscedasticity, 114
homosexual men, 229. *See also* men who have sex with men
homozygotes, 246, 273–75
hospitals, 3–4, 334
"house brand," 408
household-reported medical data, 94–96
households, census counting, 44, 45
Houston-Galveston Area Council, 307
human body internal structures, 213
human identification, 268
human immunodeficiency virus (HIV)
  infection, 227. *See also* HIV research
human mind, 211–12
humidity, relative, 175
hybrids, 274
hypothesis
  advertising experiment, 377
  in DNA profile case, 259–60
  in scientific method, 375–76, *375*
  testing, 4–5, 11–13, 17

iceberg phenomenon, 199–201, *199,* 204
"Idiot's Bayes" model, 141
improve phase of Six Sigma projects, 239–330
inbred mouse strains, 271, 273–75, 280
incubation period
  of anthrax, 201–3
  HIV infection, 207
  iceberg phenomenon, 200
  source of uncertainty, 205–6
independent errors, 368
independent segregation of alleles, 276–77
independent voters, 29

index of abundance, 113, 116
index of leading economic indicators (LEI), 368
Indian bees, 127
Indian subcontinent, 107
individual data, 314
Indonesian islands, 107
industry indicators, 368
inhalational anthrax, 197, 198, 199, 201, *202, 203*
in-house testing, 350
initial conditions
  possible situations in baseball, 396–98
  weather forecasting, 171–74
Institute of Medicine, 99
instrumental variable, 80
insurance companies, 334
Intel Corporation stock, 360–68
intent to treat (ITT) effect, 79–81
interconnected processes, 334
intercross of mouse strains, 274–75, *276,* 280
interim analysis, 239
internal corporate variables, 367–68
International Space Station, 184, *184,* 186, 193
Internet service providers, 135
interpolation, 417
intervention, controlled, 236–38
interviews, confidential, 232
inverse problem, 192
investment managers, 369
investment strategies, 369–70
investors in stock market, 359–60, 362, 370
isolating effects of interest, 214–16, 218
Ithaca, New York, 179
ITT. *See* intent to treat (ITT) effect

James, Bill, 395
Japan, 108
Java, Indonesian island, 107, 127
junk mail, 307, 309
junk mail reduction, 307–22
just-in-time maintenance, 354

Kampala, Uganda, 235
Kaplan-Meier estimator, 239–40
KDD Cup, 314, 317
Kepler, Johannes, 186
Khabarovsk, Russia, *107*
Khabarovsk Krai, 105, 108, 114
Kilgore, Sally, 71
"killer bees," 120. *See also* Africanized honey bees
Korea, 107
Kristen Gilbert, United States v., 3–18
Krueger, Alan, 84

labor-management model, 332
Lafite chateau, *408,* 409
lagged variable, 364, 369
Lagrange, Joseph-Louis, 186

La Niña, 178
Lathrop, Abbie, 271
Latour chateau, *408, 409*
Lauterbur, Paul, 213
law of segregation, 275
laws of motion, 186–87
LDLR marker, 257–58, *258,* 264, 267
leading economic indicators (LEI), 368
Lebanon County, Pennsylvania, 163–64, *163*
Leber's congenital anaurosis, 244
Lee, William Swain (judge), 29
LEI. *See* index of leading economic indicators
level of surprise, 221–23
Library of Congress, 309
life, dollar value of a, 162, 164, 166–67
life tests, accelerated, 345, 350, 351, 352–53
likelihood function, 282–83, 286
likelihood method, 204–5, *204*
likelihood of response, 317. *See also* probability
likelihood ratio
    based confidence interval procedure, 206
    in DNA profile case, 260, 264, 265, 267–68
    test, 283
likely voters, 22–25, *28*
Lindsey, George, 397–98, 400–401
linear dose-response, for radon-lung cancer,
        151, 160, 164, 165
linear regression, 178
    models, 313, 320, 362–64, 365
line transect methods, 110–11
Ling-spam corpus, 144
linkage, 276–79, 282–84
linkage analysis, 246, 247–48, 251
linkage groups, 249, 284
linkage mapping, 249, 279, 283
*Liquid Assets: The International Guide to Fine
        Wines,* 416
lists of members in population, 229–31
loci (locus) of gene, 245, 273
locomotives, 340, 352, 354
LOD, 283, 285–86, *287*
logarithms
    call scores, 302
    counting animal abundance, 115
    incubation periods, 203
    linkage, ratio of likelihoods, 283
    of odds ratio, 143
    of radon levels, 155, *155*
    response ratio, 313
    of stock prices, 363
    of wine prices, 412, *413*
logistic growth curve, 129, *129,* 130, 132
logistic regression models, 313, 320, 386
logistics of research, 236
lognormal distribution, 155, *155, 202,* 203
London auction prices of wines, *408*
longitudinal data, 94–95, 99
long lead times, weather forecasting, 178
long-run frequency distributions, 178–79

long sequences, 218
long-term trends in stock market, 367–69
Lorenz, Edward, 172–73
loss function, 159
lottery, randomized voucher evaluation, 75–76
low earth orbit, 185–87
low-income people in health care surveys, 93
lung cancer from radon exposure, 149–51,
        151, 160

mad cow disease epidemic, 207
magnetic field, 213
magnetic resonance, 213
magnetic resonance imaging (MRI), 211, 213,
        214, 218, 220
maintenance of products, 354
Mansfield, Peter, 213
manufacturers, 334, 339–40
MAPE. *See* mean absolute percentage error
mapping mouse traits, 271–89
    genetic background, 272–79
    genetic mapping, 279–88
margin of error, *24, 28*
    animal density, 115
    sampling error, 22–23
markers
    genetic, 245, 251, 257
    genetic mapping, mouse traits, 282–84,
            285–87, 288
    molecular, 247
    polymorphic, 249, 257
market basket of wines, 411–12
market indices, 366
marketing, 387
Markov chain model of baseball, 396–97,
        399–400, 402, 404, 405
Marquez, L., 144
Maryland, radon concentration, *163,* 164
Massachusetts, VA hospital, 3
Massachusetts Institute of Technology, 172
match probabilities, 256, 260–61, 264–66, *266*
Mathematica Policy Research, 75
maximum likelihood method, 204–5, *204*
mean absolute percentage error (MAPE),
        364–65, *364*
measurement error, 42
measurement-remediation decision on radon
        exposure, 161–65, *161, 163*
measurements decision, 213–14
measurement system analysis, 326
measure phase of Six Sigma projects, 325–27
medical. *See* health entries
Medical Expenditure Panel Survey (MEPS),
        90–96, 96–99
medical expenditures, 89–91, 98–99
medical imaging, 213
medical-provider survey, 94–96
medical-scanning devices, 340, 352, 354
medical trials and studies, 8, 13–15, 79

Medicare Modernization Act, 91
Meeker, W. Q., 347n1
meiosis, 246, 277, 278, *278*, 281
memory experiment. *See* short-term memory
  experiment
memory load, 211–12, 214–16, *219*, 220,
  223–24
Mendel, Gregor, 246, 272, 273
Mendelian genetics, 245–46
Mendel's laws, 246, 275–76, 276–79
men who have sex with men, 227–34
MEPS. *See* Medical Expenditure Panel Survey
messaging, 373
meteorologists, 172
meteorology, 172
method of equal proportions, 36, 56
Mexico, 121, *122, 123,* 124–25, 374
microbiological research facility, 201
microprocessor market, 360
microwave energy, 213
mind, human, 211–12
miners (uranium), radon risk, 149–51
Minner, Ruth Ann (lieutenant governor),
  29–30, *30*
Minnesota radon survey, 157
missing data, 81–83, 239–40
miss rate, 138, 139. *See also* false negative
mixed genetic profiles, 267–68
model assumption errors, 205–6
model output statistics (MOS) forecasting,
  174–75, *176*
models
  Africanized bee invasion, 129–31, *131*
  animal counts and tracks, 112–16
  anthrax outbreak, 197–209
  autoregressive, *362,* 364–65
  baseball decisions, 395, 396–97, 399–404
  Bayesian hierarchical, *156,* 157, 206
  data mining, 311–15, 317–19
  dynamical weather forecasts, 172–73, 174–77
  Eta, 175–76
  financial, 369–70
  HIV genetics, 251
  "Idiot's Bayes," 141
  labor-management, 332
  linear regression, 313, 320, 362–64, 365
  logistic regression, 313, 320, 386
  mapping mouse traits, 288
  predictive, 308, 310
  regression, 158, 362–65
  reliability, 343–44
  space debris, 187–88, 192
  stochastic logistic, 129–31, *131*
  student test scores, 72, *73*
  tree, 317–18, *318, 319*
molecular markers, 247
monitoring devices, 354
Montgomery County, Maryland, *163,* 164
monthly stock prices, 367, *368,* 370

month-to-month differences, 329
Montrose chateau, *408,* 409
moose, 108
Morbidity and Mortality Weekly Reports, 227
Morgan (distance), *284,* 286
Morgan, Thomas, 246, *284*
MOS. *See* model output statistics
mosaic chromosome, 277
Moses, Lincoln, 417
mother-to-child HIV transmission, 227–28,
  234–41
Motorola, 323
motor reliability assessment, 345–49
mouse genetics, 272
mouse genome, 273
mouse strains, 271–72, 273–75
mouse traits. *See* mapping mouse traits
moving average, 361–62, *362,* 362n2, 370
MRI. *See* magnetic resonance imaging
multilevel modeling, 157
multilocus mapping, 284
multinomial probabilities, 282
multiple comparisons, 222
multiple imputation, 83
multiple linear least-squares regression, 174
multiple sequences per level, 217–18
multiple sources (of survey data), 95–96
multiple subjects, 217
multivariable studies, 328–29
multivariate analysis, 380–82
multivariate regression, 386
Myers, David, 75

Naive Bayes approach to spam filters, 137,
  140–44, 145
Napa Valley, California, 417
NASA. *See* National Aeronautics and Space
  Administration
NASDAQ, 360
National Aeronautics and Space Administration
  (NASA), 183, 192, 341, 375
National Center for Education Statistics, 71
National Health Care Quality Report, 99
national health care surveys, 89–101
National Health Interview Survey (NHIS),
  92–93
National Institute of Health, 235
National League, 398, *398,* 400, *401*
National Residential Radon Survey, 155–58
National Uranium Resource Evaluation, 157
National Weather Service, 175
natural experiments, 415
natural population growth, 129. *See also* logistic
  growth curve
necessary, but not sufficient, condition, 349
net undercounting, 42–43
neural networks, 317, 320
neurons in brain, 214
"never reached" households, 26

nevirapine (NVP), 235, 236, *237,* 238, 240, 241
New Hampshire, House representatives, 49, *49,* 51
New Jersey, anthrax outbreak, 198, 199, 203–4, *203*
newsletters, financial, 369
"Newspaper Accuracy" case, 324
newspaper accuracy improvement, 323–32
newspaper reporters
   accuracy of newspaper, 325, 329, 330
   and prosecutor's fallacy, 265
Newton, Isaac, 186
New York, weather forecast, 175, *176,* 179
New York City private-school voucher experiment, 71, 75–78, 81–84
NHIS. *See* National Health Interview Survey
Nobel Prize, 213, 311
No Child Left Behind Act, 69
Nogales, Arizona, 125–26
"noise," statistical static, 216–20, 220–21, 362, 368
noncompliance in randomized experiments, 79–81
non-sampling error, 26–27
non-sister chromosomes, 281
nonsmokers and radon exposure, 150–51
normal distribution, 23
normally distributed errors, 368
Northampton, Massachusetts, 3
North Carolina, 35, 62
null hypothesis, 283
numerical profile, 138
nurse, murder accusation, 3–4
NVP. *See* nevirapine

obesity, 250–51
observational studies, 13–16, 17, 71–74
odds ratio, spam filters, 142–43
Ohio, school voucher program, 70, 76–78
one-number census, 43–44
"one person, one vote" principle, 36, 48, 62
one-way analysis of variance, 247
online analytical processing (OLAP), 310
open-label randomized trial, 238
opinion research, election polls, 32
Orbital Debris: A Chronology, 184
Orbital Debris Engineering Model, 192
orbital elements, 187
orbiting space debris, 183–94
orbit shape, 187
ORDEM2000 program, 192–93, *192*
O-rings on *Challenger,* 341, 353
outbreaks of anthrax
   in Russia, 201–3
   in United States, 197–99, 203–5, 207
outliers (boxplot), 316

overestimating, 366
overforecasting, 365

Pacific coastal areas, 121, 123, 125
Pakistan, 107
Paralyzed Veterans of America (PVA), 309, 314–21
parameters in model, 313
parental chromosomes, 278
parental pairs, 279
Pareto analysis, 328–29, *328*
Parker, Robert, 418
passage probability, *192*
patch in sky, 186–88
patterns in data
   data mining, 311–12
   of deaths in hospital, 6–7, *6*
Pauling, Linus, 272
pea hybrid study, 246
pediatric AIDS, 228
Penning-Rowsell, Edmund, 409
Pennsylvania, radon concentration, 163–64, *163*
People v. *Simpson,* 256–68
personal data, 314
personalized signature, 300
pest invaders. *See* Africanized bee invasion; varroa mites
Peterson, Paul, 75
phenotypes, 245, 273
phenotypic variation, 245, 272, 288
physical laws governing atmospheric behavior, 171–73
physically-based weather forecasts, 174–77
Pichon-Lalande chateau, *408,* 409
pitchers, baseball, 402
pixels in MR image, 220, 222, 223, *224*
plant breeders, 249
player evaluation, 402
player probabilities, baseball, 399, 404
plot counts of animals, 111–14, *111,* 116
Poincaré, 171
Poisson distribution, 190
polls, preelection, 19–34
polygenic control, 250, 251
Polymarker profiles, 257–59, *258,* 261, *262, 263*
Polymarker system, 257
polymorphic markers, 249, 257
Ponsor, Michael A. (judge), 10–11, 15, 16
poor people in health care surveys, 93
poplar trees, 247, *248,* 249–50
population
   of interest, 20–22, 33, 229–31
   of orbiting objects, 186–87
   of orbits, 187
   of U.S., 40, 45–48, *45–47*
population dependent QTLs, 251

population genetics, 267, 268
population size, 129–31
portfolios, 369
post-processing of weather forecast, statistical,
    174, 177
potential predictor variables, 319
poultry sales, 376
power-generating equipment, 352, 354
power-grid shutdown, 342
precipitation probabilities, 176–77, *176*
predators. *See* tigers
predictions. *See* Africanized bee invasion;
    Bordeaux wines
predictive forecasts, stock market, 360, 365,
    367, 369–70
predictive models, 308, 310
predictor variables
    data mining, 311, 312–13, 314–17, 319, 320
    radon concentration, 157–58
    time, in stock market, 364
    weather forecasting, 174–75
preelection polls, 19–34
prefrontal cortex, 222
pregnant women with AIDS, 228. *See also*
    mother-to-child HIV transmission
pre-processing steps, 219–20
prescription drug use, 91
price index for wine, 411–12, *411*
price predictions, 417–18. *See also*
    Bordeaux wines
primary sampling unit, 230–31, 234
Primorye Krai, Russia, 108, 113, *113*
priority table, 55–61
priority values, 55–62, *57–60*
private election polls, 12
private-school vouchers, 69–71, 75–77, 83–84
probabilistic model for small space debris,
    187–88
probabilities
    density dependent, 131
    mapping mouse traits, *281*, 282, *282*, *286*
    match, 256, 260–61, 264–66, *266*
    multinomial, 282
    of precipitation, 176–77, *176*
    transition, 398, 400
probability
    of collision, 186, 187, 192
    conditional, 16
    of response, 313, 317–18
    Weibull distribution plot, 348–49, *348*
    See also likelihood
probability anomalies, *179*
probability distribution
    call characteristics, 295–98, 300, 301, 304
    conditional, 296, 298
    incubation periods, 200
    for population size of AHB, 130
probability estimation, 261
process capability studies, 326, 330

process improvement, 332–33
process improvement methods. *See* Six Sigma
process input variables, 325–28
process map, 325–26, *326*
process output variables, 325–28
productivity improvement methods.
    See Six Sigma
product quality, 333
product release, 350–52
product reliability and safety assurance,
    339–58
    reliability in the news, 340–42
    space systems, 352–54
    washing machine example, 342–52
professional investment managers, 369
profile probabilities, DNA, 260–61,
    263–64, *263*
profiles, DNA, 255, 256–59
profit, 335
project charter, 324
prosecutor's fallacy, 16, 264
*Protecting the Space Shuttle from Meteoroids
    and Orbital Debris*, 183
psychographic data, 314
psychologically normal subjects, 225
public election polls, 12
public health
    control of anthrax outbreaks, 198,
        201, 207
    and HIV infections, 234
    and radon concentrations, 149, 153–54,
        159–60, 165, 166
public school system, 69
Purcell, Edward, 213
pure democracy, 39
purely statistical weather forecasting, 177–79
PVA. *See* Paralyzed Veterans of America
p-values in courtroom statistics, 7–10, 11–17

QTL. *See* quantitative trait loci
quality improvement methods. *See* Six Sigma
quality predictions. *See* Bordeaux wines
quality/quantity of wines, 422
quantifying evidence in DNA profile case,
    260–61
quantifying uncertainty, 205–6
quantitative inheritance, 249
quantitative methods for baseball decisions,
    394–95
quantitative representation of e-mail,
    139–40, 143
quantitative trait loci (QTL), 245, 247–50, 251
quantitative variables, 315
query-driven approach, 310
questionnaire, 25, 33

radar cross section (RCS), *189*, 191–92
radar detection of space debris, 188–89, *189*
radio astronomy, 188–89

radiology, 213–14
radium, 150, 157
radon, 149–50
radon concentration levels, 150–52, *155, 156,*
    160, 165
    average annual living area, 152–53, 155, 158,
        159, 162
radon concentration measurement decision,
    149–69
    advice for homeowners, 161–65
    decision analysis, 150, 159–65
    EPA's recommendation, 150, 152–53
    home radon problem, 150–51
    summary, 165–68
radon exposure, 150–51, *151,* 154–59
radon measuring tests, 152–53
rainfall and wine quality, 413–16
random digit dialing, 24, 28, 229
randomization
    in clinical trials, 238
    of e-mails, 379
randomized experiments
    versus observational studies, 13–16, 17
    school vouchers, 70–71, 74–78, *77*
    short-term memory, 218–21, *219*
    social experiment challanges, 79–84
    Turkey Mail program, 374, 376–87, 388
randomness in stock market. *See* stock market
    randomness
random sample, 23
random variation, 219–21
random walk in stock market, 365–67
range expansion of Africanized bees,
    120, 123
RCS. *See* radar cross section
reapportionment, 36
recessive alleles, 274, 285
recombinant pairs, 280
recombination, 276–79
recombination fractions, 279–83
Red Cross, 243
red deer, 105, 108, 114
redundancy of design, 353
redundant array of inexpensive disks
    (RAID), 353
red wines. *See* Bordeaux wines
refrigerators, 342, 352
"refused" households, 26
registered voters, 21
registrants on Butterball website, 377–79,
    382–84
registration lists of voters, 21
regression analysis, 247, 416
regression coefficients, 175
regression equations, 177
regression modeling, 158, 362–65
regularized logistic regression, 144
relationships in data, 311–12
relative humidity, 175

reliability, 339
    in design process, 340, 343
    of motor, assessment, 345–49
    See also product reliability and safety
        assurance
reliability budget, 344
reliability goals, 342–43, 354
reliability model, 343–44
reliability monitoring, 350–52
remediation costs of radon, 152
remediation-measurement decision on radon
        exposure, 161–65, *161, 163*
repetition, 217–18
reporters, newspaper, 325, 329, 330
resident population, 46, *46–47*
response
    in clinical trials, 235, 237
    possible outcomes in baseball, 397–99, 403
    of subject, 214–16, 224
response rate, 25–27
response variables, 313–14, 317–18
revision cycle, 325, *326*
right truncation, 202–3
Ringer, Larry, 126
Rio Grande Valley in Texas, 126
risk assessment for International Space
        Station, 186
risk behavior, unprotected sex, 229, 230
risk of collision, 183–84, 186
Rochford, Dennis, 30–31, *31*
roe deer, 108, 114
Romig, Harry, 332
Roosevelt, Franklin, 36
root beer, 407, 408
Roth, Bill (senator), 27, *28,* 29
routine maintenance, 354
RPE65 genes, 244
run chart, 327, *327*
runs scored, 397–98, 400
Russian air force, 186
Russian anthrax outbreak, 201–3
Russian Far East, 105, 107, 108, 116–17
Russian taiga forest, *106*

Sacramento, California, *163,* 164
sacrifice bunt, 402–3
sacrifice flies, 399
safety, 339–40, 351
sample requirements for health care survey,
        92–93
sample sizes
    preelection polls, 22–23, *24,* 32–33
    school choice studies, 72
sampling
    acceptance, 332–33
    in census counting, 43–44
    in clusters, 234
    space-time, 230–32, *231*
sampling error, 22–23, 205–6

sampling frame, 230–31
sampling variation, 233, 266–67
San Diego Padres baseball team, 393, 396, 404–5
S&P 500 index, 366, 369–70, 369n5
San Francisco, California, 229
San Patricio County, Texas, 128
Santa Clara, California, 360
satellites, 185
schizophrenics, 225
school choice program evaluation, 69–87
    observational data, 71–74
    randomized experiments, 74–78
    social experiment challenges, 79–84
School Choice Scholarships Foundation (SCSF), 75, 84
school vouchers, 69–71, 75–77, 83–84
scientific effort, 373–74
scientific method, 375–76, 375
scientific surprise, 5
scoring calls, 301–2
SCSF. See School Choice Scholarships Foundation
seasonal forecasts, 178
sea-surface temperatures, 178
segregation, genetic, 275–76
selection bias
    right truncation of incubation period, 202–3
    in school choice, 71, 74
    Young Men's Survey, 229–33
semiconductor industry, 360, 368
sensitivity analyses, 206
separation of church and state, 70
sequences, in experiments, 217–18
series system, 343–44, 344
service providers, 304
sex
    men who have sex with men, 227–34
    X and Y chromosomes, 273
Shewhart, Walter, 332
short lead times, weather forecasting, 177–78
short-term forecasting, stock market, 360–65
short-term memory, 211–12, 214–16, 220, 222, 225
short-term memory experiment, 211–26
    analyzing the data, 219–24
    designing the experiment, 212–19
    summary, 224–25
Siberian tiger. See Amur tiger
signature component, 296–98, 297, 300
signatures. See account signatures
significance tests, 288
sika deer, 105, 108, 109, 111, 114
simple linear regression, 362–64, 365
Simpson, Nicole Brown, 256
Simpson, O. J., 255–68
Simpson, People v., 256–68
simulation studies, 206
single hit, baseball, 399

sister chromosomes, 277–79, 278
Six Sigma, 323 37
    case study, 324
    five steps, 324–31
    history of statistics in quality, 332–35
smokers and radon exposure, 150–51, 164
smoking and health study, 13–14
smoothing techniques, 362, 370
Snee, R. D., 324
social experiments, challenges, 79–84
soil gas, 152, 158
solar activity, 186, 193
South Carolina, radon concentration, 166
space debris, 183 94
space flights, 183
space shuttle disasters, 193, 339, 340, 341–42, 353
space systems, 352–54
space-time sampling, 230–32, 231
spam, 135
spam filters, 135–47
spam folder, 136, 143
SPC. See statistical process control
SSE. See sum of squared errors
SSP. See sum of squared priority values
standard normal distribution, 365–66
State Residential Radon surveys, 154–58, 156
states, apportionment and resident populations, 46–47
statistical algorithms for spam filtering, 137, 139–40
statistical analysis of wine prices, 410, 413, 422
statistical dead heat, 27–29
statistically pure weather forecasting, 177–79
statistically significant, 78
statistical methods
    in data mining, 310
    of mapping genes, 272
statistical modeling. See models
statistical noise, 216–20, 220–21, 362, 368
statistical post-processing of weather forecast, 174, 177
statistical precision, 112
statistical principles, 216–19
statistical process control (SPC) methods, 333
statistical testing, 9–10, 9
statistical thinking, 333–35, 394
statisticians
    and advertising campaigns, 375, 387–88
    and designers, 340, 355
    and financial analysts, 361
statistics
    and quality improvement (see Six Sigma)
    use in fraud detection, 295, 304
status symbol, 421 22
stemming process, 140, 143
stimulus, 214–16, 224
St. Louis Cardinals baseball team, 393, 396, 404–5

stochastic logistic model, 129–31, *131*
stochastic process, 396
stock market randomness, 359–72
    long-term review, 367–69
    models and managers, 369–70
    random walk, 365–67
    short-term forecasting, 360–65
stop word removal, 140, 143
straight-line results, 223–24
Strat-o-matic, 395
stroke damage, 213
subject line as factor of interest, 377, 378,
    381–82
subject-matter experts, 310–12, 317, 319
subjects
    in HIVNET clinical trial, 236
    in memory experiment, 217, 218, 225
    registrants on website, 377–79, 382–84
    in Young Men's Survey, 232–34
subscription fraud, 299
SUDAAN software, 234
Sumatra, 107
sum of squared errors (SSE), 52–54, *53,* 61–62
sum of squared priority values (SSP), 61
support vector machines, 144
surprise, level of, 221–23
survey design, Young Men's Survey, 230–32
survey instrument, 25, 33
surveys
    health care, 89–101
    High School and Beyond, 71
    Medical Expenditure Panel Survey (MEPS),
        90–99
    National Health Interview Survey (NHIS),
        92–93
    preelection polls, 19–34
    radon surveys, 154–58, *156*
    Young Men's Survey (YMS), 228,
        229–34, 241
survival analysis, 239–40
Sverdlovsk, Russia anthrax outbreak, 201,
    203, 205
swarming of bees, 120, 121
systematic error, 154, 157–58
systematic variation, 219–20
system components, 343–44, 345, 349
system in series, 343–44, *344*

tables, 380
Tamaulipas Province, Mexico, 124
Taylor, Frederick, 332
tech stock boom, 366
telephone numbers, 24–25
temperature
    weather forecasting, 174–75 , 176, *176*
    and wine quality, 413–16, 417
terabytes (TB), 309
test plan, reliability assessment, 345–46
test set, 311, 319, 320

Texas, 121, 125–28, 132, 307
text categorization algorithms, 144–45
Thailand, 237–38, *237,* 240–41, *240*
Thanksgiving holiday, 376, 377
thought experiments
    counting people, 42
    spam, receipt of, 136
thresholds
    fraud detection, 303, 304
    radon, 152, 160, 165
thunderstorm probabilities, *176,* 177
tiger prey abundance, monitoring of, 105–18
tigers
    Amur tiger, 105–8, *106,* 117
    capture-recapture in India, 110
    habitat, *106,* 107–8, 116–17
    population, 107
    stripe pattern, 110
time as predictor variable, 364
time-dependent data, 361
time series plot, 361, *361*
time/space sampling, 230–32, *231*
time to failure data, 348, 353
title variable, 384
tomato plants, 247–48
total quality management (TQM), 333, 335
TQM. *See* total quality management
track rate, 113–16, *113*
tracks of animals, counting, 111–16
trading days, 363, 364
training data for spam filters, 137–40, 143, 144
training set, 311, 319, 320
trajectory orbit, 186
transactional data, 307–8, 314
transactional database, 308
transects, counting animal abundance, 113,
    114, 116
transformation, 313, 319
transition probabilities, 398, 400
transit time of AHB invasion, 122–25,
    124, 125
trap lines for bees, 121–24, *122, 123*
treatment effect estimates, 79–81, 83
tree models, 317–18, *318, 319,* 320
trees, poplar, 247, *248,* 249–50
trend model. *See* simple linear regression
true count, 42
t test, 332
Tukey, John W., 333
Turkey Mail program, 374, 376–87, *388*
twin studies, 250–51

Uganda, 228, 234–40, *237, 240,* 241
UNAIDS. *See* United Nations HIV/AIDS
    program
uncertainty
    in estimates of HIV infections, 234
    quantifying, 205–6
    of weather forecasting, 173, 180

undecided voters, 28, 29–31
underestimating, 366
underforecasting, 365
underlying model assumptions, 205
ungulate populations, 108, 116–17.
    See also deer
uninsured (health care), 91–92, 96–98, 97, 99
United Nations HIV/AIDS program
    (UNAIDS), 228, 235
United Parcel Service (UPS), 309
United States
    anthrax outbreak, 197–99, 203–5, 207
    arrival predictions of AHB invasion, 124–26
    HIV infection clinical trials, 237–41,
        237, 240
    temperature forecast, 178, 179
United States v. Kristen Gilbert, 3–18
University of Delaware, 28
University of Florida, 244
University of Minnesota, 214
unlisted telephone numbers, 24
uranium
    miners and radon risk, 149–51
    National Uranium Resource Evaluation, 157
U.S. Air Force, 185, 186
U.S. Census Bureau
    census of U.S. population, 40, 42–48, 57
    House apportionment, 42–48
    House seat allocation, 35
    source of demographic data, 308
    voting citizens, 21
U.S. Constitution, 39–41, 51
U.S. Department of Commerce, 96
U.S. Department of Education, 69
U.S. Department of Energy, 150
U.S. Department of Health and Human
    Services, 90, 99
U.S. Environmental Protection Agency (EPA),
    150, 152–53, 154–58
U.S. Federal Communications
    Commission, 293
U.S. Food and Drug Administration, 235
U.S. House of Representatives
    apportionment, 40, 41, 48, 57, 57–60
    and one-number census, 44
    seat allocation, 35–36, 62
U.S. Supreme Court
    apportionment of representatives, 35–36, 44
    guidelines on scientific testimony, 11
    school voucher ruling, 70
Utah, 35, 62, 166

VA. See Veteran's Administration
validation set, 311
value of a life in dollars, 162, 164, 166–67
variability
    animal density, 114–15
    in courtroom statistics, 7–10
    in incubation periods, 200

managing with statistics, 216–19
and process improvement, 334
in radon surveys, 153, 157
sample size factor, 23, 24
in sampling process of baseball data, 401
in wine prices, 422
variables
    accelerating, 353
    call characteristics, cell phone accounts,
        294–98, 294, 301
    categorical, 315, 383–84
    instrumental, 80
    internal corporate, 367–68
    lagged, 364, 369
    process input and output, 325–28
    quantitative, 315
    response, 313–14, 317–18
    See also predictor variables
variant Creutzfeldt-Jakob disease, 207
variation
    exists in all processes, 334
    random, 219–21
    and repetition, 217–18
    in stock market activity (see stock market
        randomness)
    systematic, 219–20
varroa mites, 126–28, 127, 132
venue-based sampling, 230
venues, 230
Veteran's Administration (VA) hospital, 3–4
vineyards, 408, 409–10
vintage, wine, 410
    overpricing, 419–22
    quality, 408–9, 411
viral load trajectories, 251, 252, 252
vocabulary list, 140, 141
volatility in stock market, 361, 367, 369, 370
voter participation, 21, 32
voters
    actual, 21
    eligible, 19–20, 21–22
    independent, 29
    likely, 22–25, 28
    registered, 21
    undecided, 28, 29–31
vouchers to private schools, 69–71, 75–77,
    83–84

Wall Street, 359
War on Spam, 135–36
warranty policies, 343, 351
washing machine reliability example, 342–52
Washington D.C.
    anthrax outbreak, 198, 199, 203–4, 203
    randomized voucher evaluation, 76–78
Watson, James, 245, 272
weather forecasting, 171–81
weather's effect on wine quality, 413–16,
    422–23

Weibull distribution probability plot,
    348–49, *348*
weighted average, 300
weighting strategy, 82–83
Welch, William (assistant U.S. attorney),
    4, 10, 11
Welder Wildlife Refuge (WWR)
    graphed data, *128, 129, 130, 131*
    study area of AHB, 128, 129,
        131, 132
"what if" results, 83–84
wind direction, 176, *176*
wind speed, 176, *176*
wine index, 411–12
wine investors, 422
Wines of Bordeaux, The, 409
winters (tiger habitat), 107, 108
wireless phone fraud. *See* cell phone fraud
    detection

WISE. *See* Women's Ischemia Syndrome
    Evaluation
within-subject comparisons, 217
Women's Ischemia Syndrome Evaluation
    (WISE), 251
World Trade Center disaster, 256
WWR. *See* Welder Wildlife Refuge
Wyoming, House representatives, 51, *52*

X chromosome, 273
X-rays, 211, 213

Yahoo! 145
Yang, Mark, 244
Y chromosome, 273
Young Men's Survey (YMS), 228, 229–34, 241

Zhu, Pei, 84
zidovudine (ZDV), 235–41, *237*